AF594956

Critical Metals in Hydrothermal Ores: Resources, Recovery, and Challenges

Critical Metals in Hydrothermal Ores: Resources, Recovery, and Challenges

Editors

Lingli Zhou
Hongrui Fan
Thomas Ulrich

MDPI • Basel • Beijing • Wuhan • Barcelona • Belgrade • Manchester • Tokyo • Cluj • Tianjin

Editors

Lingli Zhou
University College Dublin
Ireland

Hongrui Fan
Chinese Academy of Sciences
China

Thomas Ulrich
Aarhus University
Denmark

Editorial Office
MDPI
St. Alban-Anlage 66
4052 Basel, Switzerland

This is a reprint of articles from the Special Issue published online in the open access journal *Minerals* (ISSN 2075-163X) (available at: https://www.mdpi.com/journal/minerals/special_issues/CMiHO).

For citation purposes, cite each article independently as indicated on the article page online and as indicated below:

LastName, A.A.; LastName, B.B.; LastName, C.C. Article Title. *Journal Name* **Year**, *Volume Number*, Page Range.

ISBN 978-3-0365-1920-3 (Hbk)
ISBN 978-3-0365-1921-0 (PDF)

Cover image courtesy of Hongrui Fan.

Contents

About the Editors

Lingli Zhou is a research fellow in critical mineral resources. Her research interests include critical mineral resources, geochemistry, applied and analytical mineralogy, and sedimentology. Her past 15 years' research has important implication for broad geochemical and mineralogical transformation processes, from deep down to the mantle and up to the Earth's surface. She completed two PhDs from the Chinese Academy of Sciences (2014) and Aarhus University (2015), respectively, and continued her geoscience career as a postdoctoral research fellow/research fellow in universities in Ireland (2015–). She has developed profound analytical skills, using LA-ICPMS as well as extensive applications through academic and industrial collaborations.

Hongrui Fan completed his PhD degree in Petrology (specialization in ore-forming fluids) at the Institute of Geology, Chinese Academy of Sciences, in 1990. He then joined the Institute of Geology and Geophysics, Chinese Academy of Sciences, where he is currently a full Professor, leading the research group of Continental Margin Evolution and Fluid-involving Mineralization Processes. His research interests include the model and origin of a number of different types of ore deposits, particularly precious metal and REE ore deposits, and applications of fluid inclusions in metal ores, petroleum and geothermal resources. He has published more than 210 papers and two scientific books. In addition, he was a visiting fellow of University of Western Australia (1994–1995) and appointed as the regional vice-president of Society for Geology Applied to Mineral Deposits (SGA, 2008~2009).

Thomas Ulrich completed his PhD at ETH Zurich, Switzerland in 1999 and has worked at the University of Queensland and the National Australian University in Australia as a post-doctoral fellow. Following this, he was a senior researcher at Laurentian University in Sudbury, Canada, before moving to Aarhus University in 2010. He is a senior lecturer in Ore Deposit Geology, and his research focusses on the geochemistry of magmatic-hydrothermal systems. He uses in situ analytical methods such as LA ICPMS to unravel ore-forming processes from the trace element composition of mineral and fluid inclusions.

Editorial

Editorial for Special Issue "Critical Metals in Hydrothermal Ores: Resources, Recovery, and Challenges"

Lingli Zhou [1,2,3,*], Hongrui Fan [4,5,6] and Thomas Ulrich [7]

1 School of Earth Sciences, University College Dublin, Belfield, Dublin 4, Ireland
2 Irish Centre for Research in Applied Geosciences (iCRAG), University College Dublin, Belfield, Dublin 4, Ireland
3 Geological Survey Ireland, Beggars Bush, Dublin 4, Ireland
4 Key Laboratory of Mineral Resources, Institute of Geology and Geophysics, Chinese Academy of Sciences, Beijing 100029, China; fanhr@mail.iggcas.ac.cn
5 College of Earth and Planetary Sciences, University of Chinese Academy of Sciences, Beijing 100049, China
6 Innovation Academy for Earth Science, Chinese Academy of Sciences, Beijing 100029, China
7 Department of Geoscience, Aarhus University, Hoegh-Guldbergs Gade 2, 8000 Aarhus C, Denmark; thomas.ulrich@geo.au.dk
* Correspondence: lingli.zhou@ucd.ie

Citation: Zhou, L.; Fan, H.; Ulrich, T. Editorial for Special Issue "Critical Metals in Hydrothermal Ores: Resources, Recovery, and Challenges". *Minerals* **2021**, *11*, 299. https://doi.org/10.3390/min11030299

Received: 20 February 2021
Accepted: 10 March 2021
Published: 13 March 2021

The consumption of resources has rapidly increased over the last few decades, driven by the continuous growth of the global population and technological innovations. Increasing importance has been put on "Critical" raw materials, which are economically and strategically important but vulnerable to supply shortage [1]. The development of sustainable supplies of critical minerals and metals is required if society is to succeed in the decarbonisation of the global economy. While discovery of critical metal deposits is urgent, of equal importance is understanding the life cycle of critical metals already in the economy. This Special Issue includes contributions on both discovery and investigations of the life cycle of critical metals.

"Critical" raw materials are generally taken to be technology-enabling metals (e.g., cobalt, lithium and rare earth metals), which are referred to as "critical metals", "strategic metals", "E-tech elements" and "technology metals". In contrast to the major industrial, base and precious metals, the role of critical metals in the development of novel technologies, e.g., digital systems and devices, renewable energy and energy storage, electric mobility and autonomous vehicles, and aerospace exploration has increased only in past two decades as a result of the so-called the "Fourth Industrial Revolution" [2].

Criticality assessments of raw materials vary from country to country, and the list of critical raw materials is updated periodically to reflect production, market, technological developments, as well as policy priorities [1]. Although the list varies at the country level, in general there is a global concern of sustainable supply of certain metals. For instance, the United States listed 35 critical minerals that are vital to its security and economic prosperity in 2018, including aluminium (Al from bauxite), antimony (Sb), arsenic (As), barite, beryllium (Be), bismuth (Bi), caesium (Cs), chromium (Cr), cobalt (Co), fluorspar, gallium (Ga), germanium (Ge), natural graphite, hafnium (Hf), helium (He), indium (In), lithium (Li), magnesium (Mg), manganese (Mn), niobium (Nb), platinum group metals (PGMs), potash, rare earth elements (REEs), rhenium (Re), rubidium (Rb), scandium (Sc), strontium (Sr), tantalum (Ta), tellurium (Te), tin (Sn), titanium (Ti), tungsten (W), uranium (U), vanadium (V), and zirconium (Zr) [3]. As of 2020, the European Union (EU) has identified 30 critical raw materials that are essential to delivering the European Green Deal, including barite, bauxite (Al), Be, Bi, borates, Co, coking coal, fluorspar, Ga, Ge, Hf, heavy REEs (HREEs), In, Li, light REEs (LREEs), Mg, natural graphite, natural rubber, Nb, PGMs, phosphate rock, phosphorus, Sb, Sc, silicon metal, Sr, Ta, Ti, V, W [4]. China's first official policy and catalogue of "strategic minerals" was established in 2016 with the list being refined in 2018 to include four sub-groups: 1) the noble metals of Li, Be, Rb, Cs, Nb, Ta, Zr, Hf, W; 2) the rare earth metals of La, Ce, Pr, Nd, Sm, Eu, Gd, Tb, Dy, Ho, Er, Tm, Yb, Lu, Sc, Y; 3) the companion metals of Ga, Ge, Se, Cd, In, Te, Re, Tl; 4) the precious metals of PGMs, Cr, Co [5]. Mineral deposits that produce those critical metals are accordingly defined as "Critical Mineral Resources" [5]. As pointed out by Andersson [6], the difference in criticality assessments from Western countries and those from China may lie in the consideration of the factor of supply risk. Supply risk is always a key factor to determine criticality in countries (e.g., EU and US) that have a high dependency on imported raw materials, while it is a subordinate factor in Chinese assessments, as some of the strategic minerals and metals are not subject to shortage of supply, at least in the short to medium term.

Some of the critical metals defined above are typically distributed in low concentrations in Earth's crust while others like aluminium are not. A number of the critical metals, such as Bi, Co, Ga, Ge, Sb, and Te are currently produced exclusively as the by-product of the extraction of major industrial metals such as Cu, Pb, Zn and Al. The enrichment of critical metals in rocks or minerals to economically viable ores usually requires complicated and prolonged geological processes that lead to their high concentrations in a restricted domain of Earth's crust. For instance, an economic deposit of chromium must contain between 4000 and 5000 times (concentration factor) the average crustal abundance (96 ppm), that is between 38.4 and 48 wt.% Cr, to be economical. The high concentration factor may explain the apparent scarcity of certain critical metal resources (e.g., PGMs, REEs), although this parameter is usually market dependent for a metallic/non-metallic commodity. In general, a geological process of ore formation combines extraction of the constituents from magmas, rocks or ocean water, transport of the constituents in a fluid medium from the source region to the deposition site, and deposition of the constituents at favourable sites. The ore-forming processes may be broadly grouped into the following categories on the basis of different geological dynamics: (a) orthomagmatic processes, (b) sedimentary processes, (c) metamorphic processes, (d) hydrothermal processes, and (e) surficial and supergene processes [7]. The formation of mineral deposits often involves one or more of these processes. This Special Issue focuses particularly on the hydrothermal processes and its associated critical metal resources, as hydrothermal ore deposits represent a major source of metals.

The aim of this Special Issue is to advance an in-depth understanding of the resources and recovery of critical metals associated with hydrothermal processes. Such knowledge is crucial for both exploration and for exploitation of hydrothermal ores in which these critical metals are present at concentrations sufficient to allow recovery. Ten articles are included in this Special Issue, which are based on empirical studies carried out by researchers from Australia, Chile, Europe, and China, covering a wide range of critical metals, including Co, Ga, Ge, Re, REEs, In, Sb, Sn and W. These papers advance our understanding of global critical metal resources, and meanwhile, highlight the challenges for securing these elements moving forward and the necessity for responsible sourcing and sustainable use of these resources.

Hydrothermal processes involve movement of aqueous fluids derived from a variety of sources through the Earth's crust, capable of carrying and/or leaching a variety of elements. The interaction of hydrothermal fluids with wall rocks and/or the hydrosphere may result in the formation of a wide range of mineral deposit types. Two papers in this issue deal with the chemical signature of minerals that can be used to delineate the hydrothermal processes in different deposits. Cave et al. [8] investigate the geochemistry of sulphide minerals in the sediment-hosted Hilton Zn-Pb deposit, located at the late Paleoproterozoic to early Mesoproterozoic McArthur Basin of Northern Australia. Six texturally distinct generations of sphalerite were identified through field and microscopy observation, and trace element compositions, particularly Ga, Ge, and In, utilizing in situ LA-ICPMS analysis. Another mineral–chemical study was undertaken by Duan et al. [9] on garnet from the Zhuxiling W (Mo) skarn deposit in Southern China examining REE, Sn and W concentrations. The authors demonstrate that the trace element concentrations reflect varying physio-chemical conditions of the hydrothermal fluids at the time of garnet precipitation.

Carbonatite-related REE deposits comprise the main source of LREE and Nb resources in the world. Despite their economic importance, there is still much to learn about them. Liu et al. [10] use thermodynamic modelling to address the physio-chemical conditions of the ore-forming fluids and REE precipitation mechanisms at the world-class Bayan Obo REE-Nb-Fe deposit, China. The Bayan Obo deposit is a carbonatite-related type in which hydrothermal processes have played a significant role in enriching REE. This study presents a detailed paragenetic sequence of REE minerals and utilizes a variety of evidence to constrain the physical conditions (temperature and pressure) of hydrothermal fluids at the time of mineral precipitation. They conclude that multiple pulses of high temperature hydrothermal fluids were required to form the observed mineral assemblages. Focusing on a global context, the review article by Wang et al. [11] summarises the tectonic background, petrology and geochemical characteristics of carbonatite-related REE deposits, emphasising that hydrothermal processes are crucial in upgrading the REE mineralisation.

The nature and composition of the ore-forming fluids, as well as the age of mineralisation of the quartz vein-type Baiyinhan tungsten deposit, NE China, are elucidated by Wang et al. [12]. The authors employed a number of analytical techniques to analyse hydrothermal quartz, molybdenite, and wolframite. Their results indicate the importance of both hydrothermal and meteoric fluids in ore genesis.

Deep-sea pelagic sediments may be extremely enriched in REEs and Y (REY) and could represent a prominent future source of these critical metals. Zhou et al. [13] use detailed investigations of the sediments to demonstrate the importance of early hydrothermal fluids to metal enrichment.

The tectonic evolution of sedimentary basins from extension to compression has been regarded as the key factor in the formation of sandstone-type uranium deposits. Such a genetic relationship is addressed in the work of Yang et al. [14] in this Special Issue who investigate the sandstone-hosted Qianjiadian U deposit, Songliao Basin, NE China, as a case study. They suggest that weathering and alteration of mafic intrusions in the basin may have been critical in forming a reducing barrier capable of precipitating uranium.

Lemière et al. [15] evaluate the potential of soil analysis for Sb by pXRF (portable X-ray fluorescence) as a low-footprint exploration method based on a case study of the vein-type Les Brouzils ore deposit of western France. The team performed shallow-soil sampling along profiles across known veins in what is now as an agricultural area to capture endogenic geochemical anomalies. The study demonstrates that the pXRF instrument can effectively locate the Sb veins and precisely delineate Sb anomalies especially when using its multi-element capabilities.

The green mining concept promotes innovative technologies to simultaneously improve the mining sector's economic and environmental performance [16]. Responses could include extraction of additional critical elements from current production or from historic mine wastes. This Special Issue presents two such studies. One describes the development of technological innovations that allow for the extraction of critical and precious metals associated with low-grade porphyry ores (e.g., Co, Te, Au, Ag) [17]. Velásquez et al. [17] use the world-class porphyry Los Sulfatos deposit, located in the Chilean Central Andes, as a case study, and performed in situ and high-resolution geochemical characterisation of the metal-bearing sulphides by LA-ICPMS, the knowledge of which is informative for selective metal treatment. The second study demonstrates extraction of copper from lead and silver mine waste dumps by using a simple, robust, and versatile leaching process [18]. Terrones-Saeta et al. [18] investigated the mine wastes from the mining district of Linares, Spain, and carried out leaching experiments in low molarity of sulphuric acid solutions, at ambient temperature and atmospheric pressure. The results show that the Cu recovery rate can reach up to approximately 80%, together with almost total recovery rate of Zn.

On the basis of this Special Issue's findings, we outline the following major challenges and opportunities facing academic research and industrial mineral exploration:

1. An improved understanding of the geochemistry of critical metals is required. Little is known about what the levels of abundance of critical metals are in different geological reservoirs; what ligands are key, and complexes they form in different hydrothermal fluids; and what physical conditions are required for efficient transport of critical metals in hydrothermal fluids? What changes in the hydrothermal fluids can cause the deposition of critical metals? How do critical metals partition into different minerals from the hydrothermal fluids? Evolutionary and revolutionary analytical techniques will need to be developed to be able to precisely determine the critical metal contents in minerals and rocks, which are usually present at extremely low concentrations, and most importantly, to be able to differentiate the primary sources of critical metals. Experimental studies of the chemistry of critical metal transport and deposition by ore-forming fluids are also essential.
2. Refined genetic models of critical mineral deposits need to be developed. Most of the existing genetic models of ore deposits were developed for major industrial metals, while the processes responsible for the critical metal mineralisation have been neglected. Studies of critical metals at the element, the mineral and the deposit scales need to be integrated and combined with other datasets to develop refined genetic models. The results need to be interpreted in ways that are applicable for both the academic research and industry exploration.
3. Knowledge of fertility indicators of critical metal deposits is still absent. Why are some rocks more fertile than others to source/host critical metal mineralisation? How can we develop a vector tool by using mineral geochemistry for the exploration of critical metal resources? In this regard, applying machine learning to large geochemistry datasets of both barren and fertile minerals and rocks will be able to provide a new perspective on rock fertility and mineral discriminators in critical mineral exploration.
4. The ore grades are expected to decline for mineral deposits over time, which implies that significantly more materials will have to be mined and processed to produce the same amount of metal. Thus, a significant challenge for the contemporary mining industry is to move from traditional major industrial metal mining to a highly efficient multi-metallic operation. Innovation in metal-selective metallurgical processing is necessary for cost-effective simultaneous extraction of both major industrial metals and critical metals. This calls for highly interdisciplinary collaboration among geochemists, mineralogists, metallurgists, and engineers.
5. The recovery of critical metals from mining waste and unconventional sources represents a big potential for sustaining the supply of critical metals. However, the challenge lies in the fact that the traditional techniques to recover the critical metals from mine wastes are costly and energy and time consuming. Novel and advanced techniques are required to enable more sustainable and manageable extracting process of critical metals.

This Special Issue was completed while more than half of the world was under lockdown due to the COVID-19 crisis. Critical minerals and metals are widely used in the pharmaceutical industry (e.g., phosphates, platinum), telecommunications industry (e.g., indium gallium, and caesium), and the robotics industry (e.g., rare earth metals, cobalt, and gallium) all of which have been critical in combating this crisis. The editors of this issue hope that it will help shape new thinking to address contemporary societal challenges such as the pandemic and the larger climate and biodiversity challenges that lie ahead.

Funding: This research received no external funding.

Acknowledgments: The Guest Editors would like to sincerely thank all authors, reviewers, the Managing Editor Francis Wu and the editorial staff of *Minerals* for their timely efforts to successfully complete this Special Issue. We also thank Paul Sylvester, the Editor-in-Chief, for his support. Lingli Zhou is also grateful to Murray Hitzman for his constructive comments and suggestions during the drafting of this editorial.

Conflicts of Interest: The authors declare no conflict of interest.

References

1. National Research Council Minerals (NRC). *Critical Minerals, and the US. Economy*; National Academies Press: Washington, DC, USA, 2008.
2. Cugurullo, F. *Frankenstein Urbanism: Eco, Smart and Autonomous Cities, Artificial Intelligence and the End of the City*; Routledge: London, UK, 2021.
3. DoI, U.S. Final List of Critical Minerals 2018. *Fed. Regist.* **2018**, *83*, 23295–23296.
4. European Commission (EC). *Critical Raw Materials Resilience: Charting a Path towards Greater Security and Sustainability*; COM 474 Final; European Commission: Brussels, Belgium, 2020.
5. Zhai, M.G.; Wu, F.Y.; Hu, R.Z.; Jiang, S.Y.; Li, W.C.; Wang, R.C.; Wang, D.H.; Qi, T.; Qin, K.Z.; Wen, H.J. Critical Mineral Resources of China: Current Views and Challenges. *China Sci. Found.* **2019**, *2*, 106–111. (In Chinese)
6. Andersson, P. Chinese assessments of "critical" and "strategic" raw materials: Concepts, categories, policies, and implications. *Extr. Ind. Soc.* **2020**, *7*, 127–137. [CrossRef]
7. Pirajno, F. *Hydrothermal Mineral Deposits. Principles and Fundamental Concepts for the Exploration Geologist*; Springer: Berlin/Heidelberg, Germany, 1992; Volume XVIII, p. 709.
8. Cave, B.; Lilly, R.; Hong, W. The Effect of co-Crystallising Sulphides and Precipitation Mechanisms on Sphalerite Geochemistry: A Case Study from the Hilton Zn-Pb (Ag) Deposit, Australia. *Minerals* **2020**, *10*, 797. [CrossRef]
9. Duan, X.X.; Ju, Y.F.; Chen, B.; Wang, Z.Q. Garnet Geochemistry of Reduced Skarn System: Implications for Fluid Evolution and Skarn Formation of the Zhuxiling W (Mo) Deposit, China. *Minerals* **2020**, *10*, 1024. [CrossRef]
10. Liu, S.; Ding, L.; Fan, H.R. Thermodynamic Constraints on REE Mineral Paragenesis in the Bayan Obo REE-Nb-Fe Deposit, China. *Minerals* **2020**, *10*, 495. [CrossRef]
11. Wang, Z.Y.; Fan, H.R.; Zhou, L.; Yang, K.F.; She, H.D. Carbonatite-Related REE Deposits: An Overview. *Minerals* **2020**, *10*, 965. [CrossRef]
12. Wang, R.; Zeng, Q.; Zhang, Z.; Guo, Y.; Lu, J. Fluid Evolution, HO Isotope and Re-Os Age of Molybdenite from the Baiyinhan Tungsten Deposit in the Eastern Central Asian Orogenic Belt, NE China, and Its Geological Significance. *Minerals* **2020**, *10*, 664. [CrossRef]
13. Zhou, T.; Shi, X.; Huang, M.; Yu, M.; Bi, D.; Ren, X.; Yang, G.; Zhu, A. The Influence of Hydrothermal Fluids on the REY-Rich Deep-Sea Sediments in the Yupanqui Basin, Eastern South Pacific Ocean: Constraints from Bulk Sediment Geochemistry and Mineralogical Characteristics. *Minerals* **2020**, *10*, 1141. [CrossRef]
14. Yang, D.G.; Wu, J.H.; Nie, F.J.; Bonnetti, C.; Xia, F.; Yan, Z.B.; Cai, J.F.; Wang, C.D.; Wang, H.T. Petrogenetic Constraints of Early Cenozoic Mafic Rocks in the Southwest Songliao Basin, NE China: Implications for the Genesis of Sandstone-Hosted Qianjiadian Uranium Deposits. *Minerals* **2020**, *10*, 1014. [CrossRef]
15. Lemière, B.; Melleton, J.; Auger, P.; Derycke, V.; Gloaguen, E.; Bouat, L.; Mikšová, D.; Filzmoser, P.; Middleton, M. pXRF Measurements on Soil Samples for the Exploration of an Antimony Deposit: Example from the Vendean Antimony District (France). *Minerals* **2020**, *10*, 724. [CrossRef]
16. Kirkey, J. Eco Friendly Mining Trends for 2014. Available online: https://www.mining-technology.com/features/featureenvironment-friendly-mining-trends-for-2014-4168903/ (accessed on 29 January 2014).
17. Velásquez, G.; Estay, H.; Vela, I.; Salvi, S.; Pablo, M. Metal-Selective Processing from the Los Sulfatos Porphyry-Type Deposit in Chile: Co, Au, and Re Recovery Workflows Based on Advanced Geochemical Characterization. *Minerals* **2020**, *10*, 531. [CrossRef]
18. Terrones-Saeta, J.M.; Suárez-Macías, J.; Río, F.J.L.D.; Corpas-Iglesias, F.A. Study of Copper Leaching from Mining Waste in Acidic Media, at Ambient Temperature and Atmospheric Pressure. *Minerals* **2020**, *10*, 873. [CrossRef]

Review

Carbonatite-Related REE Deposits: An Overview

Zhen-Yu Wang [1,2], Hong-Rui Fan [1,2,3,*], Lingli Zhou [4], Kui-Feng Yang [1,2,3] and Hai-Dong She [1,2]

[1] Key Laboratory of Mineral Resources, Institute of Geology and Geophysics, Chinese Academy of Sciences, Beijing 100029, China; 13264722955@163.com (Z.-Y.W.); yangkuifeng@mail.iggcas.ac.cn (K.-F.Y.); shehaidong2013@163.com (H.-D.S.)

[2] College of Earth and Planetary Sciences, University of Chinese Academy of Sciences, Beijing 100049, China

[3] Innovation Academy for Earth Science, Chinese Academy of Sciences, Beijing 100029, China

[4] iCRAG and School of Earth Sciences, University College Dublin, Belfield, Dublin 4, Ireland; lingli.zhou@ucd.ie

* Correspondence: fanhr@mail.iggcas.ac.cn

Received: 24 September 2020; Accepted: 26 October 2020; Published: 28 October 2020

Abstract: The rare earth elements (REEs) have unique and diverse properties that make them function as an "industrial vitamin" and thus, many countries consider them as strategically important resources. China, responsible for more than 60% of the world's REE production, is one of the REE-rich countries in the world. Most REE (especially light rare earth elements (LREE)) deposits are closely related to carbonatite in China. Such a type of deposit may also contain appreciable amounts of industrially critical metals, such as Nb, Th and Sc. According to the genesis, the carbonatite-related REE deposits can be divided into three types: primary magmatic type, hydrothermal type and carbonatite weathering-crust type. This paper provides an overview of the carbonatite-related endogenetic REE deposits, i.e., primary magmatic type and hydrothermal type. The carbonatite-related endogenetic REE deposits are mainly distributed in continental margin depression or rift belts, e.g., Bayan Obo REE-Nb-Fe deposit, and orogenic belts on the margin of craton such as the Miaoya Nb-REE deposit. The genesis of carbonatite-related endogenetic REE deposits is still debated. It is generally believed that the carbonatite magma is originated from the low-degree partial melting of the mantle. During the evolution process, the carbonatite rocks or dykes rich in REE were formed through the immiscibility of carbonate-silicate magma and fractional crystallization of carbonate minerals from carbonatite magma. The ore-forming elements are mainly sourced from primitive mantle, with possible contribution of crustal materials that carry a large amount of REE. In the magmatic-hydrothermal system, REEs migrate in the form of complexes, and precipitate corresponding to changes of temperature, pressure, pH and composition of the fluids. A simple magmatic evolution process cannot ensure massive enrichment of REE to economic values. Fractional crystallization of carbonate minerals and immiscibility of melts and hydrothermal fluids in the hydrothermal evolution stage play an important role in upgrading the REE mineralization. Future work of experimental petrology will be fundamental to understand the partitioning behaviors of REE in magmatic-hydrothermal system through simulation of the metallogenic geological environment. Applying "comparative metallogeny" methods to investigate both REE fertile and barren carbonatites will enhance the understanding of factors controlling the fertility.

Keywords: ore genesis; fluid evolution; REE enrichment; carbonatite-related REE deposit

---

1. Introduction

Rare earth elements (REEs) are a group of 17 chemically similar metallic elements (scandium, yttrium and lanthanide series in the periodic table IIIB). Scandium and yttrium are included in the REE group considering their comparable nature to the lanthanide elements and their common occurrence in a deposit [1,2]. The REE group is usually divided into two subgroups: light rare earth elements (LREEs, La-Eu) and heavy rare earth elements (HREEs, Gd-Lu and Y). Scandium is not included in the two subgroups because of its much smaller ion radius [3,4]. Rare earth elements are not at all rare in the Earth's crust (especially LREEs), but are rather dispersed. The crustal abundance of REE is 0.017%, and the abundance of Ce, La and Nd is higher than that of W, Sn, Mo, Pb and Co. In comparison, heavy rare earth elements such as Tb and Tm are two to five times less abundant than Mo in the continental crust [5]. Under the upper mantle conditions, the trend of REE distribution into melts decreases with the decrease in ion radius from La to Lu (i.e., HREEs are generally more compatible with mantle peridotites). REEs are vital to modern technologies and society and are among the most critical of all raw materials. The demand for REE is constantly growing, driven particularly by their application in a range of modern and green technologies, including electric and conventional vehicles, communication technologies, and production and storage of renewable energy [4,6]. Supply of REE resources is largely restricted to China, along with a minor contribution from Brazil, Australia, the United States, Canada, India [7,8]. Although there is a substantial development on recycling of REE from end-products [9], the majority of demand still depends on the production from natural source of REE. The major REE deposits that are economically exploitable are restricted to a few localities in the world (Table 1), while the economic potential of a REE deposit is strongly influenced by the mineralization processes and its mineralogy [8]. The most commercially important REE deposits are associated with magmatic processes and are found in, or related to, alkaline igneous rocks and carbonatites. The supergene REE deposits also represent a large REE resource such as placer-type and ion adsorption-type (IAR) deposits. The IAR deposits distributing over a wide area of southern China are usually rich in middle and heavy REE, which meet almost all the needs of HREE such as Gd, Tb, Dy in the development of emerging industries around the world and also having extremely high economic value [10]. The reserves, production scale, and export volume of rare earth resources in China rank first in the world, which has become the center of many economic and political controversies. Rare earth resources are widely distributed in China and relatively concentrated in the individual deposits. According to statistics, rare earth resources have been found in more than 2/3 of the provinces (regions) in China, among which the Bayan Obo in Inner Mongolia, Mianning in Sichuan province, the Gannan in Jiangxi province and northern Guangdong province are the major areas [11].

Table 1. Major economically exploitable rare earth element (REE) deposits in the world.

Country	Deposit	Deposit-Type
China	Bayan Obo (REE-Nb-Fe)	Igneous carbonatite
China	Maoniuping (REE)	Carbonatite
China	HREE-enriched deposits in southern China	Weathered crust elution
Brazil	Araxá Catalão (REE)	Weathered Carbonatite
Brazil	Mrro do Ferro (Th-REE)	Carbonatite
Australia	Mount Weld (REE)	Weathered Carbonatite
United States	Mountain Pass (REE)	Carbonatite
Russia	Tomtor (REE)	Weathered Carbonatite
Russia	Lovozero (REE-Nb)	Alkaline igneous rock
India	Amba Dongar (REE)	Carbonatite
Vietnam	Mau Sai (REE)	Carbonatite
Burundi	Gakara (REE)	Carbonatite
Malawi	Kangankunde (REE)	Carbonatite
South Africa	Palabora (REE)	Carbonatite
South Africa	Steenkampskraal (REE-Th-Cu)	Alkaline igneous rock
Turkey	Aksu Diamas (REE)	Placer
Greenland	Tanbreez (REE)	Alkaline igneous rock
Sweden	Norra Kärr (REE)	Alkaline igneous rock

Worldwide, carbonatites are major sources of both REE and niobium, and are characterized by significant enrichment in the LREEs (La-Gd) over the HREEs (Tb-Lu). The Bayan Obo REE-niobium-iron deposit in Inner Mongolia, China, the world's largest REE deposit, is an important example [12,13]. Whereas, the alkaline oversaturated rocks associated with REE deposits represent one of the most economically important resources of heavy REE and yttrium. Alkaline oversaturated rocks form from magma so enriched in alkalis that they crystallize sodium- and potassium-bearing minerals (e.g., feldspathoids, alkali amphiboles). The source magma is usually mantle-derived and highly differentiated. The world-class Tanbreez REE deposit, which is hosted in the Ilímaussaq complex of southern Greenland, is a typical example [14,15].

Carbonatite usually refers to igneous rocks derived from the mantle with carbonate mineral volume > 50% and SiO_2 < 20wt.% [16,17]. Based on wt.% ratios of the major elements (MgO, CaO, FeO, MnO, Fe_2O_3, etc.), carbonatites are subdivided into magnesiocarbonatites, ferrocarbonatites and calciocarbonatites (Table 2) by the International Union of Geological Sciences (IUGS) [12]. Two unique types of carbonatite are identified to occur locally: silicocarbonatite (carbonatite with more than 20% SiO_2) existing in the Afrikanda of the Kola Peninsula (Russia), and natrocarbonatite in the Ol Doinyo Lengai volcano (Tanzania) mainly consisting of Na-K-Ca carbonates, such as nyerereite [$(Na, K)_2Ca(CO_3)_2$] and gregoryite [$(Na, K, Ca_x)_{2-x}(CO_3)$] [12,18,19]. Recent works on the different types of carbonatite in the Bayan Obo REE-Nb-Fe deposit reveal that the evolutionary sequence of carbonatitic magma is from ferroan through magnesian to calcic in composition, accompanied with an increase in LREE enrichment [20].

Table 2. Classification of carbonatites (adapted from Simandl and Paradis [12] and Gou et al. [17]).

Classification	Major Elements
Calciocarbonatite	$CaO/(CaO + MgO + FeO + Fe_2O_3 + MnO) > 0.8$
Magnesiocarbonatite	$CaO/(CaO + MgO + FeO + Fe_2O_3 + MnO) < 0.8$ and $MgO > (FeO + Fe_2O_3 + MnO)$
Ferrocarbonatite	$CaO/(CaO + MgO + FeO + Fe_2O_3 + MnO) < 0.8$ and $MgO<(FeO + Fe_2O_3 + MnO)$
Natrocarbonatite	High content of Na, K, Ca
Silicocarbonatite	SiO_2 > 20% of the whole rock

The types of REE deposits are complex and diverse, and the ones related to carbonatite–alkaline complexes are the most typical and abundant, including the Bayan Obo REE-Nb-Fe deposit in Inner Mongolia, Maoniuping REE deposit in Sichuan, Mountain Pass REE deposit in the USA and Araxá Nb-P-REE deposit in Brazil [21]. Carbonatite-related REE deposits comprising the main source of LREE and Nb resources in the world refer to REE deposits that are closely related to a set of carbonatite or alkaline rocks (usually coexisting) in genesis and space. Although carbonatite-related REE deposits are particularly crucial, there are still many debates and controversies about their geological background and genetic characteristics, and the detailed metallogenic process is not clear. Therefore, it is necessary to summarize the tectonic background, petrology and geochemical characteristics of carbonatite-related REE deposits, so as to better understand the genesis of the deposits and provide guidance for mineral exploration.

This article first summarizes the geological characteristics of carbonatite-related REE deposits, then summarizes and briefly introduces the typical carbonatite-related REE deposits in China. In combination with previous studies, we present a full discussion of the genetic characteristics and the mechanism of massive REE enrichment of carbonatite-related REE deposits. Ultimately, the existing problems are discussed and the future research directions are proposed to address the problems.

2. Types and Basic Geological Characteristics of Carbonatite-Related REE Deposits

2.1. Types of Carbonatite-Related REE Deposits

Carbonatite usually coexists with alkaline silicate igneous rocks to form carbonatite–alkaline (ring) complexes, but some appear in the form of isolated batholiths, dykes, intrusions, lava flows and pyroclastic overburdens [22]. Three types of carbonatite-related REE deposits are recognized according to the varied characteristics of mineralization, namely primary magmatic type, hydrothermal type and carbonatite weathering-crust type [23]. The former two types, collectively called carbonatite-related endogenetic REE deposits, are mainly formed in relation to carbonatite magma and its derived hydrothermal fluids, forming REE mineral accumulation and mineralization. The mineralization of primary magmatic type REE deposits mainly occur in the magmatic stage. REE minerals such as bastnäsite, monazite, allanite, xenotime, parasite occur in carbonatite or magmatic phosphorite, and the entire rock body is mineralized. REE minerals of hydrothermal type deposits are formed by hydrothermal fluids that are evolved from magmas, which are usually associated with calcite, barite, fluorite, quartz and other minerals to form major or net veins. The mineralized veins frequently interpenetrate in the contact zone and surrounding rock of carbonatite complexes. The hydrothermal REE minerals can also occur as crack fillings superimposed on the minerals that are formed during the magmatic stage. This type of deposit is usually large in scale, and the REE minerals are relatively simple, mainly bastnäsite [24]. Carbonatite weathering-crust type is secondary carbonatite REE deposits, which comprise laterite crust after long-term weathering and fluid leaching under particular environmental conditions that are hot, humid and suitable for weathering crust preservation. REEs in the weathered crust section are remobilized, enriched, and adsorbed on the surface of clay minerals such as kaolinite in the form of ions. This type of REE deposit is usually HREE-enriched due to the higher mobility of LREE during weathering [23,25].

2.2. Spatial and Temporal Distribution

Worldwide there are more than 500 proven carbonatites, only a small proportion of which is fertile with REEs (Figure 1) [26,27]. The ages of carbonatites range from the Archean to Cenozoic. The mineralization of carbonatite-related REE deposits occurs over a long period of time but is mostly concentrated in the Mesozoic. Carbonatite-related REE deposits in China are formed during the Proterozoic to Mesozoic. Among those, two major mineralization events occurred in the Mesoproterozoic and Mesozoic (Figure 2), which are represented by the Bayan Obo REE deposit in Inner Mongolia (~1.32 Ga, [20]) and the Miaoya REE deposit in Hubei province (232 Ma, [28]), respectively.

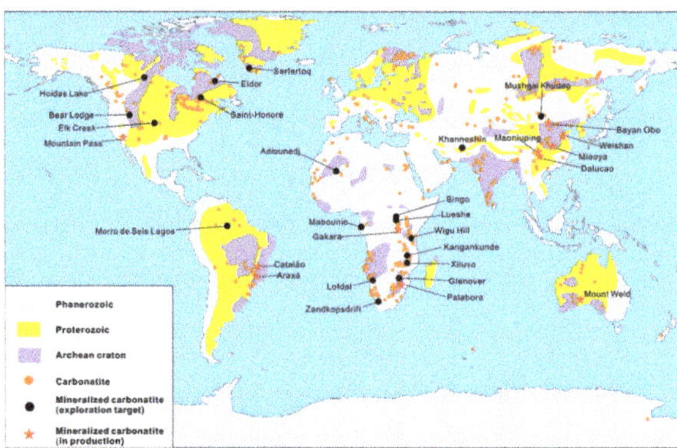

Figure 1. Distribution of carbonatites and carbonatite-related REE deposits, including deposits in production and exploration targets discussed in this paper (modified from Woolley and Kjarsgaard [26], Liu and Hou [27] and Verplanck et al. [29]).

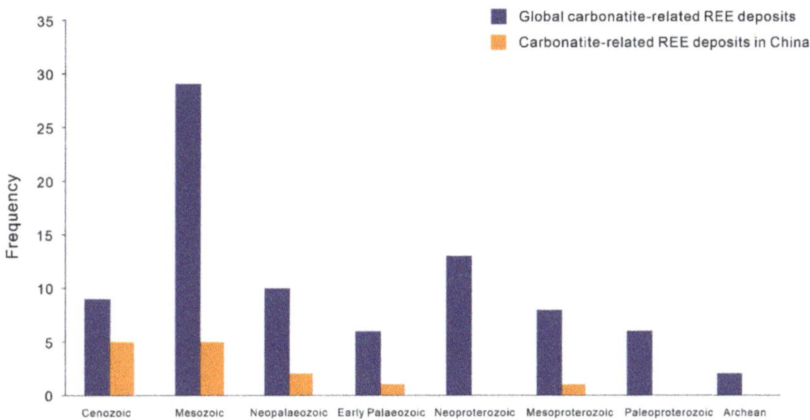

Figure 2. Age distribution histograms of major carbonatite-related REE deposits in the world and China (based on data from Woolley and Kjarsgaard [26], with minor updates).

Globally, carbonatite-related REE deposits mostly occur in continental marginal depression belts and rift belts along craton margins controlled by large-scale deep lithosphere faults (e.g., the Mountain Pass REE deposit in the USA). Subordinately, REE mineralization also occurs at stable geological structural units (platform or paraplatform) such as the Araxá Nb-P-REE deposit in Brazil and the Sarfartoq REE deposit in Greenland (Figure 1) [17,29,30]. The distribution of carbonatite-related REE deposits in China is rather restricted to continental margin rifts or orogenic belts at the margin of cratons [21]. For example, the Bayan Obo and Weishan REE deposits are located on the northern and southeastern margins of the North China Craton, respectively; the Maoniuping REE deposit is located in the rift belt of the western margin of the Yangtze Craton; the Miaoya and Shaxiongdong REE deposits in Hubei Province are located in the Qinling orogenic belt [31–35].

2.3. Ore and Orebody Characteristics

The vast majority of carbonatite-related REE deposits in China belong to endogenetic deposits including both the primary magmatic type and hydrothermal type REE deposits. Large and super large-scale REE deposits often have both magmatic and hydrothermal features, reflecting an interplay of both processes in contribution to the ore genesis [23]. The Bayan Obo REE-Nb-Fe deposit is a great example [36]. Previous studies suggest that two major mineralization events occurred at 1400 Ma and 440 Ma, respectively, at the giant Bayan Obo REE-Nb-Fe deposit, which correspond to an early emplacement of large-scale carbonatitic magma (1.4-1.3 Ga) and a late intense hydrothermal activity (440 Ma). The early carbonatitic magma is fertile in REEs, and the REEs are captured in primary REE-minerals such as bastnäsite and monazite. The late hydrothermal fluids released the REEs from its original host minerals, and upgraded the reserves, forming REE minerals such as secondary bastnäsite, parasite and cerianite, which are usually coarse-grained and appear in veins [37–40].

Carbonatite-related primary magmatic type REE deposits are mainly hosted in a set of ring-shaped complexes composed of carbonatites and alkaline rocks. The entire carbonatite body is commonly mineralized, and mineralization tends to extend into the alkaline complexes appearing as lenticular or irregular lenses [23]. The occurrence of ore bodies is usually controlled by regional deep faults existing in the extensional lithosphere. In this context, the regional deep faults act as conduits for the migration of deeply seated and fertile magma [17]. For example, the Mountain Pass REE deposit, United States, is controlled by a large deep fault that crosses the west coast of North America. The Miaoya REE deposit in China occurs in an extensional tectonic setting derived from fault zones or orogenic belts [41–43]. The ores usually are fine-grained and massive, disseminated, or striped in texture. The REE minerals are generally formed from crystallization in the late stage of magma evolution, mainly including bastnäsite, monazite, xenotime, parisite, allanite, and frequently in association with Nb-containing minerals such as pyrochlore, aeschynite, niobite. Other existing minerals include magnetite, hematite, apatite, barite, calcite, dolomite, mica and zircon [23,25].

The orebodies of hydrothermal type REE deposits are usually in the form of veins or stockwork that occur near the contact zone between orebodies and host rocks or in the host rocks. The mineralization is also controlled by regional deep faults. For example, the Maoniuping REE deposit in Sichuan Province and Weishan REE deposit in Shandong Province are typical hydrothermal type REE mineralization. They are located in the Panxi Rift Valley and Tanlu Fault Zone, respectively [44,45]. Compared with REE deposits of primary magmatic type, the REE minerals in hydrothermal type deposits are mineralogically simpler, typically containing bastnäsite and parasite. The REE minerals appear either as filling in veins or disseminated or as overgrowth on the early magmatic minerals in carbonatite [23]. The gangue minerals mainly comprise of fluorite, quartz, barite, calcite, apatite and aegirine-augite. The hydrothermal fluids exsolved from the late stage magma tend to overprint early magmatic mineralization, a process that is crucial for the massive enrichment of REEs in carbonatite-related REE deposits.

2.4. Geochemical Characteristics

The chemical composition of carbonatite varies greatly. Typical carbonatites are SiO_2 unsaturated, rich in CO_2 and CaO, and some are rich in Mg, Fe and alkalis [17,46]. Trace elements in carbonatite mainly include LREE, Nb, Ta, Th, Zr, Hf, Sr, Ba, F, and P. Carbonatite is one of the rocks with the highest REE content and the highest LREE differentiation on the Earth. It has become an essential prospecting indicator for large-scale, high-grade and LREE-enriched REE deposits. As the most characteristic and significant elements in carbonatite, LREEs are enriched to a higher level than other trace elements as main ore-forming elements in some carbonatite-related REE deposits, e.g., Bayan Obo REE-Nb-Fe deposit [17]. Niobium and Ta often appear as associated elements in the deposits and are enriched in the late petrogenic stage. The precipitation of Nb and Ta often forms pyrochlore, aeschynite, etc. Alkaline-earth metals such as Sr, Ba also show enrichment and are incorporated in carbonates. Substantial accumulation of Fe can occur to a level that is economically exploitable (e.g., Bayan Obo REE-Nb-Fe deposit) [17,40,47]. The contents of F and Cl in carbonatite are higher compared to those of other magmatic rocks, which is essential for the efficient migration and precipitation of REEs in the fluids. Occasionally, F is highly enriched in carbonatite-forming carbonatite-type fluorite deposits [48–50]. Phosphorus is also one of the most characteristic trace element in carbonatite, and often occurs in the form of apatite.

2.5. Alteration Characteristics

In the process of ascending and emplacement of carbonatitic magma, the temperature and pressure of fluids decrease, along with release of volatiles (F, Cl, P and S) and elements such as REE, Na, K, and Fe. The fluids reach the upper crust and metasomatize surrounding rocks, forming a lithological belt of fenite with asymmetric zonal distribution in the orebody, which is called fenitization (Figure 3). Fenitization is an alkali metasomatism, which is caused by alkali-rich fluids exsolved from igneous carbonatitic or alkaline magma. It is a typical and unique alteration phenomenon of carbonatite and alkaline igneous rocks [47,51–53]. McKie (1966) defined fenite as the rock produced by the metasomatism of igneous carbonatite with surrounding rocks, mainly composed of alkaline feldspar and alkaline mafic minerals [54]. During the metasomatism, a large amount of Si and substantial Al are released into the solution, while K, Na, Ca, and Fe are retained in the fenite forming new minerals. The degree of metasomatism varies according to the distance to the causative rocks, resulting in halo-like patterns of fenitization. The main minerals of fenite include Na- and K-amphiboles, Na-pyroxene, aegirine-augite, K-feldspar, albite, perthite, nepheline and pale brown mica. The accessory minerals include apatite and REE minerals such as titanite, pyrochlore, monazite and bastnäsite [12].

Figure 3. Fenitization in the wallrock quartz sandstone from the Bayan Obo REE-Nb-Fe deposit.

According to the Na_2O/K_2O ratio of whole rock, fenitization can be divided into three main types: sodic, potassic, and sodic-potassic. Potassic fenite is dominated by K-rich feldspars (orthoclase or microcline). In some cases, there are also low-Al phlogopite or biotite, which is usually developed near the upper part of intrusive calcite/dolomite carbonatite. Sodic fenite mainly consists of Na-rich amphibole, sodic pyroxene, alkali feldspar or albite. Sodic-potassic fenitization is an intermediate between the two endmembers [12,55]. In addition, sodic fenite usually occurs in the contact zones between carbonatite bodies and surrounding rocks in the early stage or deep emplacement, while potassic fenite is often found in the contact zones in the late stage or shallow emplacement [56]. Previous studies have suggested that the initial carbonatite contains a considerable amount of alkali, and the hydrothermal fluid rich in Na and K can be derived in the process of evolution and metasomatic reaction with surrounding rocks [46]. This initial carbonatite is named sodic carbonatite, such as Oldoinyo Lengai sodic carbonatite in northern Tanzania [19,57,58]. However, carbonatites exposed in nature are mostly ferroan, magnesian, and calcic carbonatites, rarely alkali in composition, although they spatially closely coexist with alkali-rich fenite. In addition to Na and K, other elements such as Fe, Zr, V, Zn, Rb and Ba are incorporated into the fluids derived from carbonatite magma [59,60]. During the separation of fluid phase, REEs, especially LREEs, prefer to partition into the liquid and vapor phases. The fenitizing fluids derived from carbonatite magma are characterized by strong enrichment of LREEs. REE-bearing minerals such as bastnäsite and monazite crystallize and accumulate from the fenitizing fluids during its auto-metasomatism to carbonatite, which favors for the formation of large REE deposits [20].

3. Typical Carbonatite-Related REE Deposits

3.1. Bayan Obo REE-Nb-Fe Deposit, China

The Bayan Obo deposit is the largest REE deposit in the world and an important Fe and Nb deposit in China (Figure 4a). It is located in the Wulanchabu grassland, Baotou City, Inner Mongolia Autonomous Region [61,62]. The ore-forming process is extremely complex, which is mainly due to the extended and intense hydrothermal alterations and transformation during the late mineralization period [63,64]. In the deposit, REE orebodies are mainly hosted in a set of dolomites in the south of

Kuangou anticline (Figure 4b). The dolomite is divided into fine-grained and coarse-grained lithofacies. Mineralization is absent in the dolomite (also known as limestone) in the northern flank of the Kuangou anticline. The contrast of mineralized and barren dolomite may reflect key factors that control the formation of the large Bayan Obo deposit. In addition, various carbonatite dykes are developed in the mining area, which are divided into three categories: dolomite type (ferroan), calcite-dolomite type (magnesian), and calcite type (calcic) based on their mineralogical composition [65]. The extensive development of carbonatite dykes has long been believed to account for the massive enrichment of REE. In addition to the mineralogical variation, the REE content is also different for the three types of carbonatite dykes. The calcic carbonatite dykes contain the highest REE content relative to the other two types (REE_2O_3 reaches up to 20%) [20,65]. This may suggest that the elemental differentiation process of different types of carbonatite dykes is crucial in controlling the massive enrichment of REE. However, the evolution mechanism of the Bayan Obo carbonatite magma is still unclear, and the magmatic affinity of the ore-bearing carbonatites is also controversial [66,67].

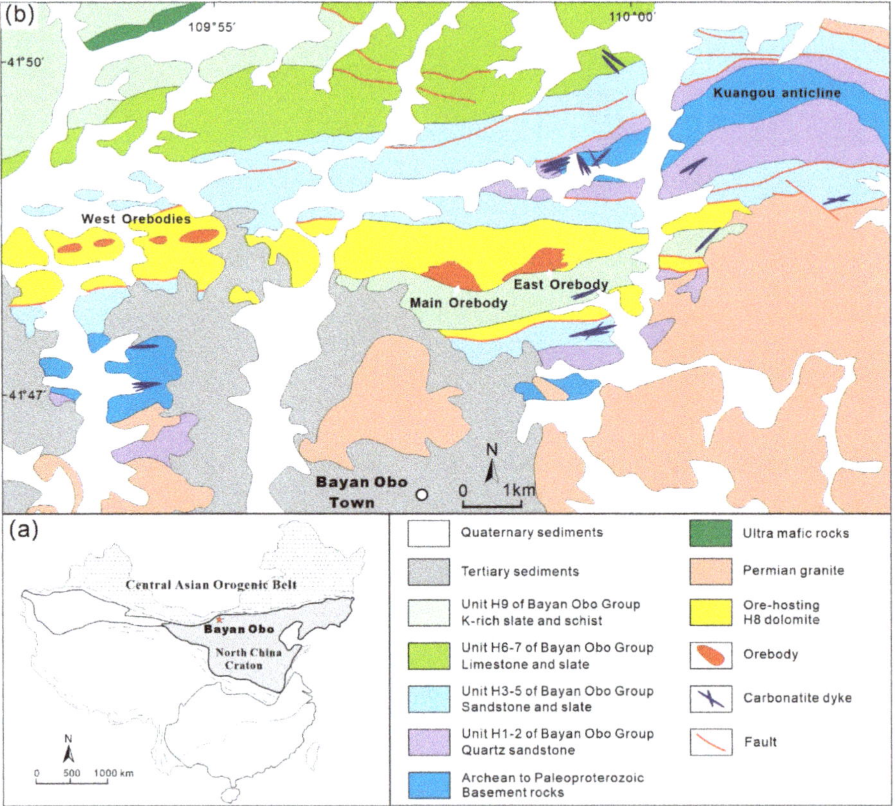

Figure 4. The location of the Bayan Obo deposit in China is marked in the interior plot (**a**). The geological setting of the Bayan Obo REE-Nb-Fe deposit (adapted from Yang et al. [65]) has been plotted in (**b**).

3.2. *Miaoya Nb-REE Deposit, China*

The Miaoya Nb-REE deposit is located in the southern margin of the Qinling Orogen, Central China. It was first discovered by the Northwest Geological Bureau in 1963 during the regional radiological survey. The deposit contains more than 40 REE-Nb orebodies with an estimated reserve of 1.21 Mt REE_2O_3, which is the largest REE deposit in the Qinling orogenic belt [68]. The Miaoya syenite–carbonatite complex hosts the entire mineralization (Figure 5). The Miaoya complex is mainly syenitic in composition and has banded structures featured by being fine-grained in the center, xenomorphic granular in the middle and porphyritic at the edge [68]. The mineralogical composition of syenite outcrops at different locations is relatively homogenous, mainly composed of K-feldspar, microcline, biotite, albite, plagioclase, quartz, and sericite, together with a small amount of zircon, monazite and kaolinite. The carbonatite occurs as stocks intruding into the alkaline syenite, and comprise both calciocarbonatite and ferrrocarbonate in composition. The main constitutional minerals are calcite, ankerite and apatite; the accessory minerals include biotite, alkaline feldspar, magnetite, ilmenite, perovskite, zircon, and Nb-rich rutile. A variety of REE-bearing minerals are formed including bastnäsite, parasite, monazite, allanite and REE-rich apatite [69]. The occurrence of syenite and carbonatite is spatially closely associated. It is speculated that they have a common source and are the products of partial melting of the mantle [41].

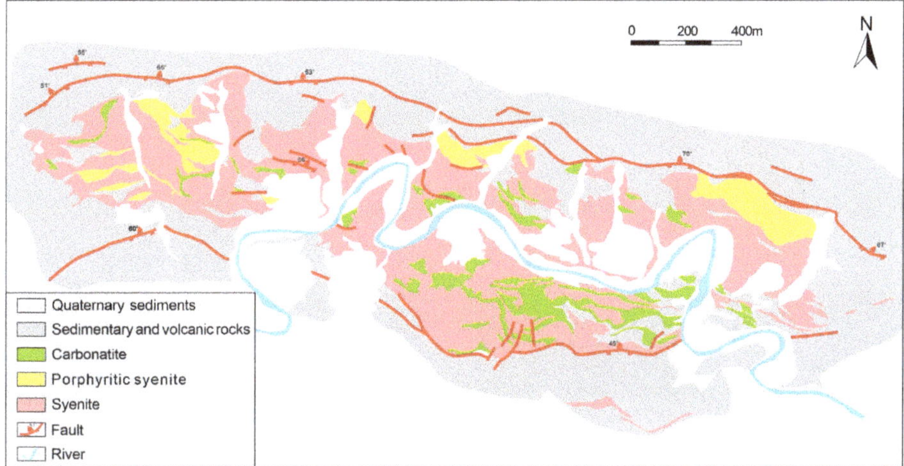

Figure 5. Simplified geological map of the Miaoya syenite–carbonatite complex (adapted from Su et al. [41]).

3.3. Maoniuping REE Deposit, China

The Maoniuping REE deposit is the second largest REE deposit in China. It is located in the east of the Tibetan Plateau and the west of the Yangtze Plate, and controlled by the Panxi rift. The Maoniuping REE deposit is also the largest and most typical REE deposit in the Mianning–Dechang REE metallogenic belt [70]. The orebodies are mainly in the form of veins, veinlets, and net veins interspersed in syenite (Figure 6). Fluoritization, carbonation and REE minerals usually occur in the contact zone between veins and surrounding rocks. The carbonatite intrudes along the center of the expanded part of syenite and is distributed in the contact zones around the orebodies. The REE minerals are mainly bastnäsite, together with minor parisite and cerianite. The gangue minerals mainly include calcite, quartz, barite, fluorite, albite, arfvedsonite, and a small amount of metal oxides and metal sulfides. The deposit is a typical hydrothermal carbonatite-related REE deposit. During the migration of hydrothermal fluids, REEs were transported by complexing with the F^-, Cl^-, $(SO_4)^{2-}$and $(CO_3)^{2-}$ ligands in the fluids. The hydrothermal fluids gradually cool down, together with crystallization and precipitation of fluorite, barite and calcite. The gradual cooling of fluids and the crystallization of fluorite, barite and calcite lead to the breakdown of REE complexes and precipitation of the main REE ore mineral bastnäsite [71].

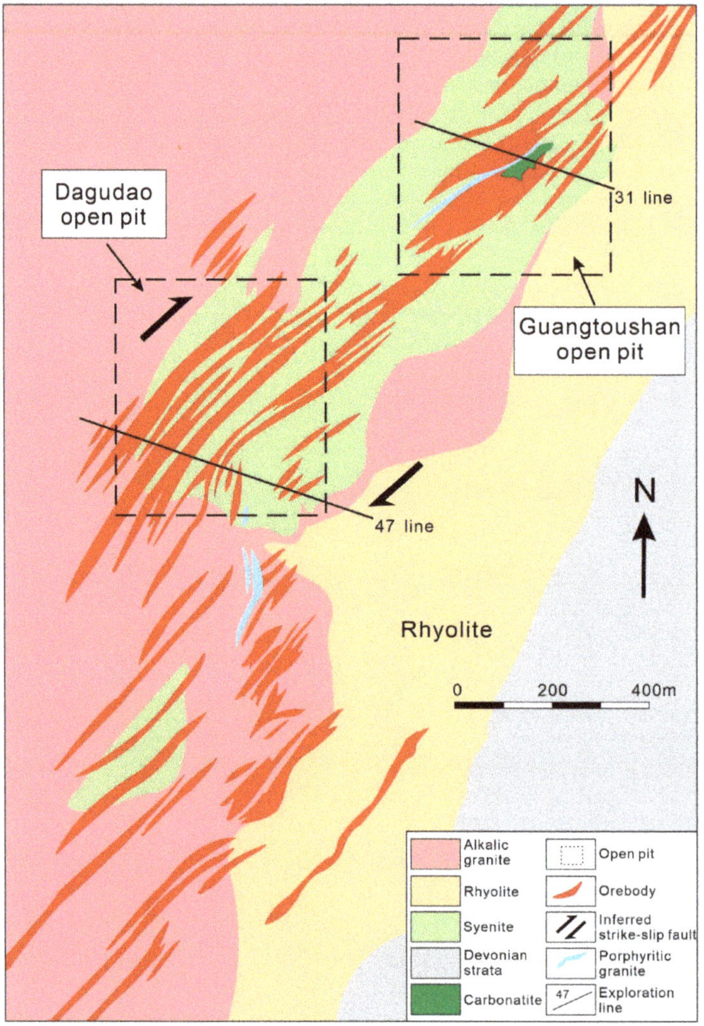

Figure 6. Simplified geological map of the Maoniuping REE deposit, showing the location of the Dagudao and Guangtoushan section (adapted from Liu et al. [71]).

3.4. Weishan REE Deposit, China

The Weishan REE deposit is one of the three major LREE bases in China. It is a typical carbonatite-related hydrothermal REE deposit, located in the Weishan County, southern Shandong Province. Tectonically it is located at the southeast margin of North China Craton and western part of the Tanlu fault zone. The carbonatite orebodies contain veins and veinlets, as well as disseminated mineralization that occur within the faults of rock mass and coexist with ijolite syenite (Figure 7) [44]. The large vein-type orebodies have the highest economic value, and the net veins and disseminated orebodies are distributed at the edge of the large veins [34]. The REE minerals mainly include bastnäsite, parisite, Ce-apatite, and a small amount of allanite, monazite, and titanite. The gangue minerals include fluorite, quartz, barite, calcite, muscovite, limonite, aegirine augite, orthoclase, albite. The deposit also contains substantial sulfides, including pyrite, galena, pyrrhotite, chalcopyrite, molybdenite.

The formation of the Weishan deposit requires two enrichment processes: generation of mineralized carbonatite and subsequent enrichment from magmatic–hydrothermal processes. The carbonate magma is sourced from enriched lithospheric mantle. Liquid immiscibility between the carbonatite melt and CO_2-rich silicate melt resulted in the separation of REE-rich carbonatite melt, from which the REE-rich magmatic–hydrothermal fluids were derived. In the late stage of hydrothermal evolution, a large number of REE minerals were precipitated. The hydrothermal fluids evolved to lower temperatures and became sulfidic, which allows for the silver mineralization to occur [34,44,72].

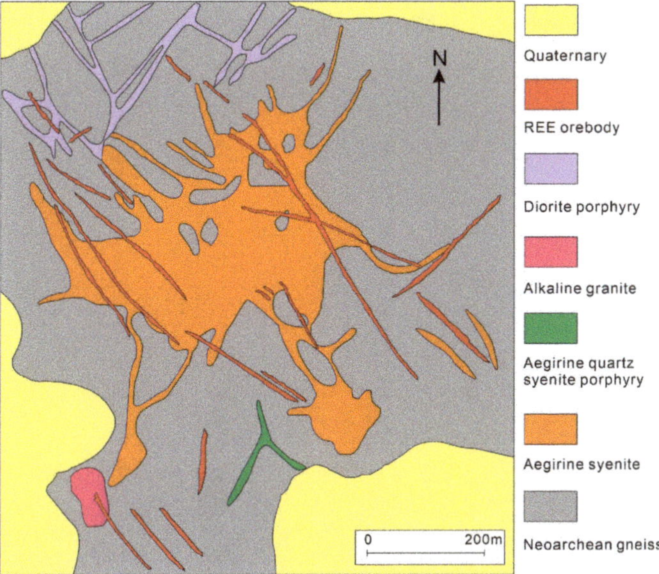

Figure 7. Simplified geological sketch map of the Weishan rare earth element (REE) deposit in Shandong Province (adapted from Jia et al. [44]).

3.5. Mountain Pass REE Deposit, USA

The Mountain Pass REE deposit, located at the southern border of California to Nevada, is the second largest REE deposit in the world following the Bayan Obo [25]. Around 16 million tons of REE oxide reserves have been proven [73,74]. The deposit is hosted in the Sulphide Queen carbonatite (Figure 8). The mineralized rocks include the shoshonite, syenite, granite, and the Sulphide Queen carbonatite. However, only the carbonatite orebodies are economically viable [75]. The orebodies intrude into the Precambrian metamorphic basement in plate or lenticular form, mainly composed of calcite, dolomite, barite, bastnäsite in mineralogical composition, and bastnäsite as the main ore mineral accounts for about 10–15%. The existing data show that cerium (Ce) is the main rare earth element in primary bastnäsite and other REE minerals in the Sulphide Queen carbonatite [73,76]. The Sulphide Queen carbonatite is distinct from other carbonatites in the world. Carbonatites are usually spatially and genetically associated with a set of sodic alkaline rocks in a ring-shaped and concentric circle, and rich in LREE, Nb, Ta, Zr, and P. However, the formation of the Sulphide Queen carbonatite is associated with a suite of ultrapotassic intrusive rocks in plate-like form, being rich in LREE and Ba and depleted in Nb and P. Previous studies suggest that the abnormal nature of Sulphide Queen carbonatite is due to the fact that the carbonatite magma and associated ultrapotassic magma have a common source, or alternatively, the carbonatite magma evolved from the ultrapotassic magma [73,75,77].

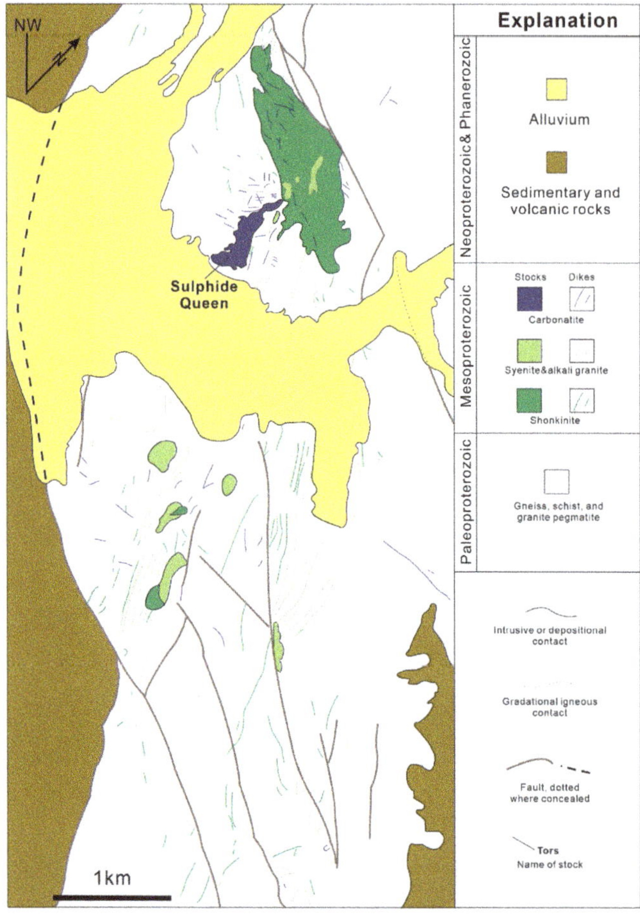

Figure 8. Generalized geological map of the Mountain Pass area (adapted from Poletti et al. [77]).

3.6. *Araxá Nb-P-REE Deposit, Brazil*

Brazil is the second largest REE-enriched country in the world, with about 22 million tons of REE reserves [78]. Among them, the Araxá carbonatite Nb-P-REE deposit is a main supplier of REE resources [78]. The deposit is located near the Araxá City, Minas Gerais State in the southeast of Brazil. It is the largest Nb deposit in the world, and produces a significant amount of P, Ti and REE. The orebody appears as a concentric ring complex, intruding into the Mesoproterozoic quartzite and schist strata (Figure 9). The core of the complex is composed of fine-grained dolomite carbonatite, which gradually transits to mica schists. The main constitutional minerals are dolomite, calcite and ankerite, and the accessory minerals are mica, apatite, alkaline amphibole, magnetite, monazite, pyrochlore, perovskite and zircon [79]. The main REE minerals are monazite (over 70%) and plumbogummite group minerals. In contrast, bastnäsite and cerianite only take up about 1% of the REE minerals. Pyrochlore is the main Nb-bearing mineral [80]. The periphery of the complex underwent intense fenitization, which is manifested in abundant alkaline feldspar, arfvedsonite and aegirine in the surrounding quartzite. Supergene enrichment in which Nb, P and REE form high-grade ores is well developed at the mining area [79].

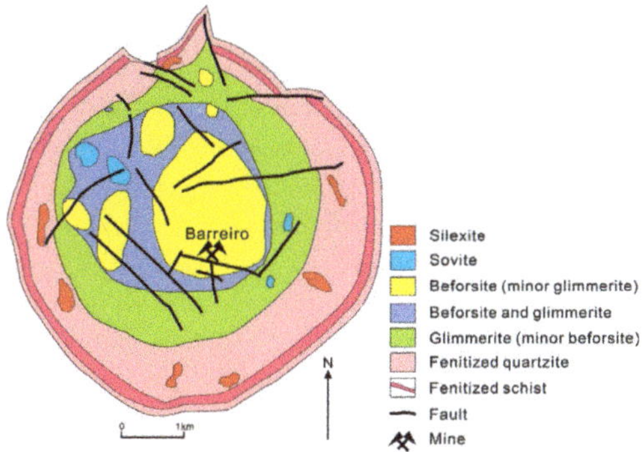

Figure 9. Simplified geological map of the alkaline–carbonatitic complex, Araxá (adapted from Li et al. [79]).

3.7. Gakara REE Deposit, Burundi

The Gakara REE deposit, located in the Republic of Burundi in east-central Africa (Figure 10), was discovered in 1936. It is one of the highest-grade REE deposits in the world, with an estimated in-situ grade of 47–67% of the total REE oxides [81,82]. The orebodies mainly occur in the wall rocks of quartzite and schist and in the form of net veins. The main constitutional minerals are bastnäsite, monazite, quartz, biotite, barite, K-feldspar and pyrite. The accessory minerals are cerianite, galena, anhydrite, rutile, molybdenite and chalcopyrite. According to previous work, the formation of the deposit includes three stages: Stage I: formation of primary coarse-grained bastnäsite, accompanied by gangue minerals such as quartz, biotite, and barite; Stage II: replacement of the primary bastnäsite by microcrystalline monazite that hosts most of the REE; microcrystalline quartz matrix was developed; silicification and brecciation occurred locally; Stage III: formation of La-rich monazite and supergene minerals such as cerianite, rhabdophane, goethite and kaolinite [81–83].

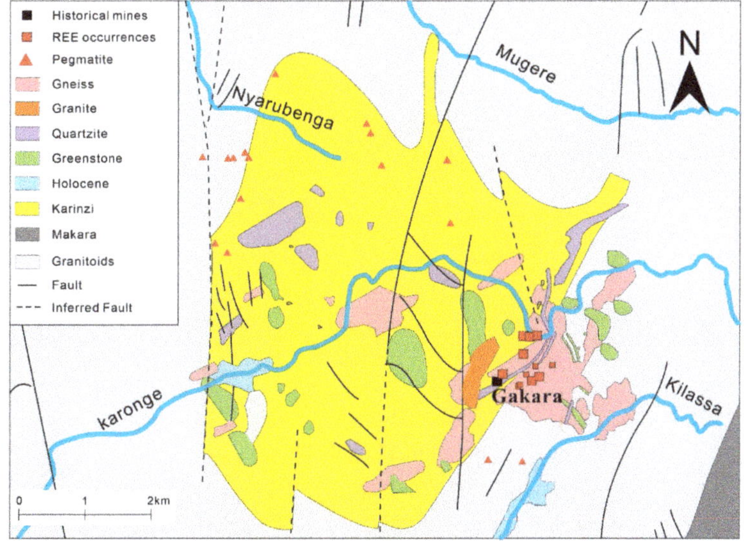

Figure 10. Simplified geological map of the Gakara REE deposit (adapted from Buyse et al. [81]).

3.8. Sarfartoq REE Deposit, Greenland

The Sarfartoq carbonatite, located in the southwest of Greenland, was discovered by the Geological Survey of Greenland in regional aerial radioactivity surveys from 1975 to 1976. Since the discovery, the Sarfartoq carbonatite has become the target of mineral exploration for resources such as diamond, Nb, U, P, REE [84]. It is located in the transitional zone between the Archean gneiss and the Paleoproterozoic Nagssugtoqidian active zone. The complex is composed of a carbonatite central core, fenitization zones and ring-like gneiss edge zones, with local intrusion of carbonate breccia veins (Figure 11). The carbonatite core area is mainly composed of rauhaugite; a small amount of sovite occurs sporadically. Extensive fenitization is developed around the core, within which a shear zone of 50–200m in width appears. Other characteristic features of the Sarfartoq REE deposit include the development of limonitization of the gneisses that is abundant in Th, U and REE. Niobium and REE mineralization, which is precipitated in the form of pyrochlore, are found in the outer fenitization zones. Hematitization and limonitization alteration are generally restricted to gneiss at the outer zones due to cataclastic deformation and hydrothermal metasomatism. Late stage of carbonatite veins, breccias and agglomerates of various shapes are developed in the alteration zone. The REE minerals in the Sarfartoq carbonatite are mainly bastnäsite, monazite and pyrochlore [14,84–86].

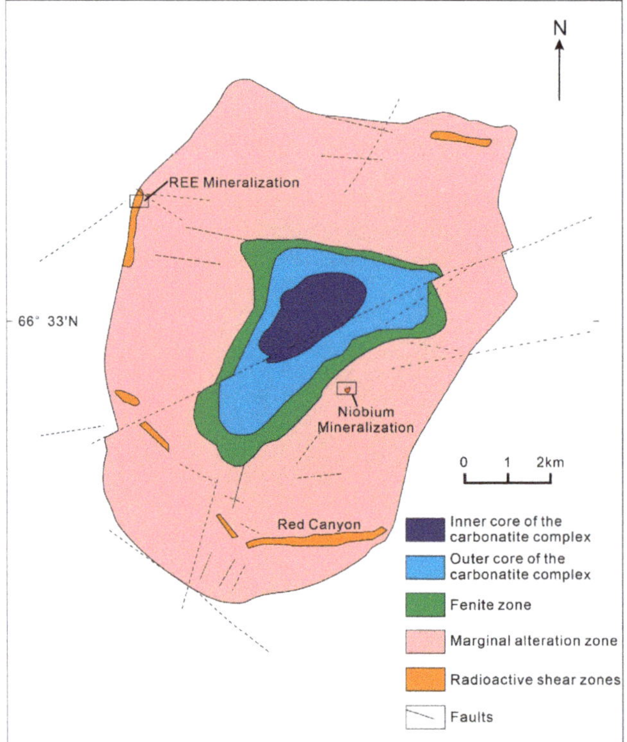

Figure 11. Geological sketch map of the Sarfartoq carbonatite complex (adapted from Bedini et al. [86]).

4. Genesis of Carbonatite-Related REE Deposits

4.1. Origin of Carbonatites

The evidence for the genesis of carbonatite mainly comes from experimental petrology, which shows that the formation of carbonatite magma is related to the mantle (lithospheric mantle or asthenosphere mantle) [87]. However, the specific evolution process of the magma is still controversial. At present, there are three main models explaining the formation of carbonatitic magma (Figure 12): (1) immiscible separation of primary carbonate–silicate magma under crustal or mantle pressure [88–92]; (2) fractional crystallization of parental carbonate–silicate magma [93]; (3) low-degree (generally < 1%) partial melting of the mantle rich in H_2O and CO_2 [88,89,94]. In addition, a recent study based on the boron (B) isotopes of global carbonatites suggests that although most carbonatites may have originated in the upper mantle, young carbonatites (<300 Ma) may contain at least part of the subducted crustal materials [95], and it is likely that the source of melts forming carbonatites is diverse. However, it is generally agreed that alkali elements (Na and K) play a significant role in the formation of calcite/dolomite carbonatite, and ferrocarbonatite intrusions, regardless of their formation mode [12].

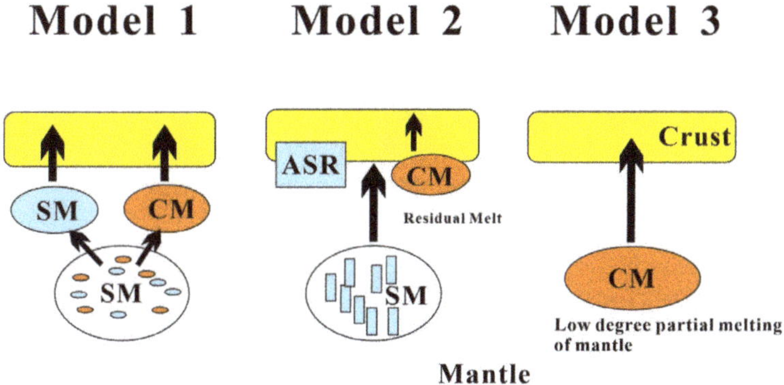

Figure 12. Lithogenesis of carbonatites. Model 1—immiscible separation of primary carbonate–silicate magma; Model 2—fractional crystallization of parental carbonate–silicate magma; Model 3—low-degree (generally < 1%) partial melting of the mantle rich in H_2O and CO_2; CM—carbonate melts; SM—silicate melts; ASR—alkaline silicate rocks (adapted from Ye et al. [87]).

4.2. Source of Ore-Forming Materials

High REE enrichment is a prerequisite for carbonatites to become a source of ore-forming materials. The genesis of carbonatites is generally recognized to be derived from the enriched mantle and related to mantle plume activities [95]. However, the source of REE in carbonatites is still controversial, including two main theories on the basis of previous research results.

The first theory suggests that the initial carbonated magma was formed by low-degree partial melting of the mantle which is rich in H_2O and CO_2, preceding the immiscible separation of primary silicate magmas that contain a large amount of REE (especially LREE) and carbonate melt. Incompatible elements such as REE preferentially partition into the carbonatite melt phase, resulting in an enrichment of REE in the melt relative to the primitive mantle [96]. For example, previous work on the Bayan Obo deposit generally suggests that the huge amount of REE in the deposit has a rich mantle source unaffected by crustal contamination, and the potential contribution of any external materials is trivial [40,97]. However, most carbonatite melts from mantle xenoliths are not enriched in incompatible elements [92,98], which is against this assumption. Enrichment of incompatible elements was otherwise discovered to exist in silicate minerals [99,100]. The carbonatite melts from mantle xenoliths have low REE content and relatively flat REE distribution patterns (i.e., low differentiation degree of LREE and HREE). At the same time, the carbonatite melt produced by partial melting of carbonated primitive mantle also has significantly lower REE content than that of carbonatite in the surface [98,101,102]. In addition, the partitioning of REE in immiscible melts is not in agreement with that of the interpretation of carbonatite formation process. The immiscibility process of carbonated silicate melts was simulated at high temperature and pressure, and it was found that REEs (especially HREE) were more inclined to enter the silicate melt [100]. Even in the case that a small quantity of volatiles (such as F, Cl, etc.) is added into carbonatite melt, the REEs will be slightly promoted into the carbonatite melt, but the overall enrichment trend of REEs in silicate melt remains unchanged [100]. Nevertheless, carbonatites are obviously more enriched in REE than the associated alkaline silicate rocks in a natural environment. Therefore, carbonatites formed by low-degree partial melting of mantle and immiscibility of carbonated silicate magma tend to be mildly enriched in REE, which is insufficient for REE mineralization to occur. Other sources of REEs are therefore required.

The second theory advocates that crustal sediments enriched in REE were subducted to the mantle, resulting in partial melting of the mantle. This process is of fundamental importance to enrich REE for the carbonatite magma, which is particularly the case for the carbonatite-related REE deposits near orogenic belts, where the subducted sediments are inevitably involved [103,104]. Kato et al. (2011) reported high concentrations of REE and Y in the deep-sea sediments at numerous sites such as the eastern South and central North Pacific [105]. Mimura et al. (2019) reported pelagic clay has high concentrations (>400 ppm) of REEs and Y in the western North Pacific Ocean, which has a significant impact on chemical composition of subducting sediments [106]. Yasukawa et al. (2014) also reported that there are remarkably high concentrations of REE and Y of potential economic significance in the deep-sea sediments of the Pacific and Indian oceans. The sediments are mainly composed of siliceous ooze with subordinate zeolitic clay, although seawater is proved to be the ultimate source of REE in the sediments [107]. This study provides a theoretical premise of reinforcing crustal sediments to be the REE source materials of carbonatite-related REE deposits. Moreover, it is found that the formation of typical carbonatite-related REE deposits in China is related to plate subduction and the associated circulation of crustal materials [30]. For example, the Bayan Obo deposit, located near the Central Asian orogenic belt, was formed during the closure of the Paleo-Asian Ocean. During the subduction of the Paleo-Asian plate, Si-rich fluid from the subducting slab was released and reacted with the calcic carbonatite pluton, from which REEs, Nb, Th, and Sr were collected. Then the fluid caused metasomatism with the overlying H8 dolomite to form a super large-scale deposit [108,109]. The Miaoya REE deposit in Hubei Province, located in the southern margin of Qinling orogenic belt, experienced complex orogenies such as continental collision, oceanic subduction, terrane accretion, and ocean basin opening and closing [110]. In the process of producing carbonatite magma, the contamination of crustal materials seems unavoidable and can contribute an additional amount of REE for the mineralization [41,111].

4.3. Nature of Ore-Forming Fluids

In carbonatite-related REE deposits (especially hydrothermal type), the ore-forming fluids play an important role in metasomatic wall-rock alteration, REE enrichment and mineralization. The ore-forming fluid derived at the late stage of carbonatite magma evolution is mainly composed of alkaline carbonate brine, which is rich in H_2O, high-density CO_2, Na, K, Ca, SO_4^{2-}, as well as REE, Nb and other ore-forming elements [45,112–117]. The fluids, which are freshly derived from carbonatite melts, are supercritical and high in temperature, pressure and solubility [45]. This fluid is also rich in alkalis, carbon and volatiles (CO_2, CH_4, C_2H_6, N_2, etc.), and has low viscosity. The nature of the fluids is ideal for establishing transport efficiency of the ore-forming elements in the fluids. The ore-forming fluids migrate along fractures at a high rate of speed and rise to the shallow surface along major structures [112,115]. In the early stage, the hydrothermal fluid is characterized by magmatic signatures, and gradually becomes mixed and diluted by meteoric water. In consequence, the fluids evolve to an aqueous, low temperature, low pressure, low salinity phase, in which the ore-forming elements are precipitated [116].

The study of fluid inclusions in carbonatites reflects the above-mentioned changes of fluid properties in different stages. Ting et al. (1994) studied fluid inclusions in the Sukulu carbonatite complex in Uganda, and found that the fluid inclusions evolve from CO_2-bearing, mid-high salinity to aqueous-rich and eventually to a CH_4-bearing composition from early to late stages, while CO_2-rich and CH_4-rich inclusions generally do not coexist [118]. Hong et al. (2014) also found a complete magmatic–hydrothermal evolution sequence from melt inclusions to melt-fluid inclusions and then to fluid inclusions for the Maoniuping REE deposit [119].

4.4. Migration and Precipitation Mechanism

REEs, as high field strength elements (HFSE), are traditionally regarded to be scarce in the form of simple ions in hydrothermal system and rarely migrate or mineralize [120]. However, it is found that REEs in certain hydrothermal fluids can be particularly enriched and migrate in the form of complexes [48,50,121]. Based on the Hard-Soft-Acid-Base (HSAB) principle, the REE ions belong to hard acid ions and are more likely to form stable complexes with hard base ions, e.g., F^-, $SO_4{}^{2-}$, Cl^-, $CO_3{}^{2-}$, $PO_4{}^{3-}$, OH^-. Previous studies suggest that REE mainly migrate in the form of chloride, sulfate and complexes in hydrothermal systems [48,122–125]. The proportions of various complexes vary with the temperature and pH of the hydrothermal fluids. In a highly acidic environment, the chlorine REE complexes (mainly $REECl^{2+}$) play a dominant role. With the increase in pH to weakly acidic or neutral conditions, the chlorine REE complexes are gradually replaced by sulfate complexes (mainly $REE(SO)_4{}^{2-}$). In low-chlorine and alkaline environments, REEs mainly migrate in the form of hydroxy complexes (mainly $REE(OH)^0{}_3$) [125–127]. In addition, temperature is critical to the transport efficiency of REEs. With the increase in fluid temperature, the complexation of REEs with chlorine and sulfate is enhanced to different extents that allows for higher capacity of REEs transportation in the fluids.

In hydrothermal systems, hard base ions complexed with REEs are not all conducive to the migration of ore-forming elements. Previous studies have found that when REEs form complexes with fluorine, carbonate and phosphate, it is more likely leading to rapid precipitation of REEs rather than migration [48,50,127]. Especially in the fluorine rich system, the REE-fluorine complex has low solubility and high-temperature instability, and the precipitation and mineralization of REEs are more likely to occur in this condition [50]. In the field investigation, we found that REEs are mainly enriched in the fluorine-enriched carbonate, which is consistent with the view that fluorine is more critical in precipitating REE than transporting REE.

The precipitation of REE in hydrothermal fluids is collectively controlled by temperature, pressure, pH, oxygen fugacity and chemical compositions [128]. The hydrothermal fluids that are initially exsolved are high in temperature and pressure, and inversely low in the pH value. The stability and solubility of REE complexes are high in the early stage of hydrothermal fluids, in which the migration of REEs is enhanced. In the process of migration, the fluid temperature and pressure decrease, leading to the separation of CO_2 and other acid gas phases. This process is accompanied with metasomatic reactions. Consequently, the pH of the hydrothermal system increases while the REE complexes (especially the fluorine complexes) become unstable and precipitated. During the metasomatism phosphorus can be introduced into the hydrothermal fluids, resulting in the precipitation of REEs in the form of REE-phosphates, e.g., xenotime and monazite [30,48,50,126]. In the late stage of fluid evolution, the temperature further decreases, and the frequent brecciations near the surface introduce meteoric water in the hydrothermal system. The fluid composition changes dramatically. Carbon dioxide is available and involved in the formation of bastnäsite that precipitates REEs [112,129]. For example, the Maoniuping REE deposit in Sichuan Province is a typical carbonatite-related hydrothermal deposit. A rich body of research has been carried out to understand the migration and precipitation mechanism of REEs in the deposit. The migration of REEs in the hydrothermal system was controlled by complexing agents such as F^-, Cl^-, $SO_4{}^{2-}$, $CO_3{}^{2-}$. At the initial stage of hydrothermal activity, REEs are mainly transported as chlorine and sulfate complexes. As the fluids advance, CO_2 became immiscible and the fluid pH increases. Corresponding to the change of pH the gangue minerals such as fluorite, celestite and barite began to precipitate, decreasing the buffering capacity of fluids and reducing the activity of fluorine complexing with REE. In consequence, a small amount of REEs was precipitated in REE minerals such as bastnäsite. During the late stage of hydrothermal activity, the fluid temperature continues to decrease, and the addition of external fluids such as meteoric water and atmospheric components such as CO_2 induces the major precipitation of gangue minerals, along with significant deposition of bastnäsite. This paragenesis is readily reflected in field observation where bastnäsite is often coexisting with fluorite, calcite and barite. To conclude, the late stage of hydrothermal evolution corresponds to the major precipitation of REE minerals, and therefore most likely represents the main mineralization stage [70,71].

4.5. REE Enrichment Mechanism

4.5.1. Fractional Crystallization of Magma

In the primary magmatic carbonatite-related REE deposits, it is common that REEs, volatiles and alkali elements are enriched in the residual magma despite the fact that primary REE minerals can also incorporate a small amount of REEs during crystallization from the carbonatite magma. This explains the close association of REE mineralization with the late stage of magmatic evolution. In particular, fractional crystallization of carbonates is important for enriching REE in the residual melts for mineralization. The carbonate minerals (such as calcite and dolomite) that are crystallized in the early stage of magmatic evolution usually have low REE content because REEs are incompatible with carbonate minerals and are therefore enriched in the residual melts. In addition to carbonate minerals, it can happen that ore minerals such as magnetite and hematite crystallize prior to the REE minerals during the fractional crystallization process. The Miaoya REE deposit in Hubei Province is a good example. Two REE mineralization events occurred during the formation of the deposit at about 420 Ma (in situ Th-Pb age of zircon from carbonatites) and 240 Ma (in situ Th-Pb age of monazite from carbonatites), respectively [28]. Zhang et al. (2019) applied monazite U-Th-Pb dating to constrain the age of the Miaoya syenite–carbonatite complexes, which yields two groups of U-Pb ages at 414 ± 11Ma (n = 5; MSWD = 0.91) and 231 ± 2Ma (n = 21; MSWD = 3.1), respectively, in addition to an age of 206 ± 4 Ma for the REE mineral bastnäsite [68]. The age of 414 Ma was obtained from homogeneous monazite grains, which is consistent with the zircon U-Pb ages of the syenite from the Miaoya syenite–carbonatite complexes (445.2 ± 2.6 Ma, MSWD = 0.66) and the carbonatite (434.3 ± 3.2 Ma, MSWD = 1.08) previously dated by Zhu et al. (2016) [33]. The two episodes of REE mineralization, in combination with field observation, are illustrated as follows. The first stage of REE mineralization is represented by the crystallization of REE minerals and phosphates (mainly monazite and fluorapatite) from early carbonatite magma, which is high in P and REE contents. The residual magma became depleted in REEs. The second stage of mineralization is characterized by fractional crystallization of calcite that endowed the residual magma with REEs (especially LREE), from which sufficient bastnäsite and a small amount of monazite were crystallized [43]. Thus, the fractional crystallization of carbonatite magma plays an important role in upgrading the REE mineralization to giant and super giant scale.

4.5.2. Immiscibility of Melts and Hydrothermal Fluids

In the carbonatite-related hydrothermal REE deposits, REEs are introduced into the carbonate fluids through immiscibility of melts and hydrothermal fluids. The REE-rich carbonate fluids migrate along fractures and precipitate under the given ambient condition, forming veins- and net veins-orebodies that are composed of REE minerals and accessory minerals such as fluorite, barite and calcite [23]. The Maoniuping REE deposit is a typical hydrothermal deposit, in which the REEs are mostly introduced and enriched in the hydrothermal evolution stage. Field observation and study of fluid inclusions in fluorite from the orebodies reveals that the evolutionary sequence of the magmatic–hydrothermal system in the Maoniuping deposit encloses three stages, namely an early magmatic stage, an intermediate high-temperature hydrothermal stage, and a late low-temperature hydrothermal stage. The carbonatite magma was initially formed by low-degree partial melting of the mantle; the immiscibility of silicate melt and carbonate melt triggered the formation of the initial ore-forming fluids. The initial ore-forming fluids, which are high in temperature, pressure and density and rich in CO_2, H_2O, H_2S, sulfates, were able to carry a large amount of REE and migrate them in gas phase. Ultimately, corresponding to the phase separation of CO_2 and aqueous fluids the temperature and pressure of the ore-forming fluids dropped, which resulted in REE precipitation and mineralization [70,71,117,119,130]. The enrichment, migration and precipitation of REEs mainly occur in the CO_2-rich fluids and at medium-high temperature, although a prerequisite for securing sufficient REEs in the early stage of fluid evolution is the liquid immiscibility between carbonatite melts and hydrothermal fluids.

5. Conclusions and Remarks

Carbonatite-related REE deposit, which provides the majority of LREE and Nb resources in the world, is a significant global deposit type. There are still many controversies about the ore genesis and enrichment mechanisms of REEs in this type of deposits. At present, we believe that the formation of carbonatites is closely related to the mantle, and there are three main mechanisms of forming REE-enriched carbonatites: (1) immiscibility of primary carbonate–silicate magma under crustal or mantle pressure; (2) fractional crystallization of parental carbonate–silicate magma; (3) low-degree (generally $< 1\%$) partial melting of the mantle which is rich in H_2O and CO_2. However, the former two mechanisms cannot explain the massive enrichment of REEs (especially LREE) as observed in the field. Therefore, we believe that the REE enrichment occurs after the formation of carbonatite magma, and most likely during the gradual evolution of the carbonatite magma by means of fractional crystallization of carbonate minerals and immiscibility of melt and hydrothermal fluids.

Considering the existing controversies, we propose the following directions for future work:

(1) At present, there is still a lack of reliable evidence to constrain the genesis of carbonatite. Experimental work needs to be designed to verify the influence of immiscibility on rock formation and REE enrichment. Meanwhile, isotope analysis, especially the non-traditional stable isotopes (such as Fe, Mg, Ba) can be carried out to examine the source of magma and its relationship with the mantle.

(2) In view of the unclear REE partitioning behaviors in a magmatic–hydrothermal system, experimental petrology work on this aspect will enhance the understanding of the distribution of REEs in different systems, such as between carbonatite melts rich in F, S, P and silicate melts, between carbonates (its symbiotic precipitates) and carbonatite melts, between carbonatite melts and fluid or gas phase. This development will provide a better understanding of REE enrichment mechanisms.

(3) Research on carbonatite-related REE deposits should expand to both REE fertile and barren carbonatites. Although carbonatite is absent in some REE deposits, the formation of those deposits is still closely related to carbonatite magma (such as the Nolans Bore P-REE-Th deposit, Australia) [131–133]. By using the method of "comparative metallogeny", a comparative study on the metallogenic characteristics of carbonatites under different or the same tectonic background should be strengthened. In combination with better knowledge of the REE partitioning behavior and the sequence of mineral formation in high temperature and pressure experiments, a critical understanding of the factors controlling the fertility of carbonatites will eventually be achieved.

Author Contributions: Conceptualization, Z.-Y.W.; resources, Z.-Y.W and H.-R.F; writing—original draft preparation, Z.-Y.W.; writing—review and editing, H.-R.F., L.Z., K.-F.Y. and H.-D.S.; funding acquisition, H.-R.F and K.-F.Y. All authors have read and agreed to the published version of the manuscript.

Funding: This study was financially supported by research grants from National Natural Science Foundation of China (Grant Nos. 41930430, 91962103), National Key R&D Program of China (grant 2017YFC0602302), the Key Research Program of the Innovation Academy for Earth Science, CAS (IGGCAS-201901) and the Zhongke Developing Science and Technology Co., Ltd. (2017H1973, ZK2018H003).

Acknowledgments: Thomas Ulrich is thanked for his constructive feedback that helped to improve an earlier draft of the manuscript. We are grateful to three anonymous reviewers for thoughtful suggestions.

Conflicts of Interest: The authors declare no conflict of interest.

References

1. Loges, A.; Migdisov, A.A.; Wagner, T.; Williams-Jones, A.E.; Markl, G. An experimental study of the aqueous solubility and speciation of Y(III) fluoride at temperatures up to 250 °C. *Geochim. Cosmochim. Acta* **2013**, *123*, 403–415. [CrossRef]
2. Binnemans, K.; Jones, P.T.; Blanpain, B.; Van Gerven, T.; Yang, Y.; Walton, A.; Buchert, M. Recycling of rare earths: A critical review. *J. Clean. Prod.* **2013**, *51*, 1–22. [CrossRef]
3. Jha, M.K.; Kumari, A.; Panda, R.; Rajesh Kumar, J.; Yoo, K.; Lee, J.Y. Review on hydrometallurgical recovery of rare earth metals. *Hydrometallurgy* **2016**, *161*, 77. [CrossRef]
4. Jordens, A.; Cheng, Y.P.; Waters, K.E. A review of the beneficiation of rare earth element bearing minerals. *Miner. Eng.* **2013**, *41*, 97–114. [CrossRef]
5. Chakhmouradian, A.R.; Wall, F. Rare Earth Elements: Minerals, Mines, Magnets (and More). *Elements* **2012**, *8*, 333–340. [CrossRef]
6. Kumari, A.; Panda, R.; Jha, M.K.; Kumar, J.R.; Lee, J.Y. Process development to recover rare earth metals from monazite mineral: A review. *Miner. Eng.* **2015**, *79*, 102–115. [CrossRef]
7. Zhang, S.J.; Zhang, L.W.; Zhang, Y.W.; Shang, L.; Li, J.B. Summarize on rare earth mineral resources and their distribution at home and abroad. *Inorg. Chem. Ind.* **2020**, *52*, 9–16, (In Chinese with English abstract).
8. Li, Z.; Hu, J.Z. World rare earth resources survey and development utilization trend. *Mod. Min.* **2017**, *33*, 97–105, (In Chinese with English abstract).
9. Li, F.Q.; Dai, T.; Wang, G.S. A review on recycling and reuse of rare earth metals. *Conserv. Util. Min. Resour.* **2019**, *39*, 84–89, (In Chinese with English abstract).
10. Wang, D.H.; Zhao, Z.; Yu, Y.; Wang, C.H.; Dai, J.J.; Sun, Y.; Zhao, T.; Li, J.K.; Huang, F.; Chen, Z.Y.; et al. A review of the achievements in the survey and study of ion-absorption type REE deposits in China. *Acta Geosci. Sin.* **2017**, *38*, 317–325, (In Chinese with English abstract).
11. Fu, T.Y.; Li, B.H.; Dong, X.Y.; Xu, L. Analysis on the distribution, classification and characteristics of rare earth deposits in China. *J. Henan Sci. Technol.* **2015**, *14*, 124–126, (In Chinese with English abstract).
12. Simandl, G.J.; Paradis, S. Carbonatites: Related ore deposits, resources, footprint, and exploration methods. *Appl. Earth Sci.* **2018**, *127*, 123–152. [CrossRef]
13. Sun, J.; Zhu, X.K.; Chen, Y.L.; Fang, N.; Li, S.Z. Is the Bayan Obo ore deposit a micrite mound? A comparison with the Sailinhudong micrite mound. *Int. Geol. Rev.* **2014**, *56*, 1720–1731. [CrossRef]
14. Liu, Q.P.; Zhao, Y.Y.; Liu, C.H. REE resources potential in Greenland and the availability evaluation in favor of China. *Geol. Bull. China* **2019**, *38*, 1386–1395, (In Chinese with English abstract).
15. Schonwandt, H.K.; Barnes, G.B.; Ulrich, T. Chapter 5—A Description of the World-Class Rare Earth Element Deposit, Tanbreez, South Greenland. In *Rare Earths Industry: Technological, Economic, and Environmental Implications*; Elsevier: Amsterdam, The Netherlands, 2016; pp. 73–85.
16. Streckeisen, A. Classification and nomenclature of volcanic rocks, lamprophyres, carbonatites and melilitic rocks IUGS Subcommission on the Systematics of Igneous Rocks. *Geol. Rundsch.* **1980**, *69*, 194–207. [CrossRef]
17. Gou, R.T.; Zeng, P.S.; Liu, S.W.; Ma, J.; Wang, J.J.; Dai, Y.J.; Wang, Z.Q. Distribution characteristics of carbonatites of the world and its metallogenic significance. *Acta Geol. Sin.* **2019**, *93*, 2348–2361, (In Chinese with English abstract).
18. Chakhmouradian, A.R.; Cooper, M.A.; Medici, L.; Abdu, Y.A.; Shelukhina, Y.S. Anzaite-(Ce), a new rare-earth mineral and structure type from the Afrikanda silicocarbonatite, Kola Peninsula, Russia. *Mineral. Mag.* **2018**, *79*, 1231–1244. [CrossRef]

19. Mattsson, H.B.; Balashova, A.; Almqvist, B.S.G.; Bosshard-Stadlin, S.A.; Weidendorfer, D. Magnetic mineralogy and rock magnetic properties of silicate and carbonatite rocks from Oldoinyo Lengai volcano (Tanzania). *J. Afr. Earth Sci.* **2018**, *142*, 193–206. [CrossRef]
20. Yang, K.F.; Fan, H.R.; Pirajno, F.; Li, X.C. The Bayan Obo (China) giant REE accumulation conundrum elucidated by intense magmatic differentiation of carbonatite. *Geology* **2019**, *47*, 1198–1202. [CrossRef]
21. Yang, Z.M.; Woolley, A. Carbonatites in China: A review. *J. Asian Earth Sci.* **2006**, *27*, 559–575. [CrossRef]
22. Linnen, R.L.; Samson, I.M.; Williams-Jones, A.E.; Chakhmouradian, A.R. Geochemistry of the Rare-Earth Element, Nb, Ta, Hf, and Zr Deposits. In *Treatise on Geochemistry*; Elsevier: Amsterdam, The Netherlands, 2014; pp. 543–568.
23. Song, W.L.; Xu, C.; Wang, L.J.; Wu, M.; Zeng, L.; Wang, L.Z.; Feng, M. Review of the metallogenesis of the endogenetic rare earth elements deposits related to carbonatite-alkaline complex. *Acta Sci. Nat. Univ. Pekin.* **2013**, *49*, 725–740, (In Chinese with English abstract).
24. Weng, Z.H.; Jowitt, S.M.; Mudd, G.M.; Haque, N. A Detailed Assessment of Global Rare Earth Element Resources: Opportunities and Challenges. *Econ. Geol.* **2015**, *110*, 1925–1952. [CrossRef]
25. Kanazawa, Y.; Kamitani, M. Rare earth minerals and resources in the world. *J. Alloy. Compd.* **2006**, *408–412*, 1339–1343. [CrossRef]
26. Woolley, A.R.; Kjarsgaard, B.A. Carbonatite occurrences of the world: Map and database. *Geol. Surv. Can. Open File* **2008**, *5796*, 1–21.
27. Liu, Y.; Hou, Z.Q. A synthesis of mineralization styles with an integrated genetic model of carbonatite-syenite-hosted REE deposits in the Cenozoic Mianning-Dechang REE metallogenic belt, the eastern Tibetan Plateau, southwestern China. *J. Asian Earth Sci.* **2017**, *137*, 35–79. [CrossRef]
28. Ying, Y.C.; Chen, W.; Lu, J.; Jiang, S.Y.; Yang, Y.H. In situ U–Th–Pb ages of the Miaoya carbonatite complex in the South Qinling orogenic belt, central China. *Lithos* **2017**, *290–291*, 159–171. [CrossRef]
29. Verplanck, P.L.; Mariano, A.N.; Mariano, A., Jr. Rare Earth Element Ore Geology of Carbonatites. In *Rare Earth and Critical Elements in Ore Deposits*; Verplanck, P.L., Hitzman, M.W., Eds.; Soc Economic Geologists, Inc.: Littleton, CO, USA, 2016; pp. 5–32.
30. Xie, Y.L.; Hou, Z.Q.; Goldfarb, R.J.; Guo, X.; Wang, L. Rare Earth Element Deposits in China. In *Rare Earth and Critical Elements in Ore Deposits*; Verplanck, P.L., Hitzman, M.W., Eds.; Soc Economic Geologists, Inc.: Littleton, CO, USA, 2016; pp. 115–136.
31. Lai, X.D.; Yang, X.Y. U-Pb Ages and Hf Isotope of Zircons from a Carbonatite Dyke in the Bayan Obo Fe-REE Deposit in Inner Mongolia: Its Geological Significance. *Acta Geol. Sin. Engl. Ed.* **2019**, *93*, 1783–1796. [CrossRef]
32. Xie, Y.L.; Li, Y.X.; Cooke, D.; Kamenetsky, V.; Chang, Z.S.; Danyushevsky, L.; Dominy, S.; Ryan, C.; Laird, J. Geochemical characteristics of carbonatite fluids at the Maoniuping REE deposit, Western Sichuan, China. In Let's Talk Ore Deposits. In Proceedings of the Eleventh Biennial SGA Meeting, Antofagasta, Chile, 26–29 September 2011; Volume 1, pp. 196–198.
33. Zhu, J.; Wang, L.X.; Peng, S.G.; Peng, L.H.; Wu, C.X.; Qiu, X.F. U-Pb zircon age, geochemical and isotopic characteristics of the Miaoya syenite and carbonatite complex, central China. *Geol. J.* **2017**, *52*, 938–954. [CrossRef]
34. Wang, C.; Liu, J.C.; Zhang, H.D.; Zhang, X.Z.; Zhang, D.M.; Xi, Z.; Wang, Z.X.; Wang, Z.J. Geochronology and mineralogy of the Weishan carbonatite in Shandong province, eastern China. *Geosci. Front.* **2019**, *10*, 769–785. [CrossRef]
35. Xu, C.; Campbell, I.H.; Allen, C.M.; Chen, Y.J.; Huang, Z.L.; Qi, L.; Zhang, G.S.; Yan, Z.F. U-Pb zircon age, geochemical and isotopic characteristics of carbonatite and syenite complexes from the Shaxiongdong, China. *Lithos* **2008**, *105*, 118–128. [CrossRef]
36. Liu, S.; Fan, H.R.; Groves, D.I.; Yang, K.F.; Yang, Z.F.; Wang, Q.W. Multiphase carbonatite-related magmatic and metasomatic processes in the genesis of the ore-hosting dolomite in the giant Bayan Obo REE-Nb-Fe deposit. *Lithos* **2020**, *354–355*, 105359. [CrossRef]
37. Yang, X.Y.; Lai, X.Y.; Pirajno, F.; Liu, Y.L.; Ling, M.X.; Sun, W.D. Genesis of the Bayan Obo Fe-REE-Nb formation in Inner Mongolia, North China Craton: A perspective review. *Precambrian Res.* **2017**, *288*, 39–71. [CrossRef]
38. Song, W.L.; Xu, C.; Smith, M.P.; Chakhmouradian, A.R.; Brenna, M.; Kynický, J.; Chen, W.; Yang, Y.H.; Deng, M.; Tang, H.Y. Genesis of the world's largest rare earth element deposit, Bayan Obo, China: Protracted mineralization evolution over ~1 b.y. *Geology* **2018**, *46*, 323–326. [CrossRef]

39. Hu, L.; Li, Y.K.; Wu, Z.J.; Bai, Y.; Wang, A.J. Two metasomatic events recorded in apatite from the ore-hosting dolomite marble and implications for genesis of the giant Bayan Obo REE deposit, Inner Mongolia, Northern China. *J. Asian Earth Sci.* **2019**, *172*, 56–65. [CrossRef]
40. Fan, H.R.; Yang, K.F.; Hu, F.F.; Liu, S.; Wang, K.Y. The giant Bayan Obo REE-Nb-Fe deposit, China: Controversy and ore genesis. *Geosci. Front.* **2016**, *7*, 335–344. [CrossRef]
41. Su, J.H.; Zhao, X.F.; Li, X.C.; Hu, W.; Chen, M.; Xiong, Y.L. Geological and geochemical characteristics of the Miaoya syenite-carbonatite complex, Central China: Implications for the origin of REE-Nb-enriched carbonatite. *Ore Geol. Rev.* **2019**, *113*, 103101. [CrossRef]
42. Ying, Y.C.; Chen, W.; Simonetti, A.; Jiang, S.Y.; Zhao, K.D. Significance of hydrothermal reworking for REE mineralization associated with carbonatite: Constraints from in situ trace element and C-Sr isotope study of calcite and apatite from the Miaoya carbonatite complex (China). *Geochim. Cosmochim. Acta* **2020**, *280*, 340–359. [CrossRef]
43. Xu, C.; Kynicky, J.; Chakhmouradian, A.R.; Campbell, I.H.; Allen, C.M. Trace-element modeling of the magmatic evolution of rare-earth-rich carbonatite from the Miaoya deposit, Central China. *Lithos* **2010**, *118*, 145–155. [CrossRef]
44. Jia, Y.H.; Liu, Y. REE Enrichment during Magmatic-Hydrothermal Processes in Carbonatite-Related REE Deposits: A Case Study of the Weishan REE Deposit, China. *Minerals* **2019**, *10*, 25. [CrossRef]
45. Hou, Z.Q.; Tian, S.H.; Xie, Y.L.; Yang, Z.; Yuan, Z.S.; Yin, S.P.; Yi, L.S.; Fei, H.C.; Zou, T.R.; Bai, G.; et al. The Himalayan Mianning-Dechang REE belt associated with carbonatite-alkaline complexes, eastern Indo-Asian collision zone, SW China. *Ore Geol. Rev.* **2009**, *36*, 65–89. [CrossRef]
46. Woolley, A.R. A discussion of carbonatite evolution and nomenclature, and the generation of sodic and potassic fenites. *Mineral. Mag.* **1982**, *46*, 13–17. [CrossRef]
47. Liu, S.; Fan, H.R.; Yang, K.F.; Hu, F.F.; Rusk, B.; Liu, X.; Li, X.C.; Yang, Z.F.; Wang, Q.W.; Wang, K.Y. Fenitization in the giant Bayan Obo REE-Nb-Fe deposit: Implication for REE mineralization. *Ore Geol. Rev.* **2018**, *94*, 290–309. [CrossRef]
48. Migdisov, A.A.; Williams-Jones, A.E. Hydrothermal transport and deposition of the rare earth elements by fluorine-bearing aqueous liquids. *Miner. Depos.* **2014**, *49*, 987–997. [CrossRef]
49. Liu, Y.; Chen, Z.Y.; Yang, Z.S.; Sun, X.; Zhu, Z.M.; Zhang, Q.C. Mineralogical and geochemical studies of brecciated ores in the Dalucao REE deposit, Sichuan Province, southwestern China. *Ore Geol. Rev.* **2015**, *70*, 613–636. [CrossRef]
50. Williams-Jones, A.E.; Migdisov, A.A.; Samson, I.M. Hydrothermal Mobilisation of the Rare Earth Elements—A Tale of "Ceria" and "Yttria". *Elements* **2012**, *8*, 355–360. [CrossRef]
51. Cooper, A.F.; Palin, J.M.; Collins, A.K. Fenitization of metabasic rocks by ferrocarbonatites at Haast River, New Zealand. *Lithos* **2016**, *244*, 109–121. [CrossRef]
52. Currie, K.L.; Ferguson, J. A Study of Fenitization Around the Alkaline Carbonatite Complex at Callander Bay, Ontario, Canada. *Can. J. Earth Sci.* **1971**, *8*, 498–517. [CrossRef]
53. Currie, K.L.; Ferguson, J. A Study of Fenitization in Mafic Rocks, with Special Reference to the Callander Bay Complex. *Can. J. Earth Sci.* **1972**, *9*, 1254–1261. [CrossRef]
54. McKie, D. Fenitization. In *Tuttle of Gittins J. (eds.) & Carbonatites*; Interscience: New York, NY, USA, 1966; pp. 261–294.
55. Le Bas, M.J. Fenites Associated with Carbonatites. *Can. Mineral.* **2008**, *46*, 915–932. [CrossRef]
56. Yang, X.M.; Yang, X.Y.; Fan, H.R.; Guo, F.; Zhang, Z.F.; Zheng, Y.F. Petrological characteristics of fenites and their geological significance. *Geol. Rev.* **2000**, *46*, 481–490, (In Chinese with English abstract).
57. Potter, N.J.; Kamenetsky, V.S.; Simonetti, A.; Goemann, K. Different types of liquid immiscibility in carbonatite magma: A case study of the Oldoinyo Lengai 1993 lava and melt inclusions. *Chem. Geol.* **2017**, *455*, 376–384. [CrossRef]
58. Mitchell, R.H. Peralkaline nephelinite-natrocarbonatite immiscibility and carbonatite assimilation at Oldoinyo Lengai, Tanzania. *Contrib. Mineral. Petrol.* **2009**, *158*, 589–598. [CrossRef]
59. Kresten, P.; Morogan, V. Fenitization at the Fen complex, southern Norway. *Lithos* **1986**, *19*, 27–42. [CrossRef]
60. Kresten, P. The chemistry of fenitization: Examples from Fen, Norway. *Chem. Geol.* **1988**, *68*, 329–349. [CrossRef]
61. Zhou, J.; Wang, X.Q.; Nie, L.S.; Jennifer, M.M.; Liu, H.L.; Zhang, B.M.; Han, Z.X. Geochemical background and dispersion pattern of the world's largest REE deposit of Bayan Obo, China. *J. Geochem. Explor.* **2020**, *215*, 106545. [CrossRef]

62. Zhang, S.H.; Zhao, Y.; Liu, Y. A precise zircon Th-Pb age of carbonatite sills from the world's largest Bayan Obo deposit: Implications for timing and genesis of REE-Nb mineralization. *Precambrian Res.* **2017**, *291*, 202–219. [CrossRef]
63. Smith, M.P.; Campbell, L.S.; Kynicky, J. A review of the genesis of the world class Bayan Obo Fe-REE-Nb deposits, Inner Mongolia, China: Multistage processes and outstanding questions. *Ore Geol. Rev.* **2015**, *64*, 459–476. [CrossRef]
64. Chen, W.; Liu, H.Y.; Lu, J.; Jiang, S.Y.; Simonetti, A.; Xu, C.; Zhang, W. The formation of the ore-bearing dolomite marble from the giant Bayan Obo REE-Nb-Fe deposit, Inner Mongolia: Insights from micron-scale geochemical data. *Miner. Depos.* **2019**, *55*, 131–146. [CrossRef]
65. Yang, K.F.; Fan, H.R.; Santosh, M.; Hu, F.F.; Wang, K.Y. Mesoproterozoic carbonatitic magmatism in the Bayan Obo deposit, Inner Mongolia, North China: Constraints for the mechanism of super accumulation of rare earth elements. *Ore Geol. Rev.* **2011**, *40*, 122–131. [CrossRef]
66. Deng, M.; Xu, C.; Song, W.L.; Tang, H.Y.; Liu, Y.; Zhang, Q.; Zhou, Y.; Feng, M.; Wei, C.W. REE mineralization in the Bayan Obo deposit, China: Evidence from mineral paragenesis. *Ore Geol. Rev.* **2017**, *91*, 100–109. [CrossRef]
67. Sun, J.; Zhu, X.K.; Chen, Y.L.; Fang, N. Iron isotopic constraints on the genesis of Bayan Obo ore deposit, Inner Mongolia, China. *Precambrian Res.* **2013**, *235*, 88–106. [CrossRef]
68. Zhang, W.; Chen, W.T.; Gao, J.F.; Chen, H.K.; Li, J.H. Two episodes of REE mineralization in the Qinling Orogenic Belt, Central China: In-situ U-Th-Pb dating of bastnäsite and monazite. *Miner. Depos.* **2019**, *54*, 1265–1280. [CrossRef]
69. Li, S. Geochemical features and petrogenesis of Miaoya carbonatites, Hubei. *Geochimica* **1980**, *1*, 345–355, (In Chinese with English abstract).
70. Zheng, X.; Liu, Y. Mechanisms of element precipitation in carbonatite-related rare-earth element deposits: Evidence from fluid inclusions in the Maoniuping deposit, Sichuan Province, southwestern China. *Ore Geol. Rev.* **2019**, *107*, 218–238. [CrossRef]
71. Liu, Y.; Chakhmouradian, A.R.; Hou, Z.Q.; Song, W.L.; Kynický, J. Development of REE mineralization in the giant Maoniuping deposit (Sichuan, China): Insights from mineralogy, fluid inclusions, and trace-element geochemistry. *Miner. Depos.* **2018**, *54*, 701–718. [CrossRef]
72. Li, J.K.; Yuan, Z.X.; Bai, G.; Chen, Y.C.; Wang, D.H.; Ying, L.J.; Zhang, J. Ore-forming fluid evolvement and its controlling to REE(Ag) mineralizing in the Weishan deposit, Shandong. *Jmineral Pet.* **2009**, *29*, 60–68, (In Chinese with English abstract).
73. Castor, S.B. The Mountain Pass Rare-Earth Carbonatite and Associated Ultrapotassic Rocks, California. *Can. Mineral.* **2008**, *46*, 779–806. [CrossRef]
74. Mariano, A.N.; Mariano, A. Rare Earth Mining and Exploration in North America. *Elements* **2012**, *8*, 369–376. [CrossRef]
75. Denton, K.M.; Ponce, D.A.; Peacock, J.R.; Miller, D.M. Geophysical characterization of a Proterozoic REE terrane at Mountain Pass, eastern Mojave Desert, California, USA. *Geosphere* **2019**, *16*, 456–471. [CrossRef]
76. Jones, A.P.; Wyllie, P.J. Low-temperature glass quenched from a synthetic, rare earth carbonatite; implications for the origin of the Mountain Pass Deposit, California. *Econ. Geol.* **1983**, *78*, 1721–1723. [CrossRef]
77. Poletti, J.E.; Cottle, J.M.; Hagen-Peter, G.A.; Lackey, J.S. Petrochronological Constraints on the Origin of the Mountain Pass Ultrapotassic and Carbonatite Intrusive Suite, California. *J. Petrol.* **2016**, *57*, 1555–1598. [CrossRef]
78. Nascimento, M.; Lemos, F.; Guimarães, R.; Sousa, C.; Soares, P. Modeling of REE and Fe Extraction from a Concentrate from Araxá (Brazil). *Minerals* **2019**, *9*, 451. [CrossRef]
79. Li, Y.K.; Chen, R.Y.; Ke, C.H.; Chen, J.; Hao, M.Z.; Li, R.P. The strategic and critical minerals associated with alkaline and alkaline-carbonatite complexes Brazil. *Acta Geol. Sin.* **2019**, *93*, 1422–1443, (In Chinese with English abstract).
80. Neumann, R.; Medeiros, E.B. Comprehensive mineralogical and technological characterisation of the Araxá (SE Brazil) complex REE (Nb-P) ore, and the fate of its processing. *Int. J. Miner. Process.* **2015**, *144*, 1–10. [CrossRef]
81. Buyse, F.; Dewaele, S.; Decrée, S.; Mees, F. Mineralogical and geochemical study of the rare earth element mineralization at Gakara (Burundi). *Ore Geol. Rev.* **2020**, *124*, 103659. [CrossRef]

82. Ntiharirizwa, S.; Boulvais, P.; Poujol, M.; Branquet, Y.; Morelli, C.; Ntungwanayo, J.; Midende, G. Geology and U-Th-Pb Dating of the Gakara REE Deposit, Burundi. *Minerals* **2018**, *8*, 394. [CrossRef]
83. Lehmann, B.; Nakai, S.i.; Höhndorf, A.; Brinckmann, J.; Dulski, P.; Hein, U.F.; Masuda, A. REE mineralization at Gakara, Burundi: Evidence for anomalous upper mantle in the western Rift Valley. *Geochim. Cosmochim. Acta* **1994**, *58*, 985–992. [CrossRef]
84. Secher, K.; Larsen, L.M. Geology and mineralogy of the Sarfartôq carbonatite complex, southern West Greenland. *Lithos* **1980**, *13*, 199–212. [CrossRef]
85. Goodenough, K.M.; Schilling, J.; Jonsson, E.; Kalvig, P.; Charles, N.; Tuduri, J.; Deady, E.A.; Sadeghi, M.; Schiellerup, H.; Müller, A.; et al. Europe's rare earth element resource potential: An overview of REE metallogenetic provinces and their geodynamic setting. *Ore Geol. Rev.* **2016**, *72*, 838–856. [CrossRef]
86. Bedini, E.; Rasmussen, T.M. Use of airborne hyperspectral and gamma-ray spectroscopy data for mineral exploration at the Sarfartoq carbonatite complex, southern West Greenland. *Geosci. J.* **2018**, *22*, 641–651. [CrossRef]
87. Ye, H.M.; Zhang, X. Advances on the carbonatite research in recent years. *Resour. Surv. Environ.* **2015**, *36*, 21–27, (In Chinese with English abstract).
88. Wallace, M.E.; Green, D.H. An experimental determination of primary carbonatite magma composition. *Nature* **1988**, *335*, 343–346. [CrossRef]
89. Wyllie, P.J.; Lee, W.J. Model System Controls on Conditions for Formation of Magnesiocarbonatite and Calciocarbonatite Magma from the Mantle. *J. Petrol.* **1998**, *39*, 1885–1893. [CrossRef]
90. Hamilton, D.L.; Freestone, I.C.; Dawson, J.B.; Donaldson, C.H. Origin of carbonatites by liquid immiscibility. *Nature* **1979**, *279*, 52–54. [CrossRef]
91. Harmer, R. The Case for Primary, Mantle-derived Carbonatite Magma. *J. Petrol.* **1998**, *39*, 1895–1903. [CrossRef]
92. Lee, W.J.; Wyllie, P.J. Processes of Crustal Carbonatite Formation by Liquid Immiscibility and Differentiation, Elucidated by Model Systems. *J. Petrol.* **1998**, *39*, 2005–2013. [CrossRef]
93. Veksler, I.; Lentz, D. Parental magma of plutonic carbonatites, carbonate-silicate immiscibility and decarbonation reactions: Evidence from melt and fluid inclusions. *Melt Incl. Plutonic Rocks* **2006**, *36*, 123–150.
94. Manthilake, M.A.G.M.; Sawada, Y.; Sakai, S. Genesis and evolution of Eppawala carbonatites, Sri Lanka. *J. Asian Earth Sci.* **2008**, *32*, 66–75. [CrossRef]
95. Hulett, S.R.W.; Simonetti, A.; Rasbury, E.T.; Hemming, N.G. Recycling of subducted crustal components into carbonatite melts revealed by boron isotopes. *Nat. Geosci.* **2016**, *9*, 904–908. [CrossRef]
96. Nelson, D.R.; Chivas, A.R.; Chappell, B.W.; McCulloch, M.T. Geochemical and isotopic systematics in carbonatites and implications for the evolution of ocean-island sources. *Geochim. Cosmochim. Acta* **1988**, *52*, 1–17. [CrossRef]
97. Zhu, X.K.; Sun, J.; Pan, C.X. Sm-Nd isotopic constraints on rare-earth mineralization in the Bayan Obo ore deposit, Inner Mongolia, China. *Ore Geol. Rev.* **2015**, *64*, 543–553. [CrossRef]
98. Ionov, D. Trace Element Composition of Mantle-derived Carbonates and Coexisting Phasesin Peridotite Xenoliths from Alkali Basalts. *J. Petrol.* **1998**, *39*, 1931–1941. [CrossRef]
99. Veksler, I.V.; Petibon, C.; Jenner, G.A.; Dorfman, A.M.; Dingwell, D.B. Trace Element Partitioning in Immiscible Silicate-Carbonate Liquid Systems: An Initial Experimental Study Using a Centrifuge Autoclave. *J. Petrol.* **1998**, *39*, 2095–2104. [CrossRef]
100. Veksler, I.V.; Dorfman, A.M.; Dulski, P.; Kamenetsky, V.S.; Danyushevsky, L.V.; Jeffries, T.; Dingwell, D.B. Partitioning of elements between silicate melt and immiscible fluoride, chloride, carbonate, phosphate and sulfate melts, with implications to the origin of natrocarbonatite. *Geochim. Cosmochim. Acta* **2012**, *79*, 20–40. [CrossRef]
101. Foley, S.F.; Yaxley, G.M.; Rosenthal, A.; Buhre, S.; Kiseeva, E.S.; Rapp, R.P.; Jacob, D.E. The composition of near-solidus melts of peridotite in the presence of CO_2 and H_2O between 40 and 60 kbar. *Lithos* **2009**, *112*, 274–283. [CrossRef]
102. Ionov, D.; Harmer, R.E. Trace element distribution in calcite-dolomite carbonatites from Spitskop: Inferences for differentiation of carbonatite magma and the origin of carbonates in mantle xenoliths. *Earth Planet. Sci. Lett.* **2002**, *198*, 495–510. [CrossRef]
103. Banerjee, A.; Chakrabarti, R. A geochemical and Nd, Sr and stable Ca isotopic study of carbonatites and associated silicate rocks from the ~65 Ma old Ambadongar carbonatite complex and the Phenai Mata igneous

complex, Gujarat, India: Implications for crustal contamination, carbonate recycling, hydrothermal alteration and source-mantle mineralogy. *Lithos* **2019**, *326–327*, 572–585.

104. Hou, Z.Q.; Tian, S.H.; Yuan, Z.X.; Xie, Y.L.; Yin, S.P.; Yi, L.S.; Fei, H.C.; Yang, Z.M. The Himalayan collision zone carbonatites in western Sichuan, SW China: Petrogenesis, mantle source and tectonic implication. *Earth Planet. Sci. Lett.* **2006**, *244*, 234–250. [CrossRef]
105. Kato, Y.; Fujinaga, K.; Nakamura, K.; Takaya, Y.; Kitamura, K.; Ohta, J.; Toda, R.; Nakashima, T.; Iwamori, H. Deep-sea mud in the Pacific Ocean as a potential resource for rare-earth elements. *Nat. Geosci.* **2011**, *4*, 535–539. [CrossRef]
106. Mimura, K.; Nakamura, K.; Yasukawa, K.; Machida, S.; Ohta, J.; Fujinaga, K.; Kato, Y. Significant impacts of pelagic clay on average chemical composition of subducting sediments: New insights from discovery of extremely rare-earth elements and yttrium-rich mud at Ocean Drilling Program Site 1149 in the western North Pacific Ocean. *J. Asian Earth Sci.* **2019**, *186*, 104059. [CrossRef]
107. Yasukawa, K.; Liu, H.; Fujinaga, K.; Machida, S.; Haraguchi, S.; Ishii, T.; Nakamura, K.; Kato, Y. Geochemistry and mineralogy of REY-rich mud in the eastern Indian Ocean. *J. Asian Earth Sci.* **2014**, *93*, 25–36. [CrossRef]
108. Ling, M.X.; Liu, Y.L.; Williams, I.S.; Teng, F.Z.; Yang, X.Y.; Ding, X.; Wei, G.J.; Xie, L.H.; Deng, W.F.; Sun, W.D. Formation of the world's largest REE deposit through protracted fluxing of carbonatite by subduction-derived fluids. *Sci. Rep.* **2013**, *3*, 1776. [CrossRef]
109. Ni, P.; Zhou, J.; Chi, Z.; Pan, J.-Y.; Li, S.-N.; Ding, J.-Y.; Han, L. Carbonatite dyke and related REE mineralization in the Bayan Obo REE ore field, North China: Evidence from geochemistry, C O isotopes and Rb Sr dating. *J. Geochem. Explor.* **2020**, *215*, 106560. [CrossRef]
110. Wu, Y.B.; Zheng, Y.F. Tectonic evolution of a composite collision orogen: An overview on the Qinling-Tongbai-Hong'an-Dabie-Sulu orogenic belt in central China. *Gondwana Res.* **2013**, *23*, 1402–1428. [CrossRef]
111. Xu, C.; Kynicky, J.; Chakhmouradian, A.R.; Li, X.H.; Song, W.L. A case example of the importance of multi-analytical approach in deciphering carbonatite petrogenesis in South Qinling orogen: Miaoya rare-metal deposit, central China. *Lithos* **2015**, *227*, 107–121. [CrossRef]
112. Fan, H.R.; Xie, Y.H.; Wang, K.Y.; Yang, X.M. Carbonatitic Fluids and REE Mineralization. *Earth Science Frontiers.* **2001**, *8*, 289–295, (In Chinese with English abstract).
113. Zhou, J.; Ni, P.; Ding, J.Y.; Zhu, X.T. Fluid inclusion study related to cabonatitic magmatic process. *Geol. J. China Univ.* **2003**, *9*, 293–301, (In Chinese with English abstract).
114. Xie, Y.L.; Xu, J.H.; Chen, W.; He, J.P.; Hou, Z.Q.; Xu, W.Y. Characteristics of carbonatite fluid in the Maoniuping REE deposit, Mianning, China. In *Mineral Deposit Research: Meeting the Global Challenge*; Springer: Berlin/Heidelberg, Germany, 2005.
115. Xie, Y.L.; Yin, S.P.; Xu, J.H.; Chen, W.; Yi, L.S. A Study on the fluids in carbonatite of the Mianning-Dechang REE metallogenic belt. *Bull. Mineral. Petrol. Geochem.* **2006**, *25*, 66–74, (In Chinese with English abstract).
116. Shu, X.C.; Liu, Y. Fluid inclusion constraints on the hydrothermal evolution of the Dalucao Carbonatite-related REE deposit, Sichuan Province, China. *Ore Geol. Rev.* **2019**, *107*, 41–57. [CrossRef]
117. Xie, Y.L.; Hou, Z.Q.; Yin, S.P.; Dominy, S.C.; Xu, J.H.; Tian, S.H.; Xu, W.Y. Continuous carbonatitic melt–fluid evolution of a REE mineralization system: Evidence from inclusions in the Maoniuping REE Deposit, Western Sichuan, China. *Ore Geol. Rev.* **2009**, *36*, 90–105. [CrossRef]
118. Ting, W.P.; Burke, E.A.J.; Rankin, A.H.; Woolley, A.R. Characterisation and petrogenetic significance of CO_2, H_2O and CH4 fluid inclusions in apatite from the Sukulu carbonatite, Uganda. *Eur. J. Mineral.* **1994**, *6*, 787–803. [CrossRef]
119. Kong, L.R.; Zhang, G.R. Ore Types and Characteristics of Melt (Fluid) Inclusions on the Maoniuping REE Deposit Mianning, Sichuan, China. *Appl. Mech. Mater.* **2014**, *675–677*, 1308–1311. [CrossRef]
120. She, H.D.; Fan, H.R.; Hu, F.F.; Yang, K.F.; Yang, Z.F.; Wang, Q.W. Migration and precipitation of rare earth elements in the hydrothermal fluids. *Acta Petrol. Sin.* **2018**, *34*, 3567–3581, (In Chinese with English abstract).
121. Migdisov, A.; Williams-Jones, A.E.; Brugger, J.; Caporuscio, F.A. Hydrothermal transport, deposition, and fractionation of the REE: Experimental data and thermodynamic calculations. *Chem. Geol.* **2016**, *439*, 13–42. [CrossRef]
122. Migdisov, A.A.; Williams-Jones, A.E. An experimental study of the solubility and speciation of neodymium (III) fluoride in F-bearing aqueous solutions. *Geochim. Cosmochim. Acta* **2007**, *71*, 3056–3069. [CrossRef]

123. Migdisov, A.A.; Williams-Jones, A.E. A spectrophotometric study of Nd(III), Sm(III) and Er(III) complexation in sulfate-bearing solutions at elevated temperatures. *Geochim. Cosmochim. Acta* **2008**, *72*, 5291–5303. [CrossRef]
124. Migdisov, A.A.; Williams-Jones, A.E.; Wagner, T. An experimental study of the solubility and speciation of the Rare Earth Elements (III) in fluoride- and chloride-bearing aqueous solutions at temperatures up to 300°C. *Geochim. Cosmochim. Acta* **2009**, *73*, 7087–7109. [CrossRef]
125. Debruyne, D.; Hulsbosch, N.; Muchez, P. Unraveling rare earth element signatures in hydrothermal carbonate minerals using a source–sink system. *Ore Geol. Rev.* **2016**, *72*, 232–252. [CrossRef]
126. Louvel, M.; Bordage, A.; Testemale, D.; Zhou, L.; Mavrogenes, J. Hydrothermal controls on the genesis of REE deposits: Insights from an in situ XAS study of Yb solubility and speciation in high temperature fluids (T < 400 °C). *Chem. Geol.* **2015**, *417*, 228–237.
127. Liu, Y.; Chen, C.; Shu, X.C.; Guo, D.X.; Li, Z.J.; Zhao, H.X.; Jia, Y.H. The formation model of the carbonatite-syenite complex REE deposits in the east of Tibetan Plateau: A case study of Dalucao REE deposit. *Acta Petrol. Sin.* **2017**, *33*, 1978–2000, (In Chinese with English abstract).
128. Verplanck, P.L. The Role of Fluids in the Formation of Rare Earth Element Deposits. *Procedia Earth Planet. Sci.* **2017**, *17*, 758–761. [CrossRef]
129. Chikanda, F.; Otake, T.; Ohtomo, Y.; Ito, A.; Yokoyama, T.D.; Sato, T. Magmatic-Hydrothermal Processes Associated with Rare Earth Element Enrichment in the Kangankunde Carbonatite Complex, Malawi. *Minerals* **2019**, *9*, 442. [CrossRef]
130. Qin, C.J.; Qiu, Y.Z.; Wen, H.J.; Xu, C. Genesis of Maoniuping REE deposit, Sichuan: Evidence from inclusions. *Acta Petrol. Sin.* **2008**, *24*, 2155–2162, (In Chinese with English abstract).
131. Anenburg, M.; Burnham, A.D.; Mavrogenes, J.A. REE Redistribution Textures in Altered Fluorapatite: Symplectites, Veins, and Phosphate-Silicate-Carbonate Assemblages from the Nolans Bore P-REE-Th Deposit, Northern Territory, Australia. *Can. Mineral.* **2018**, *56*, 331–354. [CrossRef]
132. Anenburg, M.; Mavrogenes, J.A.; Bennett, V.C. The Fluorapatite P-REE-Th Vein Deposit at Nolans Bore: Genesis by Carbonatite Metasomatism. *J. Petrol.* **2020**, *61*, egaa003. [CrossRef]
133. Anenburg, M.; Mavrogenes, J.A. Carbonatitic versus hydrothermal origin for fluorapatite REE-Th deposits: Experimental study of REE transport and crustal "antiskarn" metasomatism. *Am. J. Sci.* **2018**, *318*, 335–366. [CrossRef]

Article

Thermodynamic Constraints on REE Mineral Paragenesis in the Bayan Obo REE-Nb-Fe Deposit, China

Shang Liu [1,*], Lin Ding [1,2] and Hong-Rui Fan [3,4,5]

[1] School of Earth Sciences, Lanzhou University and Key Laboratory of Mineral Resources in Western China (Gansu Province), Lanzhou 730000, China; dinglin@itpcas.ac.cn
[2] Institute of Tibetan Plateau Research, Chinese Academy of Sciences, Beijing 100101, China
[3] Key Laboratory of Mineral Resources, Institute of Geology and Geophysics, Chinese Academy of Sciences, Beijing 100029, China; fanhr@mail.iggcas.ac.cn
[4] College of Earth and Planetary Sciences, University of Chinese Academy of Sciences, Beijing 100049, China
[5] Innovation Academy for Earth Science, Chinese Academy of Sciences, Beijing 100029, China
* Correspondence: liushang@lzu.edu.cn

Received: 6 May 2020; Accepted: 28 May 2020; Published: 29 May 2020

Abstract: Hydrothermal processes have played a significant role in rare earth element (REE) precipitation in the Bayan Obo REE-Nb-Fe deposit. The poor preservation of primary fluid inclusions and superposition or modification by multiphase hydrothermal activities have made identification of physico-chemical conditions of ore-forming fluids extremely difficult. Fortunately, with more and more reliable thermodynamic properties of aqueous REE species and REE minerals reported in recent years, a series of thermodynamic calculations are conducted in this study to provide constraints on REE precipitation in hydrothermal solutions, and provide an explanation of typical paragenesis of REE and gangue minerals at Bayan Obo. During the competition between fluocerite and monazite for LREE in the modelled solution (0.1 M HCl, 0.1 M HF and 0.1 M trichloride of light rare earth elements (LREE) from La to Sm), all LREE would eventually be hosted by monazite at a temperature over 300 °C, with continuous introduction of H_3PO_4. Additionally, monazite of heavier LREE would precipitate earlier, indicating that the Ce- and La-enriched monazite at Bayan Obo was crystallized from Ce and La pre-enriched hydrothermal fluids. The fractionation among LREE occurred before the ore-forming fluids infiltrating ore-hosting dolomite. When CO_2 (aq) was introduced to the aqueous system (model 1), bastnaesite would eventually and completely replace monazite-(Ce). Cooling of hot hydrothermal fluids (>400 °C) would significantly promote this replacement, with only about one third the cost of CO_2 for the entire replacement when temperature dropped from 430 °C to 400 °C. Sole dolomite addition (model 2) would make bastnaesite replace monazite and then be replaced by parisite. The monazite-(Ce) replaced by associated bastnaesite and apatite is an indicator of very hot hydrothermal fluids (>400 °C) and specific dolomite/fluid ratios (e.g., initial dolomite at 1 kbar: 0.049–0.068 M and 0.083–0.105 M at 400 °C and 430 °C). In hot solution (>430 °C) that continuously interacts with dolomite, apatite precipitates predating the bastnaesite, but it behaves oppositely at <400 °C. The former paragenesis is in accord with petrography observed in this study. Some mineral pairs, such as monazite-(Ce)-fluorite and monazite-(Ce)-parisite would never co-precipitate at any calculated temperature or pressure. Therefore, their association implies multiphase hydrothermal activities. Pressure variation would have rather limited influence on the paragenesis of REE minerals. However, temperature and fluid composition variation (e.g., CO_2 (aq), dolomite, H_3PO_4) would cause significantly different associations between REE and gangue minerals.

Keywords: REE precipitation; thermodynamic modelling; Bayan Obo REE-Nb-Fe deposit

1. Introduction

The Bayan Obo REE-Nb-Fe deposit is a typical carbonatite-related deposit, which indicates the rare earth element (REE) resources originated from local carbonatite magma [1,2]. However, there is still controversy about whether the abnormal REE enrichment was initiated during the intrusion of dolomitic carbonatite melt, or during hydrothermal metasomatism which overlapped a sedimentary carbonate formation [3,4]. Regardless of the origin of the ore-hosting dolomite, it is widely accepted that carbonatite-related hydrothermal processes are critical to the ultimate REE enrichment [5]. Some recent studies have even proposed that most of the REE-bearing minerals precipitated during hydrothermal processes at Bayan Obo [6,7]. There is a fundamental problem about the physico-chemical conditions of the ore-forming fluids and REE precipitation. Unfortunately, the primary fluid inclusions were poorly preserved at Bayan Obo. Although a few fluid inclusions with fluorocarbonate daughter minerals reach homogenization at over 400 °C [8], most of the fluid inclusions trapped in the megacrysts of gangue minerals from the vein-type ores display a homogenization temperature <340 °C, which is far lower than the highest temperature of the Bayan Obo carbonatite and ore-hosting dolomite calculated by dolomite–calcite geothermometry [5,9,10]. Therefore, it is difficult to analyze the Mesoproterozoic ore-forming fluids directly.

Thermodynamic modelling provides an alternative approach to the physico-chemical conditions of the ore-forming fluids and REE precipitation. Additionally, these semi-quantitative calculations are able to explain the paragenesis or sequent formation of various REE minerals qualitatively, which is vital to exploring the theoretical mechanism of Bayan Obo REE mineralization. In recent years, the fluorine and chlorine complexes of REE have been demonstrated to be the most important REE carriers during transportation, and more accurate thermodynamic properties of these species at higher temperature were also extrapolated accordingly [11]. The standard thermodynamic properties (lattice parameters, heat capacity and entropy of formation) of bastnaesite and parisite have been experimentally determined [12]. Several studies have provided general thermodynamic explanation of REE transportation and precipitation [13–15]. However, there has been no specific thermodynamic modelling based on the mineralogy and petrography of the Bayan Obo REE-Nb-Fe deposit yet. Considering that the dominating REE-bearing minerals at Bayan Obo are monazite and fluorocarbonate (bastnaesite, parisite and so on), it is time to perform thermodynamic calculations about the precipitation of REE minerals when all the requirements have been met.

2. Geological Setting

The geological setting of the Bayan Obo REE-Nb-Fe deposit is well-known and described in dozens of articles in detail [2,16,17]. There will only be a concise summary here. The Bayan Obo REE-Nb-Fe deposit is located on the north margin of the North China Craton (Figure 1b). At the age of 1.4–1.3 Ga [18], the Mesoproterozoic carbonatite intruded the Archean-Paleoproterozoic metamorphic basement rocks (2.4 Ga and 2.0 Ga of granitic gneiss and 2.0 Ga tonalite) [19–21] and the Paleoproterozoic-Mesoproterozoic metasedimentary cover, named as Lower Bayan Obo Group (1.8–1.4 Ga) [22–25]. The wall rock of some Bayan Obo carbonatite dykes suffered alkali metasomatism (fenitization) caused by hydrothermal fluids derived from these carbonatite dykes [26]. Meanwhile, the Bayan Obo ore-hosting dolomite, where most REE and Nb resources are host, also suffered extensive alkali and fluorine metasomatism. The ore-hosting dolomite is a nearly W-E-trending dolomite unit, enveloping a layer of black slate of Bayan Obo Group with similar trending (Figure 1a). Extensive alkali metasomatism also occurred in the slate in contact with the surrounding ore-hosting dolomite [27,28].

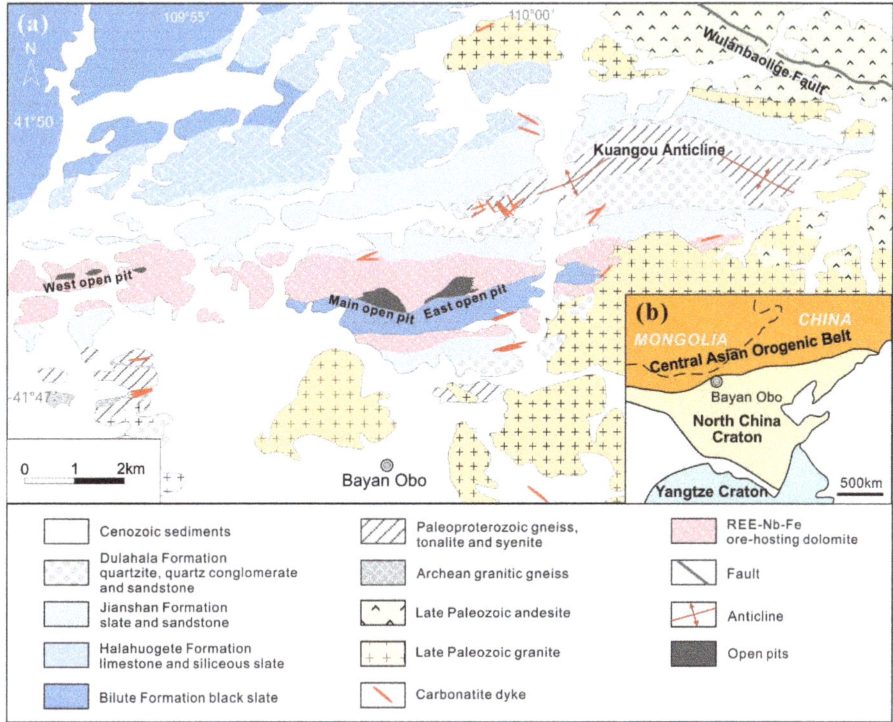

Figure 1. The geological setting of the Bayan Obo REE-Nb-Fe deposit (adapted from Yang et al. [18]) has been plotted in (**a**). The location of the Bayan Obo deposit in China is marked in the interior plot (**b**).

The carbonatite dykes at Bayan Obo were classified into magnesio-ferro-carbonatite dykes and calcio-carbonatite dykes, according to their major components. There are barely any REE-bearing minerals in the Mg-Fe-carbonatite dykes, except discrete monazite grains [7]. In contrast, the Ca-carbonatite dykes contain abundant bastnaesite and parisite, so that this kind of dyke may contain REE oxides even up to nearly 20 wt% [29]. The ore-hosting dolomite consists of coarse-grained dolomite on the margin of open pits, with only a few interstitial monazite grains, and extensively-altered fine-grained dolomite that is interior to the whole dolomite unit, containing the majority of the monazite and fluorocarbonates. Partial fine-grained dolomite was altered into diverse ores with certain grades of REE, Nb or Fe.

3. Methods

The backscatter electron imaging and energy-dispersive spectrum analyses were conducted on a FEI Nova Nano SEM 450 scanning electron microscope, at the State Key Laboratory of Continental Dynamics, Northwest University, China. The acceleration voltage was set as 20.0 kV, with a probe diameter of 4 μm. X-ray mapping was performed for monazite-bastnaesite associated thin section, because various REE minerals are indistinguishable in BSE imaging.

Thermodynamic calculations were conducted on the software GEM Selektor v.3.4, which is designed for thermodynamic modelling of an aqueous system by Gibbs energy minimization [30,31]. The database PSI/Nagra 12-07, embedded in the GEMS [32], and a third-party database named MINES 2017, packaged by Gysi and Williams-Jones [33] and Gysi [34], were adopted during the calculations. The MINES database includes revised SUPCRT92 data for aqueous species in addition to data on common ore and gangue minerals. Due to the limited data in these databases, the properties of some species were manually entered into the modified MINES 2017. The standard thermodynamic properties of fluocerite of LREE (La, Ce, Pr, Nd, Sm) were cited from Konings and Kovács [35] (S^0, Cp^0) and Migdisov et al. [11,15] (G^0, V^0). The data on phosphates of LREE ($LnPO_4$, Ln=La, Ce, Pr, Nd, Sm) were collected from Ni et al. [36] (V^0), Ushakov et al. [37] (H^0), Thiriet et al. [38] (S^0, G^0) and Popa et al. [39–41] (Cp^0). The standard thermodynamic properties of apatite-(F) ($Ca_{10}(PO_4)_6F_2$) are sourced from Jemal [42] (G^0), Cruz et al. [43,44] (H^0, V^0) and Dachs et al. [45] (Cp^0, S^0). Even though the data for bastnaesite ($Ce_{0.50}La_{0.25}Nd_{0.20}Pr_{0.05}CO_3F$), parisite, and parisite-(Ce) ($CaCe_{0.95}La_{0.60}Nd_{0.35}Pr_{0.10}(CO_3)_3F_2$) have been included in MINES by default, their properties would be summarized in Table 1, because of their significance to the calculation of REE precipitation [33]. The standard thermodynamic properties of all manually entered species and minerals entered species and minerals have been collected in Table 1. In addition, it is noticeable that the thermodynamic properties of bastnaesite and parisite were measured from natural minerals with a complex but fixed composition, while there were only properties of end members of the phosphate and fluocerite of each light lanthanide, instead of a solid solution of monazite or fluocerite, because these properties were only measured from artificially synthetic materials with simple compositions ($REEPO_4$ or $REEF_3$).

To set the initial composition of the modelled REE-bearing ore-forming fluids, the composition of orthomagmatic carbonatitic fluids trapped in fluid inclusion from Kalkfeld carbonatite (Namibia) [46] and the composition of hydrothermal fluids derived from alkali and silica-oversaturated intrusive and volcanic rocks (Capitan Pluton, U.S.) were taken into consideration [47] (Table 2). Although the LREE content of Banks et al. [47] has been adopted by Migdisov et al. [14] to calculate the transport and deposition of REE, the LREE content in fluids derived from alkali silicate magma is lower than that of carbonatitic fluids derived from Ca-carbonatite by 1 or 2 orders of magnitude [46]. The Bayan Obo REE mineralization became extensive after the carbothermal stage, when Ca-carbonatite dykes were formed by Ca-CO_2-rich carbonatite-derived fluids [7]. The fenitization occurred in both the wall rock of Ca-carbonatite dykes and the ore-hosting dolomite at Bayan Obo, and, thus, the ore-forming fluids were believed to be derived from Ca-carbonatite [48]. Therefore, as the only available composition of natural Ca-carbonatite-derived fluids, the analytical results of Kalkfeld carbonatite were chosen as the compositional benchmark of the Bayan Obo ore-forming fluids. Since the chloride of LREE has been demonstrated as the stable host of LREE at diverse pH and temperature [14,49], the LREE were introduced into the aqueous system in form of trichlorides. The initial composition of aqueous solution was set to 0.1 M HCl, 0.1 M HF, 0.1 M $LaCl_3$, 0.1 M $CeCl_3$, 0.1 M $PrCl_3$ and 0.1 M $NdCl_3$. Another 0.1 M of $SmCl_3$ and 0.001–0.1 M H_3PO_4 were added when calculating the LREE fractionation during monazite precipitation. The equal content of LREEs was chosen to assess their precipitation order, or fractionation order, avoiding influence from differentiated composition. The HCl was added to the modelled solution to help maintain an acid environment because real carbonatitic fluids may contain cations like Ca^{2+} and react with HF to form insoluble fluoride during REE transport. When monitoring the replacement of bastnaesite by monazite in the hydrothermal solution (model 1), 0.1 M of $SmCl_3$ and 0.001–0.6 M CO_2 (aq) were introduced to the aqueous system. On the other hand, there was an addition of 0.1 M $SmCl_3$, 0.1 M H_3PO_4 and 0.001–0.3 M dolomite when simulating the replacement of bastnaesite by monazite (model 2) and replacement of parisite by bastnaesite when neutralization occurred between the ore-hosting dolomite and the acid hydrothermal solution. In the above calculations, the amount of H_3PO_4, CO_2 and dolomite remained undetermined in fluid inclusions. Therefore, they were introduced to the system stepwise in rather wide ranges. The detailed species composition in the modelled solution at each step is displayed in Supplementary Tables S1–S5. P, T and X (composition of CO_2, dolomite and H_3PO_4) were set as independent variables, while the pH and pe of the aqueous system were dependent variables. The initial pH and pe states of each calculation in this study are summarized in Table S6.

Table 1. Standard thermodynamic properties of REE minerals that are added into the modified MINES database.

Species	S^0 (J/K·mol)	H^0 (J/mol)	G^0 (J/mol)	Cp^0 (J/K·mol)	V^0 (J/bar)	$Cp^0 = A0 + A1 \times T + A2/T^2 + A3/T1^{0.5} + A4 \times T^2$					Reference
						A0	A1	A2	A3	A4	
LaF_3	106.98	−1,732,132	−1,656,370	90.3	3.3000	255.870	−0.16815	−774,230	−1972.4	8.4811×10^{-5}	[11,15,35]
CeF_3	115.23	−1,719,174	−1,641,370	93.5	3.2010	103.258	−0.01299	−720,870	0	2.4688×10^{-5}	[11,15,35]
PrF_3	121.22	−1,717,746	−1,641,370	92.7	3.1530	130.599	−0.03250	−2,655,590	0	1.8169×10^{-5}	[11,15,35]
NdF_3	120.79	−1,710,367	−1,634,370	92.4	3.0930	103.387	0.00167	−1,101,170	0	1.0394×10^{-5}	[11,15,35]
SmF_3	116.50	−1,701,074	−1,624,370	91.7	3.1240	169.056	−0.07681	−4,840,760	0	0	[11,15,35]
$LaPO_4$	108.24	−1,970,700	−1,850,500	101.28	4.6013	121.128	0.03012	−2,562,500	0	0	[36–41]
$CePO_4$	119.97	−1,970,073	−1,849,800	106.43	4.5073	125.209	0.02789	−2,408,580	0	0	[36–41]
$PrPO_4$	123.24	−1,969,500	−1,850,500	106.0	4.4430	124.500	0.03037	−2,449,500	0	0	[36–41]
$NdPO_4$	125.53	−1,967,900	−1,849,600	104.8	4.3842	132.963	0.02254	−3,100,900	0	0	[36–41]
$SmPO_4$	122.49	−1,965,700	−1,846,900	105.6	4.2805	133.125	0.02347	−3,068,790	0	0	[36–41]
Bastnaesite-(Ce)	150.90	−1,808,400	−1,709,700	111.4	4.2910	9259.000	−9.07300	9,307,000	136,100	4.4600×10^{-3}	[12]
Parisite-(Ce)	391.60	−4,848,000	−4,571,500	290.8	12.2710	2494.000	−2.56400	5,943,000	−28,420	1.4790×10^{-3}	[12]
apatite-(F)	766.40	−13,598,000	−12,834,591	739.28	31.6652	1362.478	0	−12,762,680	−9243.46	1.4762×10^{9}	[42–45]

Fluocerite: Migdisov et al. [11,15] (V^0, G^0); Konings and Kovács [35] (S^0, Cp^0); Monazite: Ni et al. [36] (V^0); Ushakov et al. [37] (H^0); Thiriet et al. [38] (S^0, G^0); Popa et al. [39–41] (Cp^0); REE minerals: Gysi and Williams-Jones [12]; Apatite-(F): Jemal [42] (G^0); Cruz et al. [43,44] (H^0, V^0); Dachs et al. [45] (Cp^0, S^0).

Table 2. Composition of several vital elements in fluids derived from alkali igneous rock and carbonatite, and the initial composition of the modelled solution in this study.

Elements	Range	Banks et al. [47]		Bühn and Rankin [46]		Selected Value in This Study (mol/L)
		Content in Fluid (ppm)	Molar in 1kg Solution	Content in Fluid (ppm)	Molar in 1kg Solution	
La	max	334.7	2.4×10^{-3}	18,868	1.4×10^{-1}	0.01
	min	72.05	5.2×10^{-4}	3361	2.4×10^{-2}	
Ce	max	582.6	4.2×10^{-3}	12,797	9.1×10^{-2}	0.01
	min	83.97	6.0×10^{-4}	3802	2.7×10^{-2}	
Pr	max	57.75	4.1×10^{-4}	1223	8.7×10^{-3}	0.01
	min	6.27	4.4×10^{-5}	353	2.5×10^{-3}	
Nd	max	190.5	1.3×10^{-3}	3695	2.6×10^{-2}	0.01
	min	19.27	1.3×10^{-4}	1055	7.3×10^{-3}	
Sm	max	33.76	2.2×10^{-4}	568	3.8×10^{-3}	0.01
	min	3.43	2.3×10^{-5}	162	1.1×10^{-3}	
F	max	5245	2.8×10^{-1}	5000	2.6×10^{-2}	0.1
	min	358	1.9×10^{-2}	12,000	6.3×10^{-1}	

4. Petrographic Conditions of REE and Gangue Minerals Related to Thermodynamic Calculation

Similar to the Bayan Obo Mg-Fe carbonatite dykes, the least-altered coarse-grained ore-hosting dolomite barely contained REE minerals, except a few monazite grains (Figure 2a,b). Based on moderate ΣREE in the fresh coarse-grained ore-hosting dolomite, similar to Mg-Fe carbonatite dykes, they were regarded as a product of the magmatic stage of Bayan Obo carbonatite evolution [7]. When the ore forming fluids began to interact with dolomite grains from their boundaries, bastnaesite began to precipitate (Figure 2c,d). In extensively-altered ore-hosting dolomite or ores, it is common for abundant bastnaesite to associate with fluorite (Figure 2e,f). Occasionally, monazite would coprecipitate with fluorite in diverse types of ores (Figure 2g) or be associated with bastnaesite (Figure 2h).

There are complex paragenesis relationships among various REE minerals and gangue minerals. For example, fine-gained monazite is often surrounded and replaced by coarser-grained bastnaesite and apatite association (Figure 3). A similar phenomenon was also observed by Smith et al. [50]. The monazite gradually dissolved and then apatite was formed subsequently at the cost of phosphate anions. Meanwhile, the REE in the hydrothermal fluids precipitated in the form of bastnaesite. As a result, monazite was gradually corroded by the ore-forming fluids, while the precipitation of bastnaesite took place. It is common for bastnaesite to be associated with apatite (Figure 4a,b), even without monazite, which may be because formerly formed monazite was completely dissolved. The paragenesis of apatite and bastnaesite was often indistinguishable (Figure 4a), but occasionally it is evident that apatite precipitates predating the bastnaesite (Figure 4b). Fluorite also often coexists with bastnaesite, especially in fluorite-rich banded or massive ores (Figure 4c). It is also possible that monazite is associated with interstitial apatite and calcite, indicating the dissolution of the former (Figure 4d).

The replacement of parisite by bastnaesite and monazite was common in the Ca-carbonatite dykes and various type of ores. Megacryst of bastnaesite in the vein-type of ores was able to be altered into parisite at the margin (Figure 5a), and those bastnaesites in the fluorite-rich altered ore-hosting dolomite were replaced by parisite from grain boundaries or cleavages with the grains (Figure 5b). The parisite would corrode and replace the bastnaesite gradually, with relics of the latter inside the aggregates of parisite (Figure 5c). Occasionally, there are fine-grained monazite aggregates that were replaced by parisite from their grain boundaries. Both the parisite and apatite were crystallized interstitially (Figure 5d).

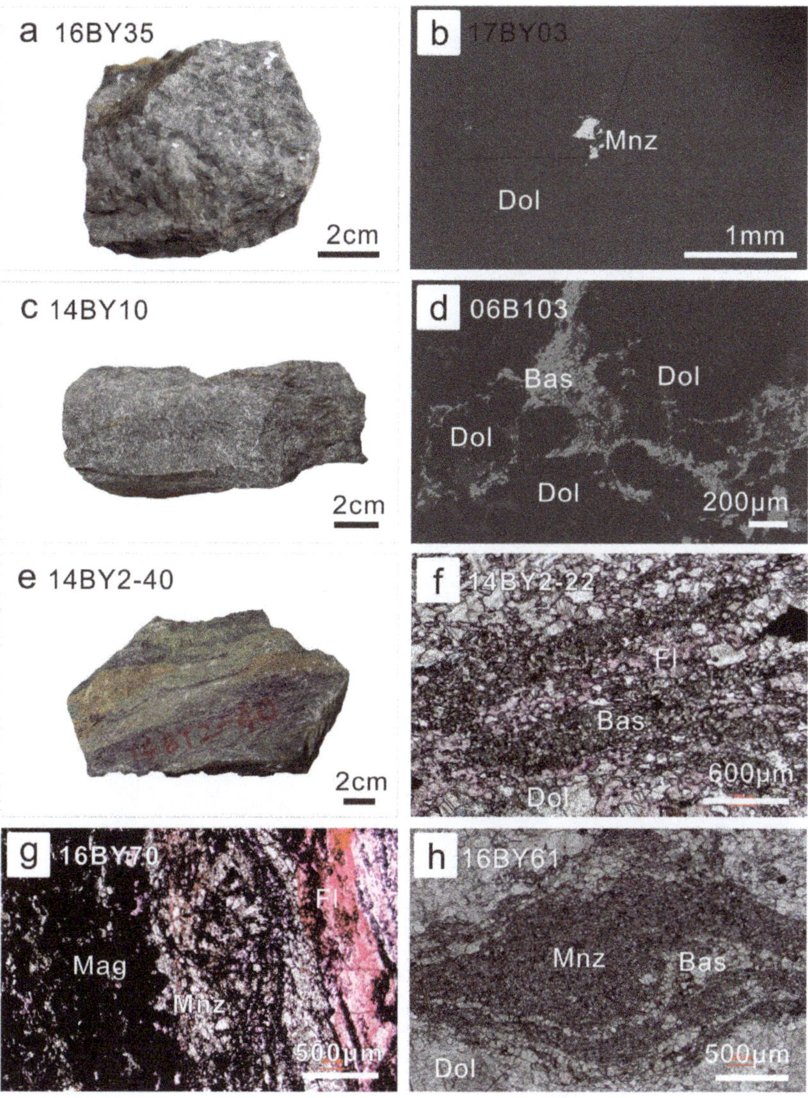

Figure 2. Petrography of ore-hosting dolomite with diverse degrees of alteration: (**a**) photography of the coarse-grained ore-hosting dolomite (sample No. 16BY35); (**b**) discrete monazite grain in the coarse-grained ore-hosting dolomite (backscatter electron (BSE) image of sample 17BY03); (**c**) photography of the slightly-altered fine-grained ore-hosting dolomite (sample No. 14BY10); (**d**) interstitial bastnaesite aggregates among relatively fine-dolomite grains (BSE of 06B103); (**e**) photography of the extensively-altered fine-grained ore-hosting dolomite (sample No. 16BY35); this sample was also called fluorite-aegirine-rich ore; (**f**) bastnaesite in the fluorite-rich ores under plane polarized light (14BY2-22); (**g**) monazite associated with fluorite and iron oxide in the aegirine-rich ores under plane polarized light (16BY70); (**h**) the association of monazite (fine-grained) and bastnaesite (coarser-grained) in the fine-grained ore-hosting dolomite (16BY61). Abbreviation of minerals: Dol—dolomite; Mnz—monazite; Bas—bastnaesite; Fl—fluorite; Mag—magnetite.

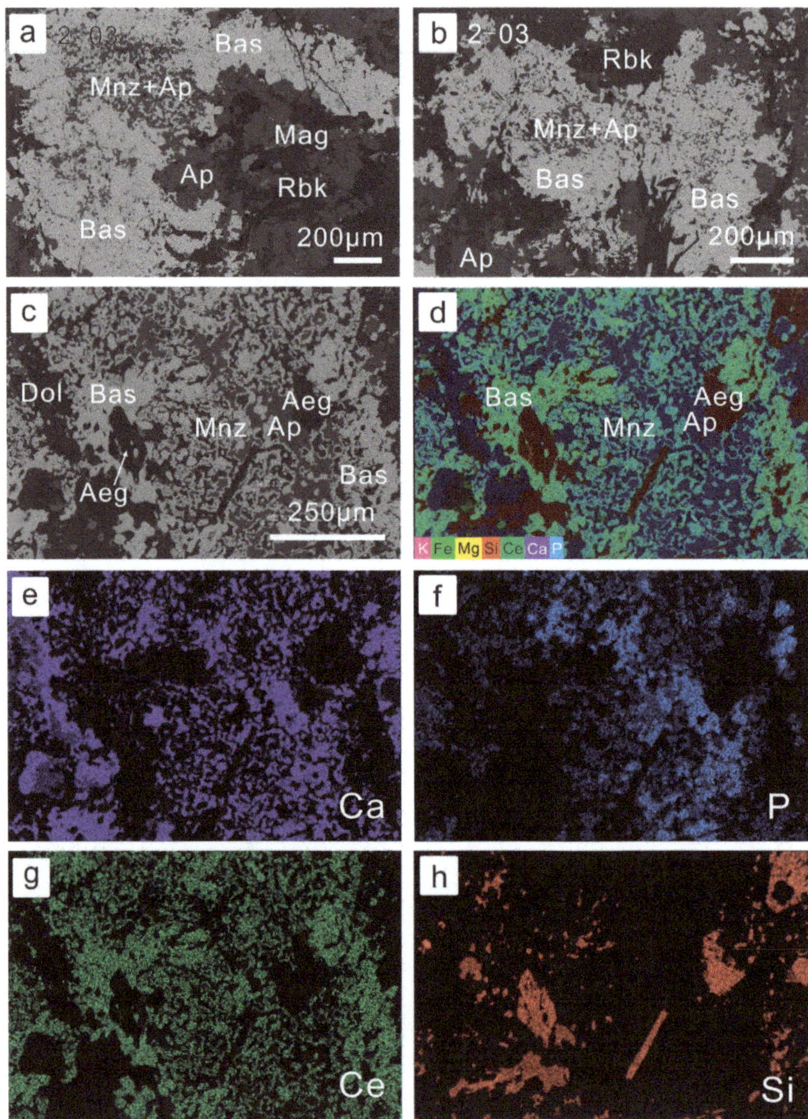

Figure 3. Petrography of replacement of bastnaesite by monazite: (**a**,**b**) monazite associated with apatite and replaced by bastnasite–apatite association (BSE image for sample 14BY-2-03, sodium-amphibole-rich ore); (**c**) coarser-grain bastnaesite replaced and associated with monazite and apatite (BSE image for sample BY-4, fluorite-rich REE-Nb ore); (**d**) compiled multi-element x-ray mapping, where monazite displays as light blue and bastnaesite has a green color; (**e**–**h**) single-element x-ray mapping of (**c**). Abbreviation of minerals: Dol—dolomite; Mnz—monazite; Bas—bastnaesite; Mag—magnetite; Rbk—riebeckite; Ap—apatite; Aeg—aegirine.

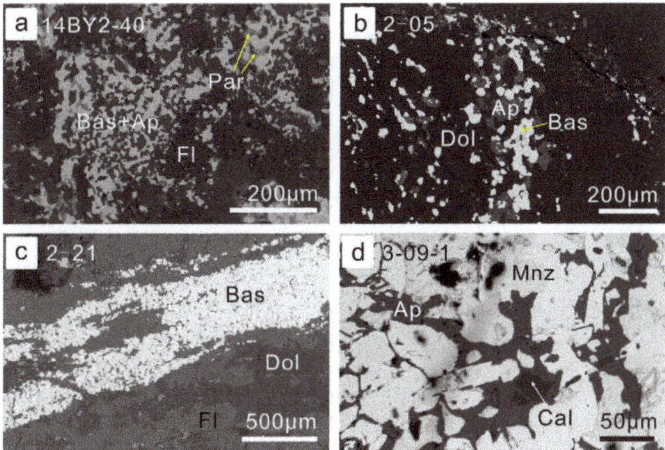

Figure 4. Petrography of association of REE minerals and gangue minerals: (**a**) the association of bastnaesite with apatite was replaced by fluorite and parisite (BSE image for sample 14BY2-40, fluorite-aegirine-rich ores); (**b**) bastnaesite association with apatite in the matrix of dolomite (BSE image for sample 14BY2-05, fine-grained ore-hosting dolomite); (**c**) banded bastnaesite aggregates in the fluorite-rich REE-Nb ores (BSE image for sample 14BY2-21); (**d**) monazite was corroded from grain boundaries and replaced by apatite and calcite (BSE image for sample 14BY3-09-1, banded REE-Fe ore). Abbreviation of minerals: Bas—bastnaesite; Ap—apatite; Par—parisite; Fl—fluorite; Dol—dolomite; Mnz—monazite; Cal—calcite. Hereinafter in the caption signatures, these abbreviations of mineral names will be used.

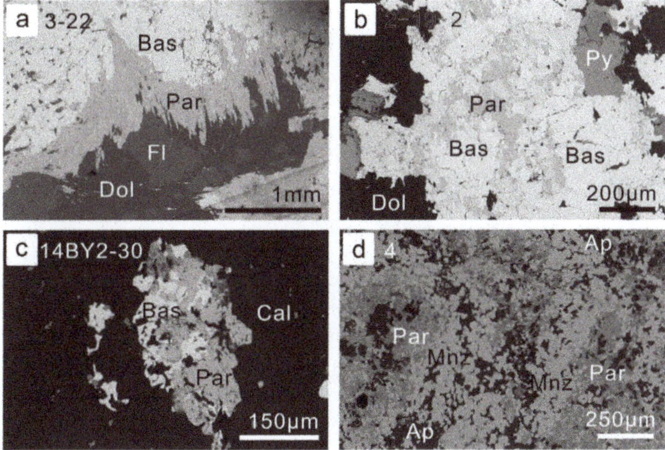

Figure 5. Petrography of replacement of parisite by bastnaesite and monazite: (**a**) bastnaesite megacryst in the vein type of ore with its margin altered into parisite (BSE image for sample 14BY3-22, vein-type of ore); (**b**) bastnaesite was replaced by parisite from the grain boundaries and cleavages within the grains (BSE image for sample 14BY2-19, fluorite-rich altered ore-hosting dolomite); (**c**) bastnaesite grains were almost replaced by parisite (BSE image for sample 14BY2-30, vein-type of ore); (**d**) fine-grained monazite aggregates were altered into parisite from grain boundaries (BSE image for sample BY-4, fluorite-rich REE-Nb ore). Abbreviation of minerals have been listed in former figures, except Par for parisite and Py for pyrite.

5. Results

5.1. LREE Fractionation during Precipitation of Monazite

In the modelled solution with 0.1 M HCl, 0.1 M HF and 0.1 M trichloride of each LREE (La, Ce, Pr, Nd, Sm), phosphoric acid was added stepwise into the aqueous solution, with 0.001 M H_3PO_4 at each step. Only LREE was taken into consideration because monazite at Bayan Obo was LREE-enriched. In these calculations, under different temperatures and pressures, there is competition between the fluocerite and phosphate of LREE in the modelled solution.

Under the pressure of 3 kbar, LREE would only precipitate as monazite at a temperature of 400 °C, in the order of heavier LREEs Sm, Pr and Nd to lighter LREEs Ce and La (Figure 6a). Fluocerites of La and Pr became stable when H_3PO_4 was instantly added to the solution at 300 °C under the same pressure (Figure 6b), but these fluocerites would be completely replaced by corresponding monazite with continuous addition of H_3PO_4 (0.021 M and 0.049 M H_3PO_4 for fluocerite-(Pr) and –(La)). At 200 °C, fluocerite of all LREE would appear when H_3PO_4 was instantly added, but only LaF_3 remained stable in the whole range of added H_3PO_4 (Figure 6c). The monazite of Sm, Nd, Pr and Ce began to dissolve and the corresponding fluocerite was crystalized when 0.001, 0.008, 0.019 and 0.030 M H_3PO_4 was added to the solution. At 100 °C, the stability of not only LaF_3 but CeF_3 also significantly increased (Figure 6d). In contrast, Sm and Nd prefers to precipitate in the form of monazite. $SmPO_4$ crystalized instantly when H_3PO_4 was added. NdF_3 dissolved and $NdPO_4$ began to form when 0.009 M H_3PO_4 was added. Pr were stored in PrF_3 until 0.033 M H_3PO_4 was introduced, and then $PrPO_4$ began to replace corresponding fluocerite.

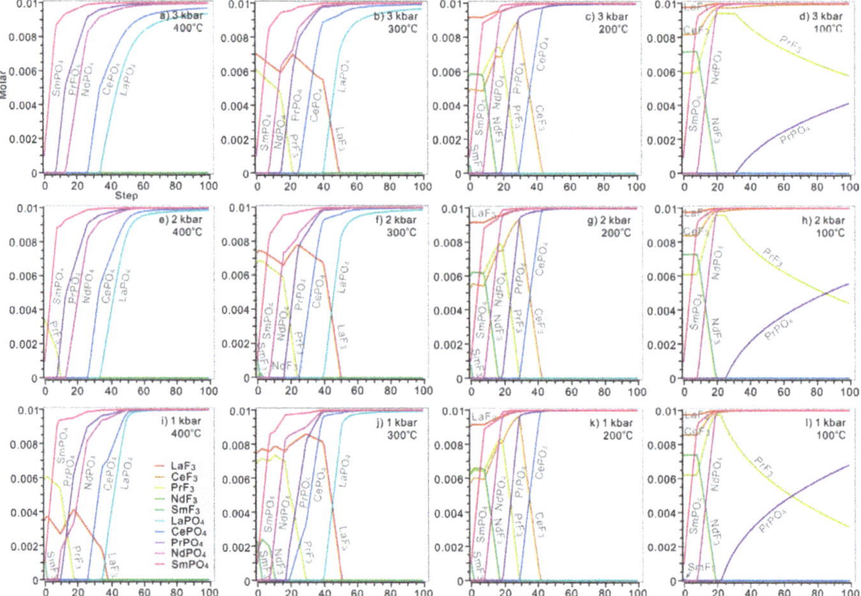

Figure 6. The precipitation of fluocerite and monazite in solution with 0.1 M HCl, 0.1 M HF, 0.1 M $LaCl_3$, 0.1 M $CeCl_3$, 0.1 M $PrCl_3$, 0.1 M $NdCl_3$, 0.1 M $SmCl_3$ and 0.001–0.1 M H_3PO_4. The H_3PO_4 was added stepwise with 0.001 M at each step. The calculation was conducted at diverse temperatures (ranging from 400 °C to 100 °C) and pressures (ranging from 3 kbar to 1 kbar). The P-T condition of each plot: (**a**) 3 kbar, 400 °C; (**b**) 3 kbar, 300 °C; (**c**) 3 kbar, 200 °C; (**d**) 3 kbar, 10 °C; (**e**) 2 kbar, 400 °C; (**f**) 2 kbar, 300 °C; (**g**) 2 kbar, 200 °C; (**h**) 2 kbar, 100 °C; (**i**) 1 kbar, 400 °C; (**j**) 1 kbar, 300 °C; (**k**) 1 kbar, 200 °C; (**l**) 1 kbar, 100 °C.

Under a lower pressure of 2 kbar, PrF_3 would exist as the only fluocerite when H_3PO_4 was introduced at 400 °C (Figure 6e). At 300 °C, the fluocerite of Sm and Nd became more unstable when a trace amount of H_3PO_4 was added (Figure 6f). The precipitation order remains the same below 200 °C (Figure 6g,h).

Under the pressure of 1 kbar, the fluocerite of La, Pr and Sm appeared once H_3PO_4 was added at 400 °C (Figure 6i). In solution below 300 °C, the precipitation order would remain similar to that of solution under lower pressure (Figure 6j–l). At 300 °C, all fluocerite of LREE completely dissolved when a certain amount of H_3PO_4 was added. LREE tends to precipitate as monazite (Figure 6j). At 200 °C, Only LaF_3 would be preserved all the way; $SmPO_4$, $NdPO_4$, $PrPO_4$ and $CePO_4$ were formed subsequently, with increasing addition of H_3PO_4 accompanying dissolution of fluocerite in the order of SmF_3, NdF_3, PrF_3 and CeF_3 (Figure 6k). La and Ce tended to be stored in corresponding fluocerite no matter how much H_3PO_4 was added at 100 °C (Figure 6l). In contrast, Sm and Nd prefer to precipitate in the form of monazite eventually. $SmPO_4$ crystalized instantly when H_3PO_4 was added. NdF_3 dissolved and $NdPO_4$ began to form when 0.01 M H_3PO_4 was added. Pr was stored in PrF_3 until 0.024 M H_3PO_4 was introduced, and then $PrPO_4$ began to replace the corresponding fluocerite.

These calculations indicate that increasing temperature or pressure would destabilize fluocerite, but the influence of pressure is rather limited. When the modelled solution was hotter than 300 °C under diverse pressure, all REE would precipitate in monazite eventually. While in a modelled solution hotter than 200 °C, no $LaPO_4$ and $CeLO_4$ would be formed, respectively (Figure 6a,b,e,f,i,j). Among all fluocerite of LREE, LaF_3 was the most stable species under 300 °C.

5.2. Replacement of Bastnaesite by Monazite

The precipitation of fluorocarbonates needs a carbonate anion in hydrothermal solution. There are two methods introducing carbonate into the aqueous system that have been considered in this study: adding CO_2 (aq) (model 1) or adding dolomite into the acid solution (model 2), which has been proposed by Williams-Jones et al. [13]. Since the natural bastnaesite, whose standard thermodynamic properties were measured by Gysi and Williams-Jones [12], contains four LREEs (La, Ce, Pr, Nd), the modelled solution is composed of 0.1 M HCl, 0.1 M HF, 0.1 M H_3PO_4 and 0.1 M trichloride of La, Ce, Pr and Nd. Sm was excluded from the system to simplify the calculation. Quantities of 0.001 M CO_2 and 0.001 M dolomite were contained in the initial solution of model 1 and model 2, respectively.

Model 1: When CO_2 (aq) was introduced to the solution, bastnaesite began to replace monazite precipitated from the solution. Under the pressure of 2 kbar, at 430 °C, the amount of CO_2 necessary to initiate the formation of bastnaesite was 0.339 M (Figure 7a). At 400 °C, monazite of LREE remained stable until over 0.09 M CO_2 was added (Figure 7b). At 300 °C, when 0.02 M CO_2 were added to the solution, bastnaesite reached its maximum amount, 0.02 M, as well (Figure 7c). At 200 °C, $LaPO_4$ would not exist and, in turn, LaF_3 stored all of the La before CO_2 was introduced. Once CO_2 was introduced, fluocerite-(La) phosphate of Ce, Pr and Nd began to dissolve, until $CePO_4$ completely dissolved and all Ce were consumed by bastnaesite ($Ce_{0.50}La_{0.25}Nd_{0.20}Pr_{0.05}CO_3F$) (Figure 7d). With decreasing temperature, complete replacement of bastnaesite by monazite requires less CO_2 (aq): 0.53 M CO_2 at 430 °C, 0.16 M CO_2 at 400 °C, 0.02 M CO_2 at 300 °C, and 0.019 M CO_2 at 200 °C. At temperatures over 400 °C, the cooling of the hydrothermal solution significantly promoted the formation of bastnaesite (Figure 7a,b), as the amount of CO_2 necessary for complete replacement decreases rapidly.

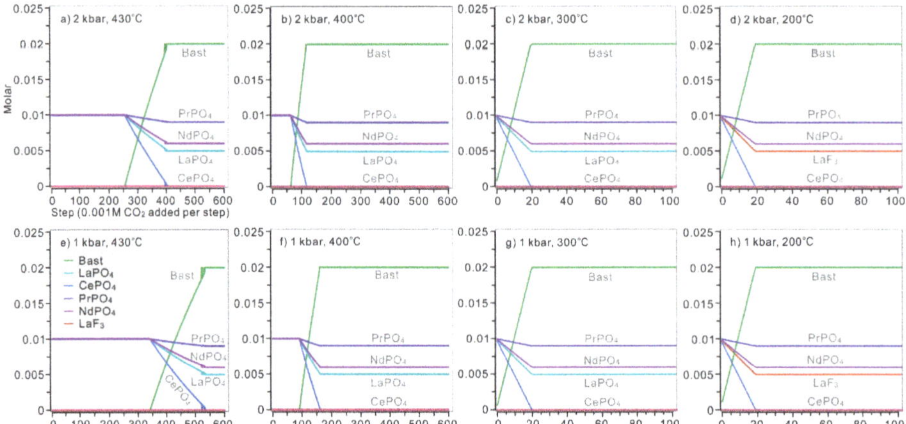

Figure 7. The replacement of bastnaesite by monazite when CO_2 (aq) was introduced stepwise (0.001 M CO_2 each step). The modelled solution contains 0.1 M HCl, 0.1 M HF, 0.1 M H_3PO_4, 0.1 M $LaCl_3$, 0.1 M $CeCl_3$, 0.1 M $PrCl_3$, 0.1 M $NdCl_3$, and 0.001–0.6 M CO_2. The calculation was conducted on diverse temperature (ranging from 430 °C to 200 °C) and pressure (ranging from 2 kbar to 1 kbar). The P-T condition of each plot: (**a**) 2 kbar, 430 °C; (**b**) 2 kbar, 400 °C; (**c**) 2 kbar, 300 °C; (**d**) 2 kbar, 200 °C; (**e**) 1 kbar, 430 °C; (**f**) 1 kbar, 400 °C; (**g**) 1 kbar, 300 °C; (**h**) 1 kbar, 200 °C.

Under the pressure of 1 kbar, the precipitation order is similar to that of the solution under the pressure of 2 kbar. At above 400 °C, it took slightly less CO_2 to initiate formation of bastnaesite compared with a solution under 1 kbar (Figure 7e,f). LaF_3 was still stable at 200 °C, storing La when bastnaesite could no longer take it (Figure 7h). The decreased pressure would hinder the replacement of bastnaesite by monazite, but the influence was rather limited in solution below 300 °C.

Model 2: If dolomite, rather than CO_2, was added into the system, the neutralization reaction between dolomite and acid model solution would produce both $CO_3{}^{2-}$ and Ca^{2+} in the system. Therefore, it is possible to monitor the precipitation of fluorite and apatite. Under the pressure of 2 kbar and temperature of 430 °C, when 0.052 M dolomite was introduced to the system, apatite began to precipitate (Figure 8a). 0.077 M dolomite was needed for the initiation of monazite dissolution and bastnaesite formation. Apatite was able to coexist with $CePO_4$ and bastnaesite when a specific amount of dolomite was introduced to the modelled solution (added dolomite was in the range of 0.076–0.096 M at 430 °C).

At 400 °C, when 0.032 M dolomite was added, the monazite began to be replaced by bastnaesite (Figure 8b). More than 0.054 M dolomite was needed for the precipitation of apatite. In addition, it is impossible for apatite to coexist with $CePO_4$ and bastnaesite because bastnaesite reached its maximum value when 0.053 M dolomite was introduced to the modelled solution, and monazite-(Ce) disappeared simultaneously.

At 300 °C, once the dolomite was introduced, bastnaesite replaced monazite instantly until $CePO_4$ was completely consumed (Figure 8c). Bastnaesite reached its maximum content after 0.01 M dolomite was added. Slightly more dolomite (0.056 M) was necessary for apatite crystallization. The replacement of bastnaesite by monazite required less dolomite in a cooler solution.

At 200 °C (Figure 8d), once dolomite was introduced into the system, LaF_3, $CePO_4$, $PrPO_4$ and $NdPO_4$ dissolved and bastnaesite began to precipitate. Bastnaesite reached its maximum content after 0.01 M dolomite was added. When 0.031 M dolomite was added, fluocerite-(La) lost stability and fluorite began to act as the major host of fluorine, and $LaPO_4$ was formed simultaneously. Apatite suddenly appeared when 0.040 M dolomite was added to the solution. Dolomite became saturated when 0.122 M dolomite was added. In general, the decreasing temperature would promote the replacement of bastnaesite by monazite with the continuous addition of dolomite.

The decreased pressure slightly influenced REE precipitation by increasing the dolomite necessary for replacement of bastnaesite by monazite (Figure 8e–h). Under the pressure of 1 kbar, only 0.045 M and 0.084 M dolomite was need for the initiation of bastnaesite formation at 400 °C and 430 °C, respectively. Under the pressure of 1 kbar, apatite was able to coexist with $CePO_4$ and bastnaesite when a specific amount of dolomite was introduced to the modelled solution (added dolomite was in the range of 0.049–0.068 M and 0.083–0.105 M at 400 °C and 430 °C).

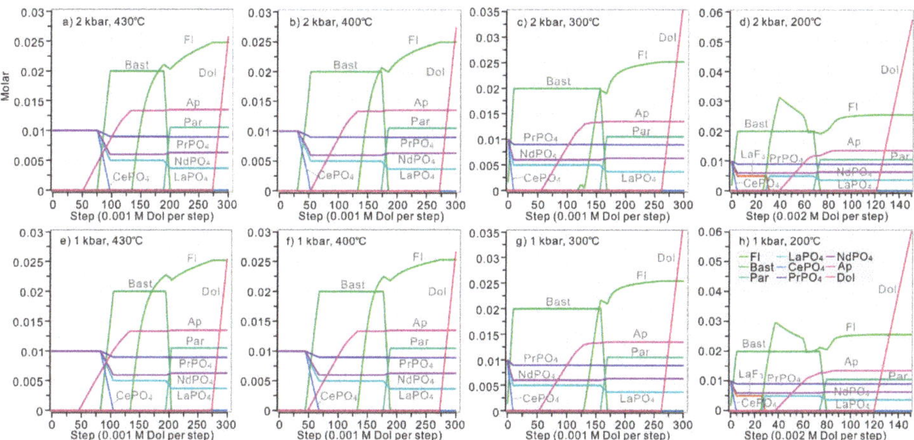

Figure 8. The replacement of bastnaesite by monazite when dolomite was introduced stepwise. Different amounts of dolomite were added in the calculations under different P and T: 0.001 M dolomite was added for each step in calculations above 300 °C, while 0.002 M dolomite was added for each step when calculating at 200 °C (Figure 8d,h). The modelled solution contains 0.1 M HCl, 0.1 M HF, 0.1 M H_3PO_4, 0.1 M $LaCl_3$, 0.1 M $CeCl_3$, 0.1 M $PrCl_3$, 0.1 M $NdCl_3$, and 0.001 M-0.3 M dolomite. The calculation was conducted at diverse temperatures (ranging from 430 °C to 200 °C) and pressures (ranging from 2 kbar to 1 kbar). The P-T condition of each plot: (**a**) 1 kbar, 200 °C; (**b**) 1 kbar, 300 °C; (**c**) 1 kbar, 400 °C; (**d**) 1 kbar, 430 °C; (**e**) 2 kbar, 200 °C; (**f**) 2 kbar, 300 °C; (**g**) 2 kbar, 400 °C; (**h**) 2 kbar, 430 °C.

5.3. Replacement of Parisite by Bastnaesite

When sufficient dolomite was added into the system, parisite would eventually replace the bastnaesite. In the last model, under the same pressure, the parisite began to replace bastnaesite with an increasing amount of dolomite at higher temperatures (Figure 8). In the modelled solution at 2 kbar, a little bit less dolomite was needed: 0.189 M, 0.173 M, 0.156 M and 0.136 M dolomite for the modelled solution at 430 °C, 400 °C, 300 °C and 200 °C, respectively (Figure 8a–d). Under the pressure of 1 kbar, the initiation of this replacement requires 0.194 M, 0.177 M, 0.159 M and 0.150 M dolomite at 430 °C, 400 °C, 300 °C and 200 °C, respectively (Figure 8e–h).

In order to evaluate the influence of temperature on the replacement of parisite by bastnaesite, a series of calculations was conducted in a continuous range of temperatures from 30 to 430 °C, with several discrete and specific amounts of added dolomite (Figure 9). When 0.001 M dolomite was introduced, there is only competition between monazite and fluocerite as the major hosts of LREE (Figure 9a). The parisite would only precipitate when over 0.10 M dolomite was added (Figure 9c–f). Specifically, when 0.10 M dolomite was introduced, parisite was able to replace bastnaesite below 100 °C (Figure 9c). When 0.15 M dolomite was added, the replacement would be initiated below 200 °C, and apatite stayed stable in the whole range of calculated temperatures (Figure 9d). When over

0.20 M dolomite was added, parisite became the dominated REE host at any calculated temperature (Figure 9e,f). Besides, dolomite turned out to be saturated when the temperature of solution dropped to 80 °C and 230 °C in the case of 0.20 M and 0.25 M dolomite addition (Figure 9e,f). In the above calculations, with several discrete initial amounts of dolomite, the replacement of bastnaesite and apatite-(F) by $CePO_4$ was possible only when 0.10 M dolomite was added into a solution at over 420 °C (Figure 9c).

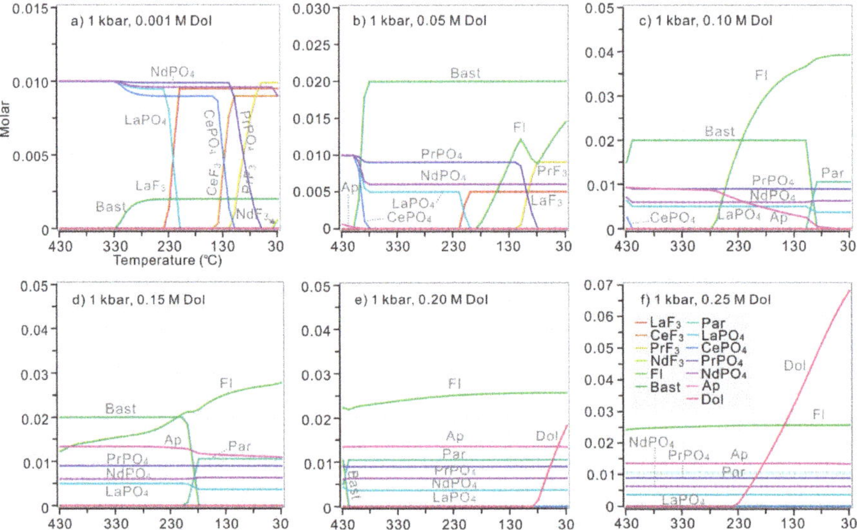

Figure 9. The precipitation of REE minerals in a continuous range of temperatures (30–430 °C) when discrete amounts of dolomite were added into a modelled solution, with 0.1 M HCl, 0.1 M HF, 0.1 M H_3PO_4, 0.1 M $LaCl_3$, 0.1 M $CeCl_3$, 0.1 M $PrCl_3$ and 0.1 M $NdCl_3$. All calculations were performed under the pressure of 1 kbar. The discrete amount of dolomite added into the initial fluids include (**a**) 0.001 M; (**b**) 0.05 M; (**c**) 0.10 M; (**d**) 0.15 M; (**e**) 0.20 M; and (**f**) 0.25 M. The temperature decreased 10 °C per step.

The influence of pressure on the replacement of parisite by bastnaesite was also verified by calculations (Figure 10). Based on the above calculations with discrete dolomite addition, the replacement of parisite by bastnaesite was possible when 0.10 M, 0.15 M and 0.20 M dolomite was introduced to the system (Figure 9c–e). Each condition was calculated under various pressures from 1 kbar to 3 kbar. In a cooling solution, the increase of pressure would generally elevate the onset temperature of the above replacement. When 0.10 M dolomite was added, parisite began to precipitate at 102 °C, 116 °C and 137 °C under the pressures of 1 kbar, 2 kbar and 3 kbar (Figure 10a). When 0.15 M and 0.20 M dolomite were added, the onset temperature of this replacement was increased from 202 °C to 225 °C and 436 °C to 441 °C if the pressure increased from 1 kbar to 2 kbar (Figure 10b,c). In general, increasing pressure promoted the replacement of parisite by bastnaesite at higher temperatures. However, the significant variation of pressure only caused replacement onset temperature to vary within 35 °C.

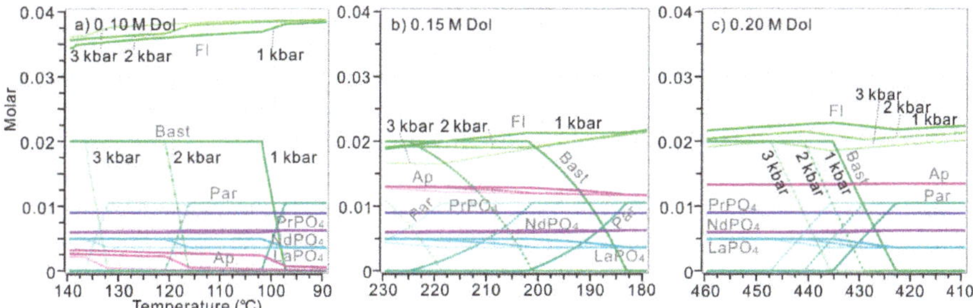

Figure 10. The replacement of parisite by bastnaesite at diverse temperatures and pressures. The initial model solution is the same as fluids in Figure 10. Discrete amounts of dolomite were introduced to the aqueous system: (**a**) 0.10 M; (**b**) 0.15 M; (**c**) 0.20 M. For each plot, the precipitation of REE minerals and gangue minerals have been calculated under the pressure from 1 kbar (solid lines) to 2 kbar (dash lines) then to 3 kbar (transparent lines with spots) and at temperature of 90–140°C for (**a**), 180–230 °C for (**b**) and 410–460 °C for (**c**). These temperature ranges were selected to cover the replacement processes. The temperature decreased 0.5 °C per step.

6. Discussion

6.1. REE Fractionation Indicated by Monazite Precipitation

According to our calculation, with increasing concentration of H_3PO_4, monazite of heavier LREE tends to crystalize from the aqueous solution earlier (Figure 6). However, monazite at Bayan Obo is characterized by Ce enrichment. For example, monazite collected from ore-hosting dolomite contains Ce (23.7–43.5 wt%, ave. 29.8 wt%, n = 12) and La (13.0–28.8 wt%, ave. 18.4 wt%), much more than the sum of other REEs [51]. This indicates that the hydrothermal fluid forming the monazite is Ce-and La-enriched, and extreme LREE enrichment was accomplished ahead of the extensive precipitation of monazite. Since monazite is one of the earliest hydrothermal minerals in the ore-hosting dolomite, the fractionation among LREEs was unlikely to be caused by fractionated crystallization during the hydrothermal process, but played as a geochemical characteristic of the initial ore-forming fluid infiltrating the ore-hosting dolomite, which conforms to the proposal that the ore-forming fluids originated from carbonatite [48]. These calculations have provided thermodynamic support. In addition, fluocerite has never been discovered and reported at Bayan Obo in the past 30 years. Thus, REE precipitation is very likely to occur when temperatures of hydrothermal fluids are over 300 °C, when LaF_3 would not be persistent in a wide range of H_3PO_4 compositions (0.001–0.1 M), and $LaPO_4$ would host all of the La eventually. Meanwhile, the concentration of H_3PO_4 in the solution should exceed 0.05 M when fluocerites remain absent under variant pressure, even though analytical measurement was unable to provide this constraint. Despite all of this, the pressure of ore-forming fluids is not easy to judge due to its limited influence on the precipitation of fluocerite and monazite in the modelled solution. On the other hand, the stability of fluocerite decreases significantly along with increasing temperature.

6.2. Replacement of Bastnaesite by Monazite

The modelled solution (model 1) contains the same amount of La, Ce, Pr and Nd. Actually, as proved above, the real ore-forming fluids are extremely La- and Ce-enriched. Therefore, $PrPO_4$ and $NdPO_4$ would be unstable in real hydrothermal fluids because limited Pr and Nd would be definitely consumed by bastnaesite. The replacement of bastnaesite by monazite would be continuous until either La or Ce was exhausted. Monazite was more likely to remain stable in hydrothermal fluids hotter than 400 °C (Figure 7a,b). In a cooling hydrothermal system hotter than 400 °C, the temperature decrease would influence the stability of monazite and promote replacement of bastnaesite to monazite more significantly. The amount of CO_2 needed for the complete replacement of bastnaesite by monazite at 400 °C was less than one third of the CO_2 needed at 430 °C. Therefore, extensive replacement of bastnaesite by monazite might begin to occur when hydrothermal fluid was cooled to about 400 °C or a little bit above, which is in accord with the complete dissolution temperature (420–480 °C) of the few discovered fluorocarbonate crystals trapped in Bayan Obo fluid inclusions [8]. However, it is still uncertain whether these few fluorocarbonate-bearing fluid inclusions are primary. Combined with the calculation of model 2, the addition of dolomite would also promote the replacement of bastnaesite by monazite. However, the neutralization reaction is not a must for the replacement if the ore-forming fluids contain sufficient CO_2 (aq) as shown in model 1, and precipitation of apatite and fluorite may need extra calcium cations in this condition. Therefore, significant pH shifting is not a fundamental requirement for the occurrence of replacement among REE minerals. However, the dolomite definitely participates in REE precipitation at Bayan Obo, as almost all REE resources are hosted in a dolomite unit.

The monazite-(Ce) would be replaced by associated apatite and bastnaesite when the hydrothermal solution was hot enough (>400 °C) (Figure 3; Figure 8a,b,e,f). Meanwhile, in a cooler modelled solution of 200 °C or 300 °C, when bastnaesite finished replacing all of the $CePO_4$, the apatite-(F) was unable to crystallize yet. As these results have shown, a certain amount of dolomite was also necessary for the coexistence of monazite-(Ce), bastnaesite and apatite-(F) (the range of dolomite necessary at 1 kbar: 0.049–0.068 M and 0.083–0.105 M at 400 °C and 430 °C; 2 kbar: 0.077 –0.097 M at 430 °C). With added dolomite less than the specific range, monazite would remain stable while bastnaesite would be unlikely to appear. When introduced dolomite exceeds the specific range, bastnaesite and apatite would remain stable while $CePO_4$ would dissolve completely. Considering that Bayan Obo monazite is Ce-enriched, the condition that monazite-(Ce) dissolves completely is inappropriate for crystallization of Bayan Obo monazite.

It is interesting that, in our calculation of model 2, monazite-(Ce) would never co-precipitate with fluorite. If there is any association of these two minerals, as in Figure 2g, according to the above results, the fluorite and monazite should crystallize subsequently, or overlap each other as products of two hydrothermal activities. Other associated relationships, such as bastnaesite with fluorite or bastnaesite with apatite, are likely to occur from the perspective of thermodynamic modelling. Generally, with continuous addition of dolomite into solution with constant temperature, bastnaesite would precipitate earlier than fluorite (Figure 8), as in the paragenesis in Figure 4c, where fluorite crystallizes from the grain boundaries of bastnaesite aggregates. For the same reason, the bastnaesite and associated apatite were replaced by fluorite in Figure 4a. Bastnaesite would be associated with apatite at any calculated temperature. At 430 °C, apatite would crystalize earlier than bastnaesite with increasing introducing of dolomite, as shown in Figure 4b. At 400 °C, bastnaesite would nearly co-precipitate with apatite. At lower temperatures, the precipitation of bastnaesite would predate the apatite. Therefore, all above petrography observed in this study supports a very-high-temperature REE precipitation.

6.3. Replacement of Parisite by Bastnaesite

Based on a series of calculations, decreasing temperature, increasing the addition of dolomite and increasing pressure are beneficial to the replacement of parisite by bastnaesite. In a real hydrothermal solution, the temperature keeps dropping when it is infiltrating and interacting with the cold ore-hosting dolomite. The replacement of parisite by bastnaesite was more likely to occur in the peripheral veinlets of the fluid passage, because this high-rock/fluid-ratio environment, as well as quenched fringe of the hydrothermal fluids, meets favorable conditions of both increasing dolomite addition and decreasing temperature. The pressure of the hydrothermal fluids is difficult to assess, as the pressure would drop in an open system but keep rather stable in a closed or semi-closed system. However, the significant variation in pressure would only affect the onset temperature of replacement slightly, according to calculations in this study. Therefore, regardless of the shift of pressure, the parisite was observed replacing the bastnaesite (or monazite occasionally as in Figure 5d) in the ore-hosting dolomite from the narrow space of their grain boundaries (Figure 5b,c). Alternatively, if the hydrothermal fluids have interacted with enough dolomite, the replacement of parisite by bastnaesite would also occur in veins or extensional fractures in the ore-hosting dolomite, instead of in a high-dolomite/fluid-ratio environment, as in the condition shown in Figure 5a.

According to the calculation in Figures 8 and 9, parisite would not directly replace monazite-(Ce) in modelled solution at any temperature. The paragenesis in Figure 5d was possibly caused by multi-phase hydrothermal activities, or parisite completely replaced interstitial bastnaesite among monazite grains.

There are mainly two REE mineralization events at Bayan Obo that have been recognized by former researchers [1–3]. The Mesoproterozoic and Paleozoic mineralization events produce similar mineral assemblages, including similar REE minerals (monazite and fluorocarbonates) and similar gangue minerals (fluorite, apatite, aegirine, sodium amphibole, phlogopite). The Paleozoic mineralization also contains some unique gangue minerals, including barite and various sulfides. Besides mineral composition, these two mineralization events caused different occurrences of the above minerals: disseminated grains, banded or massive aggregates in the Mesoproterozoic ores and altered ore-hosting dolomite, and megacrysts or veinlets of the above minerals in the vein-breccia types of ores. The similar mineral assemblage of two mineralization events indicates that their corresponding ore-forming fluids contained similar bulk composition (except sulfur-bearing species in the Paleozoic ore-forming fluids). According to former research on the Bayan Obo Sm-Nd isotopic system, there is no external REE introduced to the Paleozoic hydrothermal fluids [1,52]. This implies that the Paleozoic REE ores were product of local dissolution–reprecipitation. Therefore, thermodynamic calculations in this study are also able to provide an explanation of the Paleozoic REE precipitation process (Figure 5a,b), as if the vein-type of ores could be regarded as the product of a geochemically closed-system, except for the addition of dolomite or CO_2 (aq). The variation of pressure would not influence the paragenesis of REE minerals significantly. Therefore, modelling in this study could be applied to aqueous systems at diverse depths.

7. Conclusions

(1) During the precipitation of monazite in REE-rich fluids, heavier LREE tends to precipitate earlier and be fractionated from the fluids earlier. The Ce- and La-enriched monazite at Bayan Obo was caused by La and Ce pre-enriched hydrothermal fluids, supporting the carbonatitic fluids as ore-forming fluids from the perspective of thermodynamic calculation.

(2) With continuous introduction of sole CO_2 (aq), bastnaesite would eventually replace $CePO_4$ entirely. If dolomite is introduced solely to the acid hydrothermal fluids, bastnaesite replaces all monazite-(Ce) and is then replaced by parisite. The drop in temperature in hot hydrothermal fluids (>400 °C) would significantly promote the replacement of parisite by bastnaesite by decreasing CO_2 or dolomite demands.

(3) Based on our calculation of the modelled solution, the association of bastnaesite, monazite-(Ce) and apatite-(F) is a characteristic texture of hydrothermal fluids over 400 °C, which provides a constraint on the temperature of Bayan Obo REE mineralization. It is possible that post-magmatic hydrothermal fluids (carbonatitic fluids), or hot-magma-induced local recrystallization triggered the Bayan Obo REE mineralization.

(4) The association of REE minerals and gangue minerals have the potential to provide further constraints on the physico-chemical conditions of Bayan Obo hydrothermal fluids. Based on this study, apatite is often associated with bastnaesite at any temperature, but only crystallizes earlier, at temperatures as high as ~430 °C. The association of monazite-(Ce)-fluorite and monazite-parisite indicates multi-phase hydrothermal activities, as they were unlikely to co-precipitate in the modelled solution at any calculated temperature from 200 °C to 430 °C.

(5) Pressure variation has a rather limited influence on the paragenesis of REE minerals, whereas temperature and composition variation cause significantly different associations of REE minerals and gangue minerals.

Supplementary Materials: The following are available online at http://www.mdpi.com/2075-163X/10/6/495/s1, Table S1: The composition of REE-bearing minerals (fluocerite and monazite) in solution with 0.1 M HCl, 0.1 M HF, 0.1 M $LaCl_3$, 0.1 M $CeCl_3$, 0.1 M $PrCl_3$, 0.1 M $NdCl_3$, 0.1 M of $SmCl_3$ and 0.001–0.1 M H3PO4. The H_3PO_4 was added stepwise with 0.001 M of each step. Table S2: The composition of REE minerals and gangue minerals when CO_2 (aq) was introduced stepwise (0.001 M CO_2 each step). The modelled solution contains 0.1 M HCl, 0.1 M HF, 0.1 M H_3PO_4, 0.1 M $LaCl_3$, 0.1 M $CeCl_3$, 0.1 M $PrCl_3$, 0.1 M $NdCl_3$, and 0.001 M–0.6 M CO_2. Table S3: The composition of REE minerals and gangue minerals when dolomite was introduced stepwise (0.002 M for each step of 1 kbar-200 °C and 2 kbar–200 °C; 0.001 M for the rest of plots). The modelled solution contains 0.1 M HCl, 0.1 M HF, 0.1 M H_3PO_4, 0.1 M $LaCl_3$, 0.1 M $CeCl_3$, 0.1 M $PrCl_3$, 0.1 M $NdCl_3$, and 0.001 M–0.3 M dolomite. Table S4: The composition of REE minerals and gangue minerals in a continuous range of temperature (30–430 °C) when discrete amount of dolomite was added into a modelled solution, with 0.1 M HCl, 0.1 M HF, 0.1 M H_3PO_4, 0.1 M $LaCl_3$, 0.1 M $CeCl_3$, 0.1 M $PrCl_3$, 0.1 M $NdCl_3$. The temperature increased 10 °C per step. Table S5: The composition of REE minerals and gangue minerals have been calculated under the pressure from 1 kbar to 3 kbar when 0.10 M, 0.15 M and 0.20 M dolomite was added into the system (the same initial composition as in table S4) and replacement of parisite to bastnaesite occurred. The temperature increased 0.5 °C per step. Table S6: The initial pH and per of calculated solutions in this study. (Automatically calculated by GEMS).

Author Contributions: Conceptualization, S.L.; resources, S.L., L.D. and H.-R.F.; writing—original draft preparation, S.L.; writing—review and editing, L.D. and H.-R.F.; funding acquisition, S.L. and L.D. All authors have read and agreed to the published version of the manuscript.

Funding: This research was financially supported by the National Natural Science Foundation of China (Grant No. 41902066; 41930430) and the Fundamental Research Funds for the Central Universities of China (Grant No. Lzujbky-2019-pd06; 2022019zr106).

Acknowledgments: We are grateful to Kai-Yi Wang, Yong-Gang Zhao and Wen-Liang Ma for their kindly contribution during the field work. Pang Yun-Long was appreciated for his help during the operation of SEM.

Conflicts of Interest: The authors declare no conflict of interest.

References

1. Zhu, X.K.; Sun, J.; Pan, C.X. Sm–Nd isotopic constraints on rare-earth mineralization in the Bayan Obo ore deposit, Inner Mongolia, China. *Ore Geol. Rev.* **2015**, *64*, 543–553. [CrossRef]
2. Fan, H.R.; Yang, K.F.; Hu, F.F.; Liu, S.; Wang, K.Y. The giant Bayan Obo REE-Nb-Fe deposit, China: Controversy and ore genesis. *Geosci. Front.* **2016**, *7*, 335–344.
3. Yang, X.Y.; Lai, X.D.; Pirajno, F.; Liu, Y.L.; Ling, M.X.; Sun, W.D. The Bayan Obo (China) giant REE accumulation conundrum elucidated by intense magmatic differentiation of carbonatite. *Precambrian Res.* **2017**, *288*, 39–71. [CrossRef]
4. Yang, K.F.; Fan, H.R.; Pirajno, F.; Li, X.C. The Bayan Obo (China) giant REE accumulation conundrum elucidated by intense magmatic differentiation of carbonatite. *Geology* **2019**, *47*, 1198–1202.
5. Smith, M.P.; Henderson, P. Preliminary Fluid Inclusion Constraints on Fluid Evolution in the Bayan Obo Fe-REE-Nb Deposit, Inner Mongolia, China. *Econ. Geol.* **2000**, *95*, 1371–1388. [CrossRef]
6. Chen, W.; Liu, Y.L.; Lu, J.; Jiang, S.Y.; Simonetti, A.; Xu, C.; Zhang, W. The formation of the ore-bearing dolomite marble from the giant Bayan Obo REE-Nb-Fe deposit, Inner Mongolia: Insights from micron-scale geochemical data. *Miner. Deposita* **2020**, *55*, 131–146.
7. Liu, S.; Fan, H.R.; Groves, D.I.; Yang, K.F.; Yang, Z.F.; Wang, Q.W. Multiphase carbonatite-related magmatic and metasomatic processes in the genesis of the ore-hosting dolomite in the giant Bayan Obo REE-Nb-Fe deposit. *Lithos* **2020**, *354–355*, 105359. [CrossRef]
8. Fan, H.R.; Xie, Y.H.; Wang, K.Y.; Tao, K.J. REE Daughter Minerals Trapped in Fluid Inclusions in the Giant Bayan Obo REE-Nb-Fe Deposit, Inner Mongolia, China. *Int. Geol. Rev.* **2004**, *46*, 638–645.
9. Qin, C.J.; Qiu, Y.Z.; Zhou, G.F.; Wang, Z.G.; Zhang, T.R.; Xiao, G.R. Fluid inclusion study of carbonatite dykes/veins and ore-hosted dolostone at the Bayan Obo deposit. *Acta Petrol. Sin.* **2007**, *23*, 161–168.
10. Wang, K.Y.; Fan, H.R.; Yang, K.F.; Hu, F.F.; Wu, C.M.; Hu, F.Y. Calcite-dolomite geothermometry of Bayan Obo carbonatites. *Acta Petrol. Sin.* **2010**, *26*, 1141–1149, (In Chinese with English abstract).

11. Migdisov, A.A.; Williams-Jones, A.E.; Wagner, T. An experimental study of the solubility and speciation of the Rare Earth Elements (III) in fluoride- and chloride-bearing aqueous solutions at temperatures up to 300 °C. *Geochim. Cosmochim. Acta* **2009**, *73*, 7087–7109.
12. Gysi, A.P.; Williams-Jones, A.E. The thermodynamic properties of bastnäsite-(Ce) and parisite-(Ce). *Chem. Geol.* **2015**, *392*, 87–101. [CrossRef]
13. Williams-Jones, A.E.; Migdisov, A.A.; Samson, I.M. Hydrothermal mobilization of the rare earth elements-a tale of "Ceria" and "Yttria". *Elements* **2012**, *8*, 355–360. [CrossRef]
14. Migdisov, A.A.; Williams-Jones, A.E. Hydrothermal transport and deposition of the rare earth elements by fluorine-bearing aqueous liquids. *Mineral. Deposita* **2014**, *49*, 987–997. [CrossRef]
15. Migdisov, A.A.; Williams-Jones, A.E.; Brugger, J.; Caporiscio, F.A. Hydrothermal transport, deposition, and fractionation of the REE: Experimental data and thermodynamic calculations. *Chem. Geol.* **2016**, *439*, 13–42.
16. Drew, L.J.; Meng, Q.R.; Sun, W.J. The Bayan Obo iron-rare-earth-niobium deposits, Inner Mongolia, China. *Lithos* **1990**, *26*, 43–65. [CrossRef]
17. Chao, E.C.T.; Back, J.M.; Minkin, J.A. Host-rock controlled epigenetic, hydrothermal metasomatic origin of the Bayan Obo REE-Fe-Nb ore deposit, Inner Mongolia, P.R.C. *Appl. Geochem.* **1992**, *7*, 443–458. [CrossRef]
18. Yang, K.F.; Fan, H.R.; Santosh, M.; Hu, F.F.; Wang, K.Y. Mesoproterozoic mafic and carbonatitic dykes from the northern margin of the North China Craton: Implications for the final breakup of Columbia supercontinent. *Tectonophysics* **2011**, *498*, 1–10.
19. Wang, K.Y.; Fan, H.R.; Xie, Y.H.; Li, H.M. The zircon U-Pb dating of basement complex of the giant Bayan Obo REE-Fe-Nb deposit. *Chin. Sci. B* **2001**, *46*, 1390–1394. (In Chinese)
20. Fan, H.R.; Yang, K.F.; Hu, F.F.; Wang, K.Y.; Zhai, M.G. Zircon geochronology of basement rocks from the Bayan Obo area, Inner Mongolia, and tectonic implications. *Acta Petrol. Sin.* **2010**, *26*, 1342–1350. (In Chinese with English abstract)
21. Liu, J.; Li, Y.; Ling, M.X.; Sun, W.D. Chronology and geological significance of the basement rock of the giant Bayan Obo REE-Nb-Fe ore deposit. *Geochimica* **2011**, *40*, 209–222. (In Chinese with English abstract)
22. Zhong, Y.; Zhai, M.G.; Peng, P.; Santosh, M.; Ma, X.D. Detrital zircon U–Pb dating and whole-rock geochemistry from the clastic rocks in the northern marginal basin of the North China Craton: Constraints on depositional age and provenance of the Bayan Obo Group. *Precambrian Res.* **2015**, *258*, 133–145. [CrossRef]
23. Liu, Y.F.; Bagas, L.; Nie, F.J.; Jiang, S.H.; Li, C. Re–Os system of black schist from the Mesoproterozoic Bayan Obo Group, Central Inner Mongolia, China and its geological implications. *Lithos* **2016**, *261*, 296–306. [CrossRef]
24. Liu, C.H.; Zhao, G.C.; Liu, F.L.; Shi, J.R. Detrital zircon U-Pb and Hf isotopic and whole-rock geochemical study of the Bayan Obo Group, northern margin of the North China Craton: Implications for Rodinia reconstruction. *Precambrian Res.* **2017**, *303*, 372–391. [CrossRef]
25. Zhou, Z.G.; Hu, M.M.; Wu, C.; Wang, G.S.; Liu, C.F.; Cai, A.R.; Jiang, T. Coupled U–Pb dating and Hf isotopic analysis of detrital zircons from Bayan Obo Group in Inner Mongolia: Constraints on the evolution of the Bayan Obo rift belt. *Geol. J.* **2017**, *53*, 1–16. [CrossRef]
26. Le Bas, M.J.; Keller, J.; Tao, K.J.; Wall, F.; Williams, C.T.; Zhang, P.S. Carbonatite dykes at Bayan Obo, Inner Mongolia, China. *Mineral. Petrol.* **1992**, *46*, 195–228. [CrossRef]
27. Xiao, R.G.; Fei, H.C.; Wang, A.J.; Yang, F.; Yan, K. Formation and geochemistry of the ore-bearing alkaline volcanic rocks in the Bayan Obo REE-Nb-Fe deposit, Inner Mongolia, China. *Acta Geol. Sin.* **2012**, *86*, 735–751. (In Chinese with English abstract)
28. Wang, K.Y.; Fang, A.M.; Zhang, J.E.; Yu, L.J.; Dong, C.; Zan, J.F.; Hao, M.Z. Genetic relationship between fenitized ores and hosting dolomite carbonatite of the Bayan Obo REE deposit, Inner Mongolia, China. *J. Asian Earth Sci.* **2019**, *174*, 189–204. [CrossRef]
29. Yang, X.M.; Yang, X.Y.; Chen, T.H.; Zhang, P.S.; Le Bas, M.J.; Henderson, P. Geochemical constraints on the genesis of the Bayan Obo Fe-Nb-REE deposit in Inner Mongolia, China. *J. Chin. Rare Earths Soc.* **1999**, *17*, 289–295. (In Chinese with English abstract) [CrossRef]
30. Wagner, T.; Kulik, D.A.; Hingerl, F.F.; Dmytrieva, S.V. GEM-Selektor geochemical modeling package: TSolMod library and data interface for multicomponent phase models. *Can. Mineral.* **2012**, *50*, 1173–1195. [CrossRef]

31. Kulik, D.A.; Wagner, T.; Dmytrieva, S.V.; Kosakowski, G.; Hingerl, F.F.; Chudnenko, K.V.; Berner, U. GEM-Selektor geochemical modeling package: Revised algorithm and GEMS3K numerical kernel for coupled simulation codes. *Comput. Geosci.* **2013**, *17*, 1–24. [CrossRef]
32. Thoenen, T.; Hummel, W.; Berner, U.; Curti, E. *The PSI/Nagra Chemical Thermodynamic Database 12/07*; Paul Scherrer Institut, Villigen PSI: Villigen, Switzerland, 2014; pp. 1–445.
33. Gysi, A.P.; Williams-Jones, A.E. Hydrothermal mobilization of pegmatite-hosted Zr and REE at Strange Lake, Canada: A reaction path model. *Geochim. Cosmochim. Acta* **2013**, *122*, 324–352. [CrossRef]
34. Gysi, A.P. Numerical simulations of CO_2 sequestration in basaltic rock formations: Challenges for optimizing mineral-fluid reactions. *Pure Appl. Chem.* **2017**, *89*, 581–596. [CrossRef]
35. Konings, R.J.M.; Kovács, A. Thermodynamic properties of the lanthanide (III) halides. *Handb. Phys. Chem. Rare Earths* **2003**, *33*, 147–247.
36. Ni, Y.X.; Hughes, M.; Mariano, A.N. Crystal chemistry of the monazite and xenotime. *Am. Mineral.* **1995**, *80*, 21–26. [CrossRef]
37. Ushakov, S.V.; Helean, K.B.; Navrotsky, A.; Boatner, L.A. Thermochemistry of rare-earth orthophosphates. *J. Mater. Res.* **2001**, *16*, 2623–2633. [CrossRef]
38. Thiriet, C.; Konings, R.J.M.; Jovorský, P.; Magnani, N.; Wastin, F. The low temperature heat capacity of $LaPO_4$ and $GdPO_4$, the thermodynamic functions of the monazite-type $LnPO_4$ series. *J. Chem. Thermodyn.* **2005**, *37*, 131–139. [CrossRef]
39. Popa, K.; Sedmidubský, D.; Beneš, O.; Thiriet, C.; Konings, R.J.M. The high-temperature heat capacity of $LnPO_4$ (Ln = La, Ce, Gd) by drop calorimetry. *J. Chem. Thermodyn.* **2006**, *38*, 825–829. [CrossRef]
40. Popa, K.; Jutier, F.; Wastin, F.; Konings, R.J.M. The heat capacity of $NdPO_4$. *J. Chem. Thermodyn.* **2006**, *38*, 1306–1311. [CrossRef]
41. Popa, K.; Konings, R.J.M. High-temperature heat capacities of $EuPO_4$ and $SmPO_4$ synthetic monazites. *Thermochim. Acta* **2006**, *445*, 49–52. [CrossRef]
42. Jemal, M. Thermochemistry and relative stability of apatite phosphates. *Phosphorus Res. B* **2004**, *15*, 119–124.
43. Cruz, F.J.A.L.; Minas da Piedade, M.E.; Calado, J.C.G. Standard molar enthalpies of formation of hydroxy-, chlor-, and bromapatite. *J. Chem. Thermodyn.* **2005**, *37*, 1061–1070.
44. Cruz, F.J.A.L.; Canongia Lopes, J.N.; Calado, J.C.G.; Minas da Piedade, M.E. A Molecular Dynamics Study of the Thermodynamic Properties of Calcium Apatites. 1. Hexagonal Phases. *J. Phys. Chem. B* **2005**, *109*, 24473–24479. [CrossRef] [PubMed]
45. Dachs, E.; Harlov, D.; Benisek, A. Excess heat capacity and entropy of mixing along the chlorapatite–fluorapatite binary join. *Phys. Chem. Miner.* **2010**, *37*, 665–676. [CrossRef]
46. Bühn, B.; Rankin, A.H. Composition of natural, volatile-rich Na-Ca-REE-Sr carbonatitic fluids trapper in fluid inclusions. *Geochim. Cosmochim. Acta* **1999**, *63*, 3781–3797. [CrossRef]
47. Banks, D.A.; Yardley, B.W.D.; Campbell, A.R.; Jarvis, K.E. REE composition of an aqueous magmatic fluid: A fluid inclusion study from the Capitan Pluton, New Mexico, U.S.A. *Chem. Geol.* **1994**, *113*, 259–272. [CrossRef]
48. Liu, S.; Fan, H.R.; Yang, K.F.; Hu, F.F.; Rush, B.; Liu, X.; Li, X.C.; Yang, Z.F.; Wang, Q.W.; Wang, K.Y. Fenitization in the Giant Bayan Obo REE-Nb-Fe Deposit: Implication for REE Mineralization. *Ore Geol. Rev.* **2018**, *94*, 290–309. [CrossRef]
49. Krneta, S.; Ciobanu, C.L.; Cook, N.J.; Ehrig, K.J. Numerical Modeling of REE Fractionation Patterns in Fluorapatite from the Olympic Dam Deposit (South Australia). *Minerals* **2018**, *8*, 342. [CrossRef]
50. Smith, M.P.; Henderson, P.; Zhang, P.S. Reaction relationships in the Bayan Obo Fe-REE-Nb deposit Inner Mongolia, China: Implications for the relative stability of rare-earth element phosphates and fluorocarbonates. *Contrib. Mineral. Petrol.* **1999**, *134*, 294–310.

51. Liu, Y.L.; Chen, J.F.; Li, H.M.; Qian, H.; Xiao, G.W.; Zhang, T.R. Single-grain U-Th-Pb-Sm-Nd dating of monazite from dolomite type ore of the Bayan Obo deposit. *Acta Petrol. Sin.* **2005**, *21*, 881–888. (In Chinese with English abstract)
52. Liu, S.; Fan, H.R.; Yang, K.F.; Hu, F.F.; Wang, K.Y.; Chen, F.K.; Yang, Y.H.; Yang, Z.F.; Wang, Q.W. Mesoproterozoic and Paleozoic hydrothermal metasomatism in the giant Bayan Obo REE-Nb-Fe deposit: Constrains from trace elements and Sr-Nd isotope of fluorite and preliminary thermodynamic calculation. *Precambrian Res.* **2018**, *311*, 228–246. [CrossRef]

Article

The Influence of Hydrothermal Fluids on the REY-Rich Deep-Sea Sediments in the Yupanqui Basin, Eastern South Pacific Ocean: Constraints from Bulk Sediment Geochemistry and Mineralogical Characteristics

Tiancheng Zhou [1,2], Xuefa Shi [1,2,*], Mu Huang [1,2], Miao Yu [1,2], Dongjie Bi [1,2], Xiangwen Ren [1,2], Gang Yang [1,2] and Aimei Zhu [1,2]

1 Key Laboratory of Marine Geology and Metallogeny, First Institute of Oceanography, Ministry of Natural Resources, Qingdao 266061, China; ztc1989@163.com (T.Z.); huangmu@fio.org.cn (M.H.); myu@fio.org.cn (M.Y.); dongjiebi@fio.org.cn (D.B.); renxiangwen@fio.org.cn (X.R.); yanggang@fio.org.cn (G.Y.); zhuaimei@fio.org.cn (A.Z.)

2 Laboratory for Marine Geology, Qingdao National Laboratory for Marine Science and Technology, Qingdao 266061, China

* Correspondence: xfshi@fio.org.cn; Tel.: +86-0532-88967491

Received: 15 October 2020; Accepted: 15 December 2020; Published: 19 December 2020

Abstract: Rare earth elements (REEs) and yttrium (Y), together known as REY, are extremely enriched in deep-sea pelagic sediments, attracting much attention as a promising new REY resource. To understand the influence of hydrothermal processes on the enrichment of REY in deep-sea sediments from the eastern South Pacific Ocean, we conducted detailed lithological, bulk sediment geochemical, and in situ mineral geochemical analyses on gravity core sample S021GC17 from the Yupanqui Basin of eastern South Pacific. The REY-rich muds of S021GC17 are dark-brown to black zeolitic clays with REY contents (ΣREY) ranging from 1057 to 1882 ppm (average 1329 ppm). The REY-rich muds display heavy rare earth elements (HREE) enriched patterns, with obvious depletions in Ce, and positive anomalies of Eu in Post-Archean Australian Shale (PAAS)-normalized REE diagrams. In contrast, the muds of S021GC17 show light rare earth elements (LREE) enriched patterns and positive anomalies of Ce and Eu in the seawater-normalized REE diagrams. Total REY abundances in the core show positive correlations with CaO, P_2O_5, Fe_2O_3, and MnO concentrations. In situ analyses of trace element contents by laser ablation-inductively coupled plasma–mass spectrometry (LA-ICP-MS) demonstrate that bioapatite fossils contain high REY concentrations (998 to 22,497 ppm, average 9123 ppm), indicating that they are the primary carriers of REY. The in situ Nd isotope values of bioapatites are higher than the average values of seawater in Pacific Ocean. Fe–Mn micronodules are divided into hydrogenetic and diagenetic types, which have average REY concentrations of 1586 and 567 ppm, respectively. The high contents of Fe-Mn-Ba-Co-Mo, the positive correlations between ΣREY and Fe-Mn, the ratios of Fe/Ti and Al/(Al + Fe + Mn), and the LREE-enriched patterns in the REY-rich muds, combined with high Nd isotope values shown by bioapatite fossils, strongly indicate that the hydrothermal fluids have played an important role in the formation of the REY-rich sediments in the eastern South Pacific Ocean.

Keywords: deep-sea sediment; eastern South Pacific; bioapatite; hydrothermal fluids; LA-(MC)-ICP-MS

1. Introduction

Rare earth elements and yttrium (REY) are critical materials in many present-day industrial products, including applications in the new energy, electronics, and medical fields. Recently, deep-sea sediments with high concentrations of REY, called "REY-rich mud", have attracted attention as a promising new resource of REY [1–3].

REY-rich mud is defined as a deep-sea sediment that contains more than 700 ppm of total REY [4], which is higher than the grade of ion-adsorption-type ore deposits in southern China [5,6]. REY-rich muds from the eastern South Pacific and the central North Pacific have been firstly reported as potential resources with significant amounts of REY [1]. Furthermore, thick layers of REY-rich muds with comparable ΣREY contents have been found in the Indian Ocean, and they share similar geochemical and mineralogical features with those in the Pacific Ocean [7,8]. In 2016, extremely high REY-rich muds containing >7000 ppm of ΣREY were reported within the Japanese exclusive economic zone (EEZ) around Minamitorishima Island in the western North Pacific Ocean [9,10]. Most recently, our group discovered large areas of REY-rich muds in the eastern South Pacific Ocean in 2018 [4].

The influence of hydrothermal fluids on the chemistry of REY-rich muds is debated. In the eastern South Pacific Ocean, the formation of REY-rich muds appears to be strongly associated with hydrothermal activity along the East Pacific Rise (EPR). Kato et al. [1] proposed that hydrothermal iron oxyhydroxides that precipitate from hydrothermal plumes are the main carrier of REY in the Pacific. Moreover, the far-field spreading of hydrothermal plumes from the EPR is consistent with the spatial distribution of REY-rich muds in the Pacific Ocean [11–13]. However, most of the other REY-rich muds reported in the central North Pacific, the western Pacific, and the Indian Ocean, appear to be unrelated to hydrothermal fluids [7–9]. Recent studies using both SEM and laser ablation–inductively coupled plasma–mass spectrometry (LA-ICP-MS) have shown that apatite is certainly the host mineral of REY in the REY-rich muds of the Indian and western North Pacific Oceans, where the hydrothermal influence is minor [7–9,14–16]. Moreover, even in strongly hydrothermal-affected sediments around the EPR, apatite is also the major host of La, according to a preliminary X-ray adsorption fine structure (XAFS) analysis [17]. Thus, the influence of hydrothermal fluids on the REY enrichment mechanism of the deep-sea sediments remains controversial.

The eastern South Pacific Ocean contains large volumes of REY-rich muds, which have ∑REY contents of 1000–2230 ppm with total heavy rare earth elements concentrations (∑HREE) of 200–430 ppm [1]. Although bulk sediment analyses of REY-rich muds have been conducted on several samples from this area [1], the host minerals of the REY and their relative abundances remain poorly known. Moreover, no systematic study about the influence of hydrothermal fluids on the REY-rich muds has been conducted.

In this paper we present the results of bulk sediment elemental analyses and data of in situ electron microprobe analysis (EMPA) and laser ablation inductively coupled plasma mass spectrometry (LA-ICP-MS) elemental analyses of bioapatite fossils, Fe–Mn micronodules, and phillipsite included in the REY-rich muds. We also used in situ laser ablation multiple collector inductively coupled plasma mass spectrometry (LA-MC-ICP-MS) Sr-Nd isotope analyses of the bioapatite fossils and multiple collector thermal ionization mass spectrometer (MC-TIMS) Sr-Nd isotope analyses of the muds to characterize the isotopic features of these sediments. These data were used to identify the main mineral carriers of REY and discuss the influence of hydrothermal fluids in the REY enrichment process.

2. Regional Geology

The eastern South Pacific Ocean features a large-scale aseismic ridge called the East Pacific Rise (EPR), which generally rises 2000–4000 m above the ocean floor. The EPR is a typical fast to superfast-spreading ridge with a full spreading rate of 80–200 mm/year [18,19]. The EPR trends roughly N-S through the eastern South Pacific near 110 °W, and it features numerous transform faults on both sides [20]. Hydrothermal and volcanic activity are intense along the EPR, generating small- and large-scale hydrothermal plumes rich in δ^3He, in addition to Mn, Fe, and methane (Figure 1) [21].

Resing et al. [22] proposed that dissolved Fe and Mn from the EPR can be transported thousands of kilometers from the venting site in the eastern South Pacific Ocean. Fitzsimmons et al. [23] calculated that the Fe content in the seawater 2000 km from the EPR is 1–1.5 nmol/kg, which is significantly higher than that in the Pacific Ocean (0.4–0.9 nmol/kg).

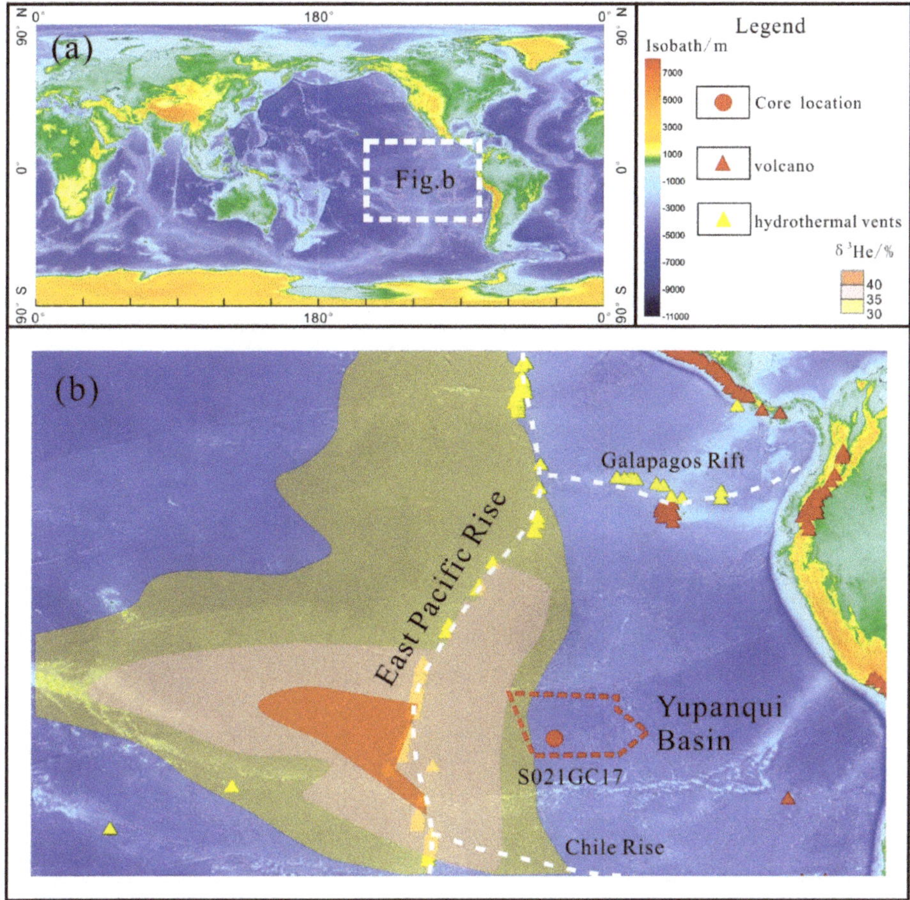

Figure 1. (**a**). The area of eastern South Pacific Ocean in the map of the world. (**b**). Location of sampling site S021GC17 in the Yupanqui Basin, eastern South Pacific Ocean (base image from Google Earth).

The eastern South Pacific Ocean has a relatively complex system of seamounts and ridges, and these divide the area into multiple deep-sea basins [24]. To the east of EPR, the Roggeveen and Chile basins are bounded to the north by the Salay Gomez Ridge (SGR) near 25 °S, which is part of the Nazca Plate that extends towards the coast of Chile [20]. The Nazca Plate is surrounded by the East Pacific Rise to the west, Galapagos rift to the north, and Chile–Peru ridges to the south [25], and the Yupanqui Basin lies in the western part of the Nazca Plate (Figure 1b). The age of the basement in the Yupanqui Basin is <20 Ma [26].

The eastern South Pacific Ocean has rich water masses, which bring in dissolved oxygen and cause sediment decomposition [27]. The North Pacific water masses flow generally towards the Southern Ocean and meet the Antarctic Circumpolar Current (ACC), upstream of Drake Passage. Subantarctic Surface Water (SASW) is found to the north of the Subantarctic Front, and is carried northward in the eastern South Pacific Ocean along the rim of the Subtropical Gyre [27]. Low salinity Antarctic Intermediate Water (AAIW) occurs at depths between 400 and 1000 m [26,28].

3. Samples and Analytical Methods

3.1. Lithological Characteristics

The gravity piston core S021GC17 was collected from the Yupanqui Basin in the eastern South Pacific Ocean. The water depth was about 4212 m and the gravity core penetrated 3.04 m below the seafloor. Samples of core (2 cm thick) were taken at intervals of 10 cm from top to bottom, and the samples were numbered sequentially from S021GC17 0-2 to S021GC17 302-304. The REY-rich muds of S021GC17 are homogeneous dark-brown to black zeolitic clays (Figure 2a) characterized by high contents of clay minerals (mainly Kaolinite and illite) (50%), phillipsite (15%), Fe–Mn micronodules (15%), and biogenic apatite (5%), in addition to terrigenous detrital materials (quartz and feldspar) (10%) and carbonate (5%) (Figures 2b,c and 3). The proportions of the different constituents were estimated roughly by microscope observation. The biogenic apatite (fish teeth fossils) and Fe-Mn oxide in the muds showed no peaks on X-ray diffraction analyses (Figure 3), indicating an amorphous or low crystallinity habit. The selected 31 samples were sealed immediately in clean polyethylene bags and placed in 4 °C storage.

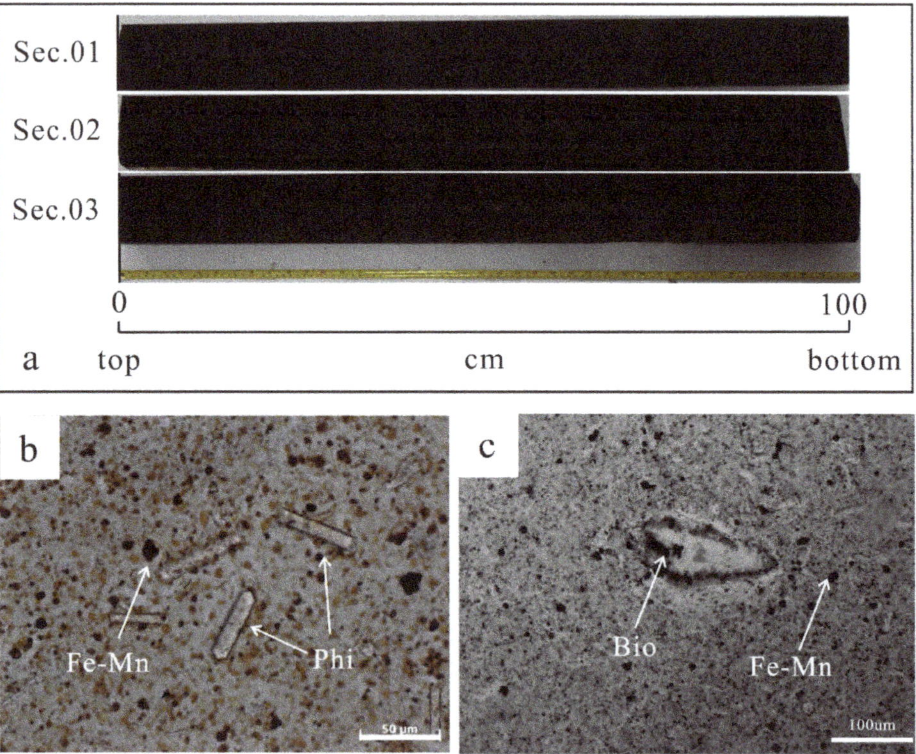

Figure 2. (**a**) The core material from the S021GC17 site. (**b**,**c**) Photomicrographs of sediment samples from the core. Phi = phillipsite; Bio = bioapatite (fish teeth).

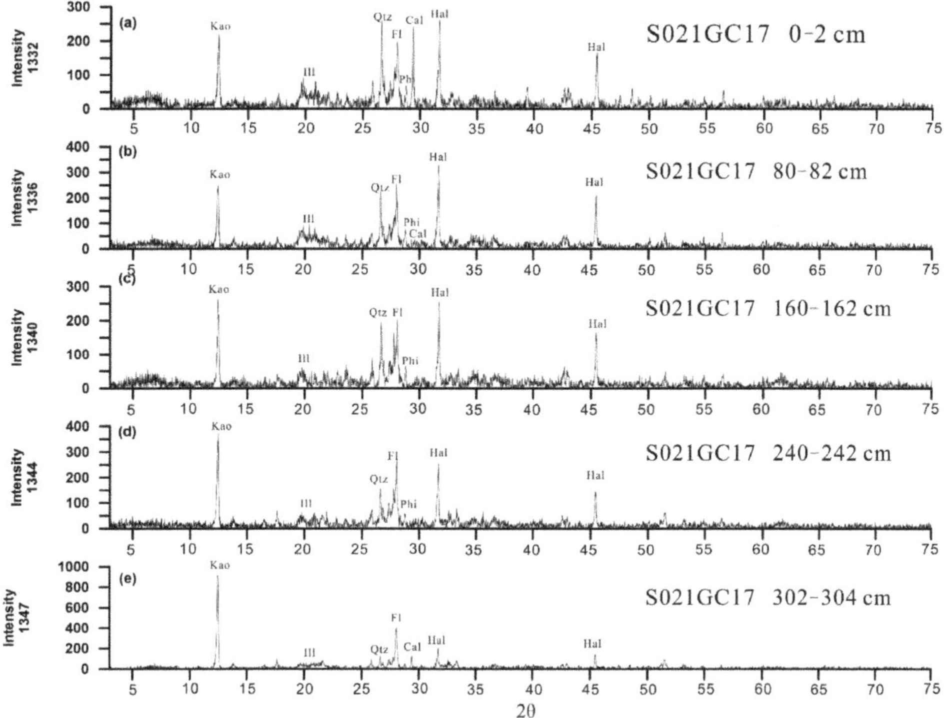

Figure 3. (**a**–**e**). X-ray powder diffraction patterns of bulk sediment samples from different depths. Kao = kaolinite; Ill = illite; Phi = phillipsite; Qtz = quartz; Fl = feldspar; Cal = calcite; Hal = halite.

3.2. Bulk Sediment X-ray Diffraction (XRD)

X-ray diffraction (XRD) analysis can effectively identify the constituents of the sediments. The bulk mineralogy of the samples was assessed using X-ray diffraction (XRD; InXitu Terra, Mountain View, CA, USA) at the Key Laboratory of Marine Geology and Metallogeny, First Institute of Oceanography, Ministry of Natural Resources (MNR), Qingdao, China. Five samples of REY-rich muds were prepared by salt washing, drying, crushing, and filling. The accuracy of the goniometer was better than 0.05° (2θ), instrument resolution better than 60%, and overall stability better than ±1%. Whole patterns from 0° to 75° 2θ were scanned at 2°/min for qualitative analysis of major constituent minerals. Mineral identification was performed using MDI Jade 5.0 software.

3.3. Bulk Sediment Major-Trace Element Analyses

Bulk-sediment major and trace element analyses were conducted at the Key Laboratory of Marine Geology and Metallogeny, First Institute of Oceanography, MNR, Qingdao, China. The samples were freeze-dried, powdered to 200 mesh. Each 0.05 g sample was weighed, placed in a polytetra fluoroethylene digestion tank, dissolved twice in HF-HNO_3 (1:1), and dried again at 190 °C for 48 h. A total of 3 mL of 50% HNO_3 was then added to each sample, and the samples were dried at 150 °C for at least 8 h, removed, and analyzed. Element concentrations were analyzed by inductively coupled plasma–optical emission spectrometry (ICP-OES; for Al, Ca, Fe, K, Mg, Mn, Na, P, Ti) and inductively coupled plasma–mass spectrometry (ICP-MS; for REY, Sc, Cr, Co, Ni, and Cu). Several samples were analyzed in replicate to determine the precision of the measurements, and the elemental composition of the GSD-9 reference standard was measured to confirm the accuracy of the analyses, which was better than 2% for most elements.

3.4. Bulk Sediment Sr-Nd Isotopes

Sr-Nd isotopic signatures can provide important information on the source materials in deep-sea sediments. Bulk sediment Sr-Nd isotope analyses were conducted at the same lab in the First Institute of Oceanography, MNR. Rb, Sr, Sm, and Nd concentrations were measured using a Thermo Fisher Scientific Triton Plus multi-collector thermal ionization mass spectrometer (Thermo Fisher Scientific, Waltham, MA, USA). The isotope ratios were corrected for mass fractionation by normalizing to $^{88}Sr/^{86}Sr = 8.375209$ and $^{146}Nd/^{144}Nd = 0.7219$. The international standards NBS-987 and JNdi-1 were employed to evaluate instrument stability during data collection. The measured international standard NBS-987 yielded $^{87}Sr/^{86}Sr = 0.710229 \pm 0.000012$ (2σ) and the JNdi-1 standard yielded $^{143}Nd/^{144}Nd = 0.512110 \pm 0.00012$ (2σ) [29].

3.5. EMPA Major-Element Analyses

Using the results of the bulk sediment analyses as a guide, typical bioapatite fossils (Figure 4a–c), Fe-Mn micronodules (Figure 4d,g), and phillipsites (Figure 4h,i) were hand-picked using a binocular microscope and cleaned with deionized water. The major-element compositions of minerals were determined by electron microprobe (EMPA-JXA-8230, Jeol Ltd., Tokyo, Japan) at the Key Laboratory of Marine Geology and Metallogeny, First Institute of Oceanography, MNR, China. The microprobe was operated at an accelerating voltage of 15 kV and a beam current of 20 nA, with an electron beam defocused to a 1–3 μm spot. Count times were 20 s for all constituents. Standard specimens used for calibration were Fe_2O_3 (for Fe), (Mn, Ca)SiO_3 (for Mn), (Ca, Na)Al(Al, Si)Si_2O_8 (for Ca and Al), $NaAlSi_3O_8$ (for Na, Si and O), $KAlSi_3O_8$ (for K), PrP_5O_{14} (for P), and $(Mg, Fe)_2SiO_4$ (for Mg), and the analyzing crystals used were LIFH (for Fe and Mn), PETH (for K, Ca and P), TAP (for Na, Mg, Al and Si), and LDE1H (for O). While performing the electron microprobe analysis (EMPA), backscattered electron (BSE) images were obtained for representative constituents.

3.6. LA-ICP-MS REY and Trace Element Analyses

In situ analysis of REY and trace elements in bioapatite fossils, Fe-Mn micronodules, and phillipsites extracted from the REY-rich muds was conducted at Beijing Createch Testing International (Beijing, China). Laser ablation was performed using an ESI NWR 193 nm excimer laser ablation system (Omaha, NE, USA). An Analytik Jena Plasma Quant mass spectrometer (Jena, Germany) was used to acquire ion-signal intensities. The spot diameter of the laser beam was 35 μm and the ablation frequency was 10 Hz. Helium was used as a carrier gas. Each spot analysis consisted of ~20 s of background acquisition, ~45 s of measurement, and ~20 s of cleaning. Elemental abundances were calibrated against multiple synthetic reference glasses (NIST SRM610, NIST SRM612, BHVO-2G, BCR-2G, and BIR-1G). The suitable internal standards were also carried out for data calibration (Ca was used as an internal standard for bioapatites; Mn for Fe-Mn micronodules; and Al for phillipsites), which corresponded to the Ca, Mn, and Al contents measured by EMPA of each spot (refer to Supplementary Tables). The following isotope suite was analyzed: ^{139}La, ^{140}Ce, ^{141}Pr, ^{146}Nd, ^{147}Sm, ^{153}Eu, ^{157}Gd, ^{159}Tb, ^{163}Dy, ^{89}Y, ^{165}Ho, ^{166}Er, ^{169}Tm, ^{172}Yb, ^{175}Lu,^{137}Ba, ^{88}Sr, ^{59}Co, ^{60}Ni, ^{65}Cu. The linear correlation between Ba and Eu contents is weak, suggesting that the mass interference of BaO on the Eu measurement is very minor.

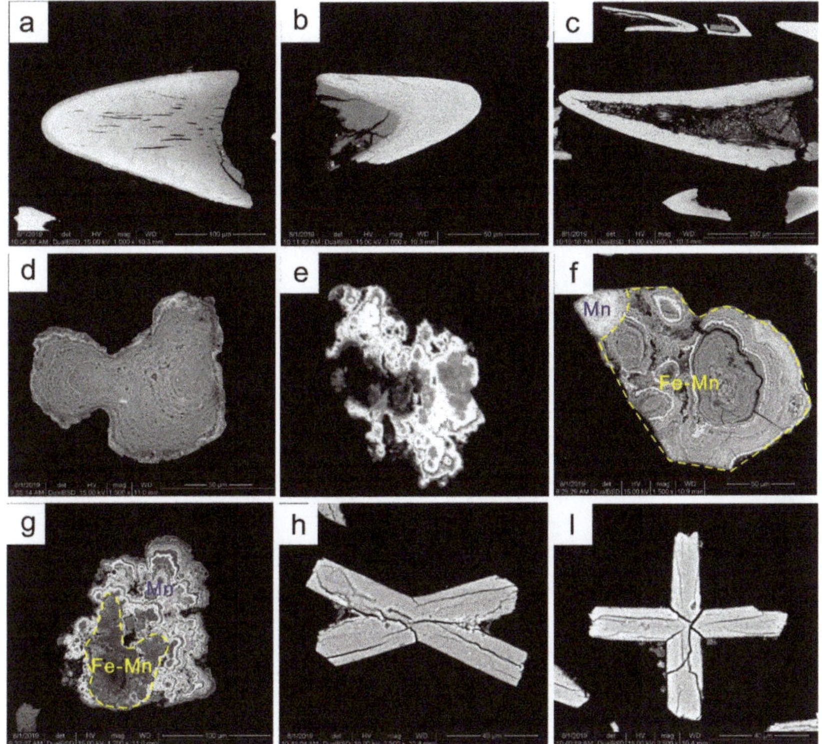

Figure 4. SEM images of the main constituents in the samples. (**a–c**) Bioapatite (fish teeth). (**d**) Fe–Mn micronodules. (**e**) Mn micronodules. (**f**,**g**) Mn micronodules in the outer rim of Fe–Mn micronodules. (**h**,**i**) Phillipsite.

3.7. LA-MC-ICP-MS Sr-Nd Isotope Analyses

In situ Sr-Nd isotope analyses of bioapatite in the REY-rich muds were conducted at Beijing Createch Testing International (Beijing, China). Laser sampling was performed using a Resolution SE 193 nm excimer laser ablation system. The analytical system involved a Neptune MC-ICPMS. The spot diameter of the laser beam ranged between 35 and 50 μm and the ablation frequency was 10 Hz. The aerosol ablated by the laser was transported to the mass spectrometer using helium as a carrier gas. The isotope ratios were corrected for mass fractionation by normalizing to $^{88}Sr/^{86}Sr = 8.375209$ and $^{147}Sm/^{149}Sm = 1.0868$. The apatite standard sample Durango was employed to evaluate the MC-ICP-MS instrument stability during data collection, yielding $^{87}Sr/^{86}Sr = 0.70599 \pm 0.00045$ (2σ) and $^{143}Nd/^{144}Nd = 0.512478 \pm 0.000029$ (2σ) [30].

4. Results

4.1. Bulk Sediment Major and Trace Element Analyses

Results of the bulk sediment major and trace element analyses are presented in Table 1a. The REY-rich muds are characterized by high contents of Fe_2O_3 (12.70–16.33 wt.%) and Al_2O_3 (8.53–11.73 wt.%). Other major elemental oxide contents are CaO (2.75–10.45 wt.%), MnO (3.85–5.30 wt.%), Na_2O (5.27–6.46 wt.%), K_2O (2.30–2.61 wt.%), P_2O_5 (1.30–3.46 wt.%), and TiO_2 (0.45–0.60 wt.%). The Ba contents range from 9187 to 13,410 ppm. The contents of Co and Mo range from 346 to 479 ppm and 49 to 104 ppm, respectively.

Table 1a. Bulk sediment geochemical data for the rare earth elements and yttrium (REY)-rich deep-sea sediments from the Yupanqui Basin.

Samples	S021GC17-0-2	S021GC17-10-12	S021GC17-20-22	S021GC17-30-32	S021GC17-40-42	S021GC17-50-52	S021GC17-60-62	S021GC17-70-72	S021GC17-80-82	S021GC17-90-92	S021GC17-100-102	S021GC17-110-112	S021GC17-120-122	S021GC17-130-132	S021GC17-140-142	S021GC17-150-152
Al_2O_3	11.03	11.49	11.73	11.59	11.50	11.56	11.50	11.48	11.54	11.26	11.27	11.27	11.30	11.21	11.15	11.05
CaO	5.37	3.37	2.80	2.75	2.78	2.84	2.84	2.80	2.85	2.89	3.01	2.90	2.87	2.86	2.87	2.88
$^{T}Fe_2O_3$	12.70	13.29	13.66	13.57	13.60	13.79	13.73	13.70	13.92	13.99	14.10	14.29	14.30	14.41	14.50	14.46
MnO	3.85	4.04	4.23	4.21	4.19	4.22	4.21	4.22	4.31	4.40	4.47	4.56	4.60	4.67	4.70	4.67
K_2O	2.30	2.39	2.46	2.44	2.45	2.49	2.47	2.47	2.50	2.47	2.46	2.47	2.47	2.47	2.47	2.47
MgO	2.99	3.08	3.18	3.12	3.16	3.18	3.16	3.13	3.19	3.15	3.14	3.16	3.15	3.17	3.18	3.18
Na_2O	6.04	6.04	6.37	6.20	6.35	6.38	6.38	6.25	6.46	6.32	6.13	6.16	6.12	6.08	6.08	6.13
P_2O_5	1.30	1.33	1.37	1.40	1.43	1.46	1.46	1.43	1.44	1.46	1.51	1.51	1.52	1.53	1.56	1.61
TiO_2	0.56	0.58	0.60	0.59	0.59	0.59	0.59	0.59	0.59	0.58	0.58	0.59	0.59	0.59	0.59	0.58
Ba	11888	12560	12537	12589	12402	12270	12464	12114	11732	12252	11766	11903	12453	12209	12466	11754
Co	345.7	359.7	372.1	375.1	374.6	371.0	373.4	378.2	379.9	408.1	392.8	413.3	408.3	418.2	429.4	425.8
Mo	48.91	58.42	65.44	72.63	74.60	76.11	79.91	80.20	81.19	85.50	83.90	86.35	88.93	90.46	93.96	95.54
La	170.51	169.98	175.01	178.66	172.71	169.55	179.99	177.03	174.40	180.98	179.91	183.35	181.84	179.07	181.63	182.72
Ce	177.62	178.07	182.42	179.26	174.55	177.91	179.71	171.65	171.94	176.04	165.81	170.02	170.22	169.39	169.32	174.09
Pr	37.80	39.01	39.92	40.71	40.12	39.74	41.45	39.83	40.09	42.88	42.29	43.34	42.52	42.45	42.99	42.58
Nd	169.07	173.56	180.48	180.28	178.66	177.16	183.94	177.41	182.73	189.69	184.93	191.85	188.83	190.23	191.04	189.31
Sm	34.96	36.19	37.18	36.97	36.95	36.94	38.61	38.09	37.74	39.95	39.80	40.86	40.08	40.35	41.00	40.52
Eu	14.63	15.49	15.72	15.12	15.21	15.07	15.77	15.31	14.81	16.05	15.82	16.00	15.85	15.51	15.69	15.30
Gd	35.00	36.17	37.98	37.76	37.80	37.28	38.59	38.24	37.81	40.85	39.92	41.64	41.13	41.16	40.68	40.09
Tb	5.60	5.82	5.96	6.01	5.97	5.94	6.14	6.06	6.11	6.50	6.44	6.58	6.57	6.69	6.60	6.67
Dy	45.70	47.00	49.20	49.13	49.19	49.19	50.67	50.08	50.88	53.95	53.11	54.47	53.57	53.86	54.44	54.01
Ho	9.36	9.82	10.25	10.34	10.25	10.44	10.47	10.48	10.64	11.14	11.41	11.37	11.32	11.29	11.49	11.50
Er	24.12	24.93	25.91	26.02	26.07	25.90	26.60	26.20	26.39	28.08	28.09	28.28	28.19	28.64	28.91	28.63
Tm	4.07	4.23	4.34	4.46	4.46	4.38	4.57	4.43	4.50	4.84	4.81	4.90	4.80	4.90	4.92	4.97
Yb	24.56	25.63	26.22	26.48	26.78	26.67	27.56	26.81	27.56	29.01	28.77	29.51	28.85	28.90	29.49	29.11
Lu	4.07	4.25	4.46	4.47	4.53	4.56	4.55	4.61	4.60	4.96	4.92	5.02	4.77	4.91	4.87	4.83
Y	299.97	306.37	324.59	331.60	331.57	328.29	333.46	339.43	344.18	366.40	355.96	368.94	352.49	355.52	367.59	370.01
$\sum$REY	1057	1077	1120	1127	1115	1109	1142	1126	1134	1191	1162	1196	1171	1173	1191	1194
δCe	0.52	0.52	0.51	0.50	0.50	0.51	0.49	0.48	0.49	0.47	0.45	0.45	0.46	0.46	0.45	0.47
δEu	1.27	1.29	1.27	1.23	1.23	1.23	1.24	1.21	1.19	1.20	1.20	1.17	1.18	1.15	1.16	1.15
Y/Ho	32.05	31.19	31.66	32.08	32.36	31.44	31.84	32.40	32.35	32.89	31.20	32.45	31.14	31.49	31.98	32.16
La/Sm	4.88	4.70	4.71	4.83	4.67	4.59	4.66	4.65	4.62	4.53	4.52	4.49	4.54	4.44	4.43	4.51
Sm/Nd	0.21	0.21	0.21	0.21	0.21	0.21	0.21	0.21	0.21	0.21	0.22	0.21	0.21	0.21	0.21	0.21
Sm/Yb	1.42	1.41	1.42	1.40	1.38	1.38	1.40	1.42	1.37	1.38	1.38	1.38	1.39	1.40	1.39	1.39
La/Yb	6.94	6.63	6.67	6.75	6.45	6.36	6.53	6.60	6.33	6.24	6.25	6.21	6.30	6.20	6.16	6.28

Table 1b. Bulk sediment geochemical data for the rare earth elements and yttrium (REY)-rich deep-sea sediments from the Yupanqui Basin.

Samples	S021GC17-160-162	S021GC17-170-172	S021GC17-180-182	S021GC17-190-192	S021GC17-200-202	S021GC17-210-212	S021GC17-220-222	S021GC17-230-232	S021GC17-240-242	S021GC17-250-252	S021GC17-260-262	S021GC17-270-272	S021GC17-280-282	S021GC17-290-292	S021GC17-302-304
Al_2O_3	11.11	11.17	11.02	11.11	11.19	10.98	10.88	10.46	10.31	10.03	10.10	9.76	8.66	8.53	9.20
CaO	2.94	3.01	3.02	3.15	3.19	3.44	3.62	3.59	3.71	3.87	4.33	4.77	8.82	10.45	6.85
$^{T}Fe_2O_3$	14.45	14.42	14.14	14.23	14.34	14.20	14.91	14.73	14.64	14.46	14.66	15.37	14.29	15.27	16.33
MnO	4.73	4.70	4.69	4.76	4.79	4.75	4.95	5.00	4.97	4.88	5.02	5.30	4.98	4.83	5.11
K_2O	2.50	2.55	2.53	2.59	2.61	2.60	2.56	2.50	2.47	2.43	2.52	2.52	2.32	2.52	2.55
MgO	3.22	3.24	3.19	3.21	3.23	3.20	3.27	3.22	3.14	3.04	3.00	2.99	2.65	2.48	2.48
Na_2O	6.28	6.34	6.09	6.06	6.08	6.11	6.08	5.90	5.90	5.88	6.02	6.00	5.43	5.27	5.43
P_2O_5	1.63	1.68	1.76	1.88	1.90	2.10	2.24	2.27	2.36	2.45	2.71	2.96	3.17	3.28	3.46
TiO_2	0.58	0.58	0.56	0.57	0.57	0.56	0.57	0.55	0.54	0.52	0.52	0.51	0.45	0.45	0.51
Ba	11716	11777	12667	12903	13410	10556	9491	9700	9187	9250	9774	12284	12509	12958	10607
Co	427.7	419.4	418.1	433.8	418.1	437.0	464.2	479.1	450.0	447.8	465.1	457.0	428.0	392.5	413.5
Mo	94.34	88.96	92.05	97.76	96.80	93.03	96.94	104.10	96.51	90.22	83.71	81.92	82.92	85.92	92.73
La	191.13	192.78	194.05	207.26	207.10	217.25	228.83	236.51	254.99	260.20	278.97	286.29	300.11	294.06	311.45
Ce	166.55	163.58	162.66	164.63	164.93	155.86	159.95	158.44	156.99	148.39	155.09	140.62	134.17	136.86	151.96
Pr	44.08	43.79	44.61	45.88	46.30	47.58	50.06	49.00	51.57	49.95	53.34	51.73	52.74	53.50	57.97
Nd	197.25	198.93	200.21	213.74	208.73	218.59	225.26	228.34	238.30	232.70	250.10	239.60	247.63	247.42	266.70
Sm	42.67	41.93	42.37	44.88	45.16	45.61	46.78	46.26	47.39	47.63	49.64	48.13	47.88	49.36	53.57
Eu	16.35	15.87	16.49	17.22	17.29	16.39	15.48	15.64	15.90	15.72	16.99	17.80	18.40	19.14	18.47
Gd	42.22	42.28	42.31	45.55	43.86	46.34	47.17	47.14	49.35	49.55	52.27	52.59	52.74	54.20	57.26
Tb	6.92	6.90	7.07	7.49	7.34	7.69	8.03	8.11	8.43	8.48	9.05	8.95	9.00	9.07	9.93
Dy	56.76	56.76	57.63	61.36	61.63	65.19	67.98	67.82	71.76	71.86	77.72	77.91	78.92	79.59	85.97
Ho	12.10	12.21	12.31	13.18	13.13	14.39	14.85	15.07	16.06	16.26	17.93	18.11	18.32	18.10	19.60
Er	29.99	30.04	31.29	33.59	33.59	36.05	37.77	38.14	40.95	41.44	46.01	46.50	47.23	46.75	50.65
Tm	5.13	5.15	5.37	5.70	5.72	6.27	6.56	6.70	7.13	7.25	8.01	8.12	8.41	8.28	8.88
Yb	30.11	30.30	31.35	33.59	33.29	36.61	38.12	39.16	41.47	42.20	47.07	47.94	48.52	48.98	52.28
Lu	5.14	5.12	5.24	5.62	5.57	6.12	6.31	6.49	7.11	7.17	8.02	8.13	8.42	8.45	8.98
Y	387.15	391.30	407.42	444.14	430.11	494.34	530.46	537.16	561.31	585.36	684.69	701.84	735.22	703.36	728.37
∑REY	1234	1237	1260	1344	1324	1414	1484	1500	1569	1584	1755	1754	1808	1777	1882
δCe	0.43	0.42	0.41	0.40	0.40	0.36	0.35	0.34	0.32	0.30	0.29	0.26	0.24	0.25	0.26
δEu	1.16	1.14	1.18	1.15	1.17	1.08	1.00	1.01	1.00	0.98	1.01	1.08	1.11	1.13	1.01
Y/Ho	32.00	32.06	33.11	33.69	32.75	34.35	35.72	35.64	34.94	35.99	38.19	38.75	40.12	38.86	37.16
La/Sm	4.48	4.60	4.58	4.62	4.59	4.76	4.89	5.11	5.38	5.46	5.62	5.95	6.27	5.96	5.81
Sm/Nd	0.22	0.21	0.21	0.21	0.22	0.21	0.21	0.20	0.20	0.20	0.20	0.20	0.19	0.20	0.20
Sm/Yb	1.42	1.38	1.35	1.34	1.36	1.25	1.23	1.18	1.14	1.13	1.05	1.00	0.99	1.01	1.02
La/Yb	6.35	6.36	6.19	6.17	6.22	5.93	6.00	6.04	6.15	6.17	5.93	5.97	6.19	6.00	5.96

$^{T}Fe_2O_3$: Total contents of Fe_2O_3.

The total REY concentrations in samples from the Yupanqui Basin ranged from 1057 to 1882 ppm (average 1329 ppm), gradually increasing with depth. The CaO, P_2O_5, Fe_2O_3, and MnO contents obtained for the REY-rich muds show positive correlations with depth, whereas the Al_2O_3, TiO_2, and K_2O contents either show no correlation or negative correlations with depth (Figure 4). The $\sum$REY co-vary positively with CaO, P_2O_5, Fe_2O_3, and MnO, and negatively with Al_2O_3, TiO_2, K_2O, δCe ($Ce_N/(La_N \times Pr_N)^{0.5}$), and δEu ($Eu_N/(Sm_N \times Gd_N)^{0.5}$) (Figure 5). There is no change in the REY patterns of the sediments with the increasing depth. The REY-rich muds have HREE-enriched patterns on Post-Archean Australian Shale (PAAS)-normalized REE diagrams. All of the samples show depletions in Ce, and positive anomalies in Eu and Y (Figure 6a). In contrast, on seawater-normalized REE diagrams the samples show light rare earth elements (LREE) enriched patterns and positive anomalies in Ce and Eu (Figure 7a).

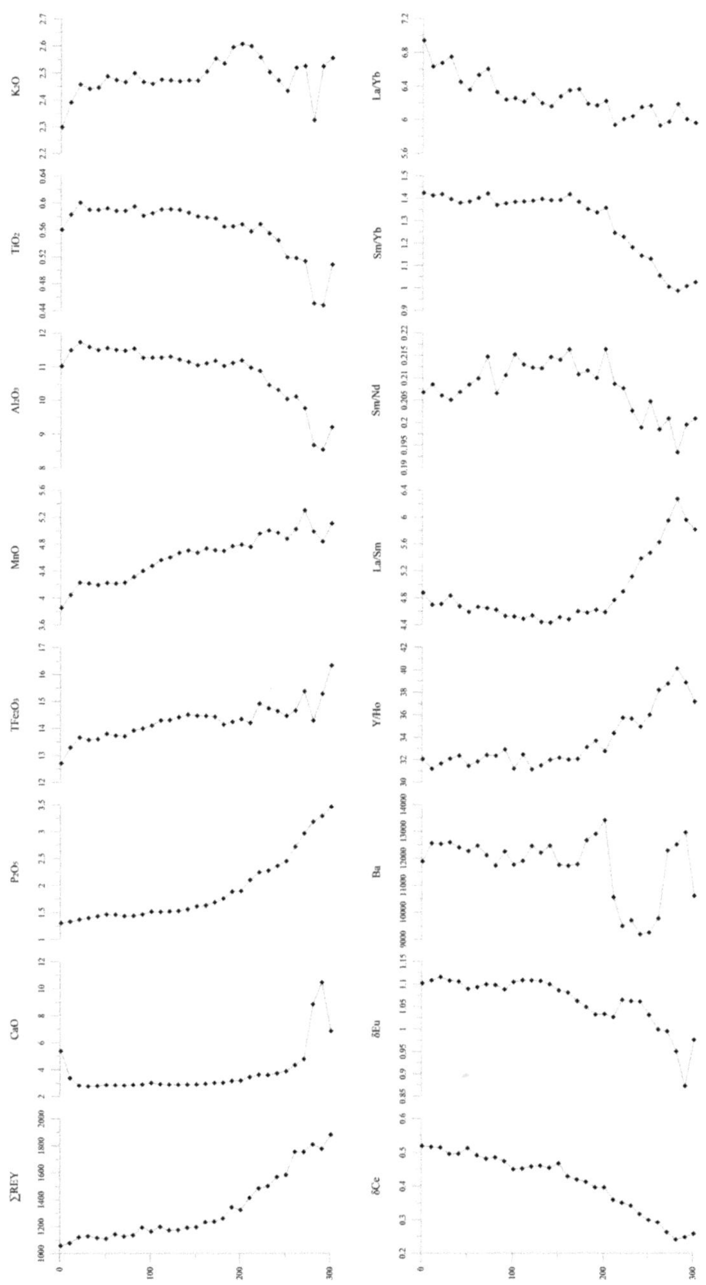

Figure 5. Profiles of REY content and other characteristic parameters of the bulk sediment samples from core S021GC17.

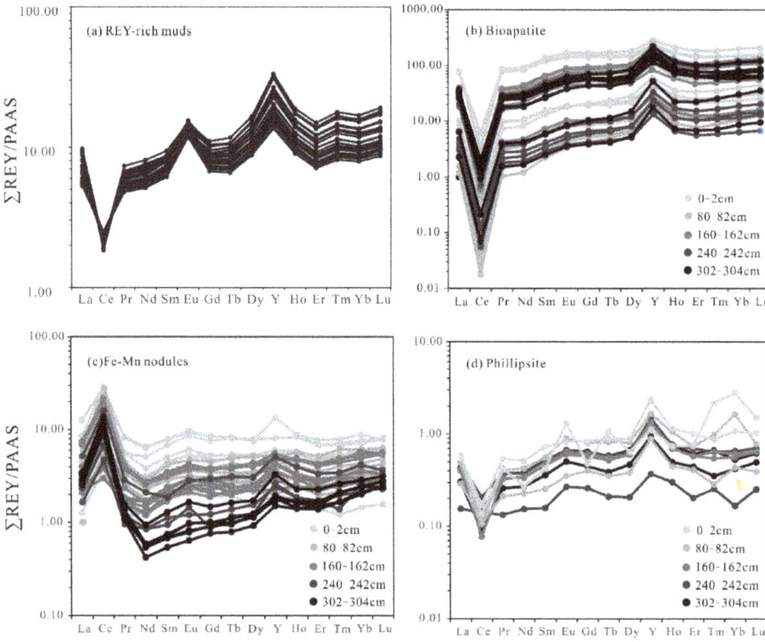

Figure 6. Post-Archean Australian Shale (PAAS)-normalized REY patterns for (**a**) REY-rich muds, (**b**) bioapatite, (**c**) Fe-Mn micronodules, and (**d**) phillipsite. Normalizing values are from [31].

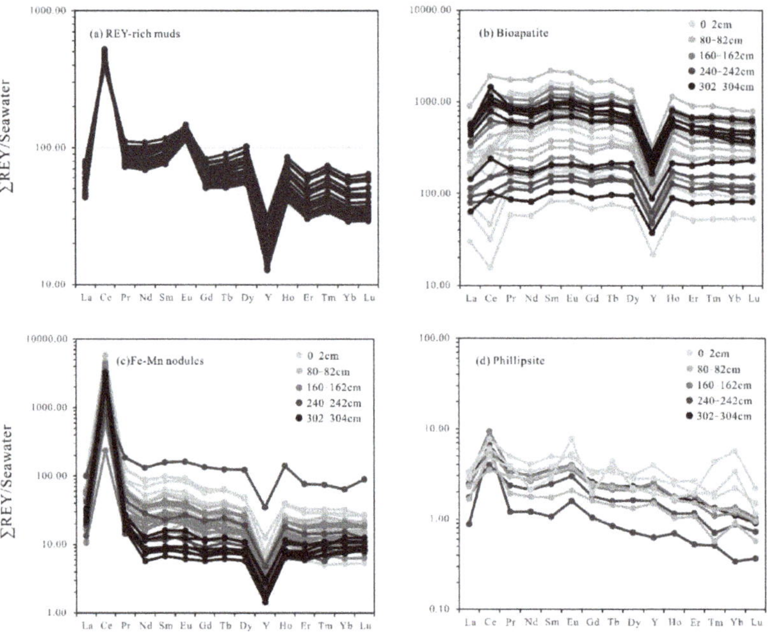

Figure 7. Seawater-normalized REY patterns for (**a**) REY-rich muds, (**b**) bioapatite, (**c**) Fe–Mn micronodules, and (**d**) phillipsite. Normalizing values are from [32].

4.2. Bulk Sediment Sr-Nd Isotope Compositions

Considering the compositional regularity of sediments, we selected five samples for Sr-Nd isotope analyses at equal intervals. The results of Sr-Nd isotope analyses of the REY-rich muds (at depths of 0–2, 80–82, 160–162, 240–242, and 302–304 cm) are presented in Table 2.

Table 2. Sr-Nd isotope values for the REY-rich deep-sea sediments and bioapatites in the Yupanqui Basin

Samples	Samples No.	Depths (cm)	$^{87}Sr/^{86}Sr$	Error (2 s)	$^{143}Nd/^{144}Nd$	Error (2 s)	εNd (0)
Whole rock	S021GC17-1	0–2	0.70951	0.00000644	0.51245	0.00000209	−3.667305
	S021GC17-2	80–82	0.70919	0.00000526	0.51246	0.00000237	−3.530757
	S021GC17-3	160–162	0.71039	0.00000681	0.51247	0.00000199	−3.355194
	S021GC17-4	240–242	0.71081	0.00000661	0.51248	0.00000164	−3.004069
	S021GC17-5	302–304	0.71084	0.00000796	0.51249	0.00000225	−2.945548
Bioapatite	18777-1-1	0–2	0.71050	0.000134288	0.51274	0.000032	1.945343344
	18777-1-2	0–2	0.70975	0.000231394	0.51248	0.000032	−3.14039429
	18777-1-3	0–2	0.71048	0.000686548	0.51240	0.000066	−4.57182653
	18777-1-4	0–2	0.70989	0.000292885	0.51240	0.000041	−4.676482
	18777-2-1	80–82	0.71113	0.000226153	0.51250	0.000033	−2.76070548
	18777-2-2	80–82	0.71097	0.000298258	0.51246	0.000041	−3.53428332
	18777-3-1	160–162	0.71089	0.000282621	0.51252	0.000048	−2.35227029
	18777-3-3	160–162	0.71056	0.000346566	0.51252	0.000044	−2.3695646
	18777-3-4	160–162	0.71091	0.000213473	0.51246	0.000036	−3.5659117
	18777-4-1	240–242	0.71158	0.000316567	0.51251	0.000064	−2.43800703
	18777-4-2	240–242	0.70914	0.000312533	0.51266	0.000063	0.458839981
	18777-5-1	302–304	0.70985	0.000162341	0.51255	0.000046	−1.62802313
	18777-5-2	302–304	0.71008	0.000293541	0.51258	0.000065	−1.15135076
	18777-5-3	302–304	0.70961	0.000106755	0.51246	0.000036	−3.47546196
	18777-5-4	302–304	0.70891	0.00020346	0.51255	0.000059	−1.76812022
	18777-5-5	302–304	0.71100	0.000557245	0.51268	0.000076	0.770968358
	18777-5-6	302–304	0.70941	0.000176067	0.51250	0.000050	−2.68444508

The initial $^{87}Sr/^{86}Sr$ ratios for the REY-rich muds at different depths vary from 0.70919 to 0.71084, and their initial $^{143}Nd/^{144}Nd$ values range from 0.512450 to 0.512487. The calculated εNd(0) values range from −3.667305 to −2.945548. The εNd(0) values of sediments increase gradually with depth, which is consistent with the trend displayed by $\sum$REY abundances.

4.3. EMPA and LA-ICP-MS Major and Trace Element Analyses

Major element, trace element, and REY compositions of the bioapatite fossils, Fe–Mn micronodules, and phillipsite extracted from the REY-rich muds at depths of 0–2, 80–82, 160–162, 240–242, and 302–304 cm in the core, were measured using EMPA and LA-ICP-MS, and the data are listed in Supplementary Tables S1–S3.

The bioapatite fossils are composed mainly of Ca (29.82–37.27 wt.%), P (14.55–18.57 wt.%), and O (28.09–45.04 wt.%). The total REY contents measured in bioapatites, measured with in situ LA-ICP-MS spot analyses, range from 998 to 22,497 ppm (average 9123 ppm), and do not show any correlation with depths. The REE patterns of bioapatites from different depths are quite uniform. All the samples show clear depletions in Ce, and positive anomalies of Y in PAAS-normalized REE patterns, in addition to variation in Ce anomalies and negative Y anomalies in seawater-normalized patterns (Figures 6b and 7b).

The Fe-Mn micronodules in the REY-rich sediments from the Yupanqui Basin can be divided into two types according to their composition: type I Fe-Mn micronodules (Figure 3d) and type II Mn micronodules (Figure 3e). The main constituents of the Fe–Mn micronodules are Fe (5.90–14.57 wt.%), Mn (11.32–31.73 wt.%), and O (10.48–22.98 wt.%). The total REY contents of the Fe–Mn micronodules, measured by in situ LA-ICP-MS spot analyses, range from 263 to 3153 ppm (average 1586 ppm). The Mn micronodules are composed mainly of Mn (22.64–42.24 wt.%) and O (17.88–35.96 wt.%), and the contents of Fe are low (0.18–3.93 wt.%). The REY content in Mn micronodules ranges from 258 to 826 ppm (average 567 ppm). The Fe-Mn micronodules have the similar REY patterns to those of the Mn micronodules, with clear Ce positive anomalies in both the PAAS- and seawater-normalized REE patterns (Figures 6c and 7c).

The main constituents of phillipsite are Si (21.75–30.15 wt.%), Al (7.34–10.95 wt.%), and O (38.82–48.04 wt.%), with minor amounts of K (1.62–3.05 wt.%) and Na (0.92–1.60 wt.%). The REY content in phillipsite ranges from 35 to 127 ppm (average 80 ppm) and the REY patterns are irregular (Figures 6d and 7d).

4.4. Bioapatite LA-MC-ICP-MS Sr-Nd Isotope Compositions

Results of the Sr-Nd isotope analyses of bioapatite fossils from different depths (0–2, 80–82, 160–162, 240–242, and 302–304 cm) are presented in Table 2. The $^{87}Sr/^{86}Sr$ ratios for the bioapatites vary from 0.708912 to 0.711583, and their $^{143}Nd/^{144}Nd$ values from 0.512398 to 0.512738. The calculated εNd (0) values range from –4.676482 to 1.945343 (average –2.173041), and show no correlation with depth.

5. Discussion

5.1. Main Carriers of REY

5.1.1. Bioapatites

Previous studies concluded that biogenic Ca phosphates, such as bioapatite, are the main carriers of REY in deep-sea sediments and that biogenic phosphate occurs mainly in the form of fish teeth [33,34]. Bioapatite has a strong adsorption capacity for REY complexes, and is the main carrier of REYs in the REY-rich muds of the Indian Ocean and the western and central North Pacific [7,8,14,16]. Kashiwabara et al. [17] reported XAFS evidence that apatite is the main host phase of La in the Pacific REY-rich mud. Kon et al. [14] conducted in situ geochemical analyses of the bioapatite fossils from the Minami-Torishima area in the western Pacific, and reported that bioapatite fossils are the main REY-hosting phase with the ∑REY range from 9300 to 32,000 ppm. Liao et al. [16] reported the average contents of 6182 ppm of REY in bioapatites from muds of the central North Pacific.

The CaO and P_2O_5 contents of REY-rich muds in the Yupanqui Basin show positive correlations with ∑REY (Figure 5), which suggests that a Ca-phosphate phase is the dominant carrier of REY. The average REY content of bioapatites fossils from the Yupanqui Basin is 9123 ppm (with 8921 ppm exclusion of Ce), which confirms that bioapatite is the main REY carrier. The bioapatites are typically fish teeth, and their REY concentrations decrease significantly from the tooth roots to the tips, suggesting that the REYs enter through the roots and diffuse to the tips (Figure 8a,c). This occurs because the dense enamel that coats the bioapatite hinders REY incorporation. REY can only diffuse through the breach at the root, leading to the observed REY gradients [16].

We proposed the assumption that the content of P in the sediments was entirely supplied by bioapatites. Thus, it can be calculated that the percentage of bioapatite in the sediments is 5.1%. Considering that the average REY content of the bioapatites is ~9000 ppm and ∑REY in the bulk sediment is ~1000 ppm, it can be roughly estimated that the 5% bioapatite accounts for ~45% of the REY budget in the bulk sediments. Our results show that bioapatite is the main carrier of REY, even in the eastern South Pacific where significant REY accumulation is related to hydrothermal Fe–Mn (oxyhydr) oxides from the EPR.

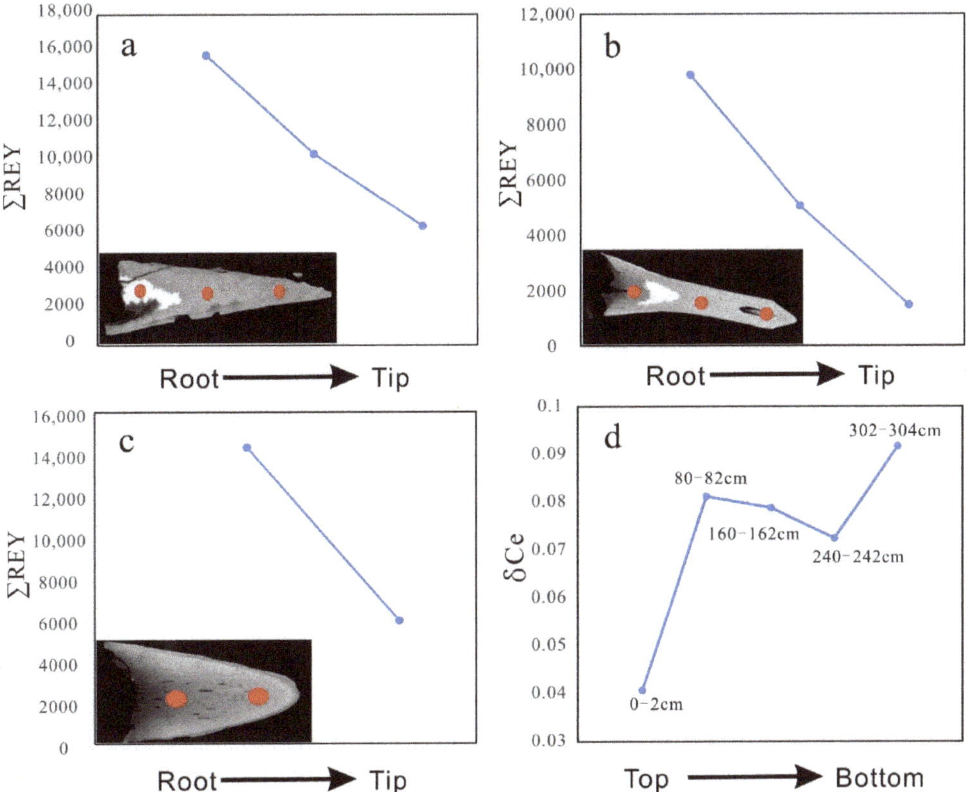

Figure 8. In situ compositional analyses showing the REY and δCe gradients in the bioapatites. (**a**–**c**) The contents gradients of REY in the bioapatite fossils. (**d**) δCe gradients in the bioapatite fossils in different depths.

5.1.2. Fe-Mn Micronodules

The ∑REY contents of our REY-rich samples not only correlate with CaO and P_2O_5, but also other elements that are not present in bioapatite (e.g., Fe_2O_3 and MnO) (Figure 5), which suggests that Fe- and Mn-bearing phases also contribute to the formation of the REY-rich muds. The Fe-Mn micronodules in sediments formed as a result of hydrogenetic, diagenetic, and hydrothermal plume processes, and can also act as REY carriers into the sediments [1,11,13,35,36]. Fe-Mn micronodules have a low degree of crystallinity and are active scavengers of REYs [1,37–39]. Menendez et al. [40] reported an average REY content of hydrogenetic Fe–Mn micronodules in deep-sea sediments of the Atlantic Ocean of 3620 ppm. Fe-Mn micronodules in REY-rich deep-sea muds recovered from the central North Pacific Ocean contain REY concentrations ranging from 439 to 1654 ppm [16].

To clarify the genetic type of micronodules, their elemental contents of Mn, Fe, Co, Ni, and Cu were plotted on a ternary discrimination diagram (Figure 9a) [41]. Most of the Fe-Mn micronodules fall into the hydrogenetic and mixed-type fields, whereas the Mn micronodules plot in the diagenetic field, which indicates the co-existence of different genetic mechanisms. Most of the micronodules plot in the transition area between the hydrogenetic and diagenetic fields, and the diagenetic-type (Fe)-Mn micronodules cluster mainly in the oxic diagenesis area on the 100 × (Zr + Y + Ce) vs. 15 × (Cu + Ni) vs. (Fe + Mn)/4 diagram proposed by Josso et al. [42] (Figure 9b). The average REY content of Fe–Mn micronodules is 1586 ppm (with 490 ppm exclusion of Ce), and the average REY content of

Mn micronodules is 567 ppm (with 222 ppm exclusion of Ce), which indicates that hydrogenetic micronodules can release REY during the early stages of diagenesis. Given the high proportion of Fe–Mn micronodules in the sediments (15%), they contribute as much as 22.5% of the REY content of the sediments. Considering Ce is the least valuable element in REY, the economical values of micronodules in the sediments represent a significant discount.

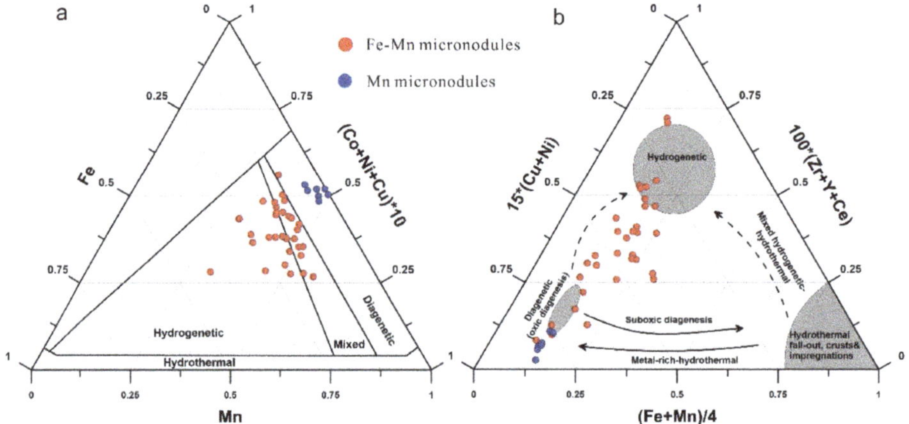

Figure 9. Discrimination diagrams for genetic classification of the Fe-Mn and Mn micronodules in core S021GC17. (**a**) Ternary Mn vs (Co + Ni + Cu) × 10 vs. Fe discrimination diagram, modified after [37]. (**b**) 100 × (Zr + Y + Ce) vs. 15 × (Cu + Ni) vs. (Fe + Mn)/4 diagram, proposed by [38].

Diagenesis can lead to the remobilization and inter-elemental fractionation of REYs, and these effects often occur in conjunction with redox reactions in the pore waters of sediments [32]. Previous researchers have suggested that REEs were associated primarily with Fe-Mn oxides in the sediments, then subsequently released to pore water during early diagenesis, and ultimately incorporated by bioapatite [43–45]. Variations in the Ce anomalies can be induced by the transfer of REEs from Mn-Fe (oxyhydr)oxides to bioapatite during early diagenesis [45]. Hence, it can be inferred that the REE patterns with positive Ce anomalies in the Fe-Mn micronodules could overprint the primary composition of the bioapatite. The Ce anomalies in our samples of bioapatite show a general increase with depth of burial (Figure 8d), which implies that the uptake of REEs was due to their release from the Fe-Mn micronodules. During early diagenesis, hydrogenetic Fe-Mn micronodules released REYs and associated metals into the pore water, forming the Mn micronodules (Figure 4f,g). Although the complete process remains an open question, the role of pore water in the formation of the REY-rich muds of the eastern South Pacific cannot be neglected.

5.1.3. Phillipsite

In a pelagic environment, phillipsite that forms by the alteration of volcanic ash is the one of the most important constituents in the REY-rich muds. Bernat [33] reported that REE contents in fish teeth fossils are two to three orders of magnitude higher than those in phillipsite. Dubinin [46] measured the REE concentrations in phillipsite samples from the Southern Basin of the Pacific, and suggested that phillipsite does not accumulate REY. Kon et al. [14] reported REY contents of 60–170 ppm in phillipsites by in situ mineral analyses.

The REY contents of phillipsites in our samples, measured by in situ LA-ICP-MS spot analyses, range from 35 to 127 ppm (average 80 ppm), showing that phillipsite is not the main carrier. The requisite for the formation of phillipsite in a marine sediment is an environment with a sufficiently low sedimentation rate, which is similar to the formation of pelagic clays with high REY contents [47].

The growth of large phillipsite grains may reflect a low sedimentation rate and long-term exposure of crystals on the seafloor [48]. Hence, phillipsite in the REY-rich muds of the eastern South Pacific is an indicator of low sedimentation rates.

5.2. *The Influence of Hydrothermal Fluids*

Previous researchers suggested that seawater was the main source of the REY in REY-rich sediments, based on the similar REE patterns of seawater and REY-rich muds [7,34,49]. The REY patterns in Figure 6a,b for the muds and the bioapatites do indeed show characteristics similar to those of seawater with their obvious depletions in Ce and positive anomalies of Y, and this indicates the REYs were mainly derived mainly from seawater. This is confirmed by the Sr–Nd isotopic compositions of the bulk muds, which are consistent with the isotopic characteristics of seawater in the Pacific (Figure 10). However, the high contents of Fe, Mn, Ba, Co, and Mo in the REY-rich muds from the Yupanqui basin conceivably imply the influence of other fluids (Table 1a).

It is well known that the East Pacific Rise (EPR) is a fast-spreading mid-ocean ridge [50] that is accompanied by intense hydrothermal activity due to high rates of magma production [11,51]. The dominance of Ce (IV) in the eastern South Pacific could reflect the larger contribution of hydrothermal Fe-Mn oxyhydroxides in this area compared with the central North Pacific [52]. The ^{3}He anomalies in seawater (Figure 1b) show that hydrothermal activity can spread far from a mid-ocean ridge, which is consistent with the distribution of REY-rich muds in the Pacific Ocean [1,11–13]. Hydrothermal fluids are enriched in iron (Fe) and manganese (Mn) by more than 10^6 relative to ambient deep ocean concentrations. Recent studies have also confirmed that Fe-Mn derived from hydrothermal fluid vents can be transported over distances of hundreds or thousands of kilometers [23,53,54]. The modern marine environment is generally in an oxic state, as seen in the pronounced negative Ce anomaly of seawater and bioapatite. It is reasonable to suggest that hydrothermal fluids containing considerable amounts of Fe–Mn oxyhydroxide particulates were spread eastwards from the EPR. The oxidized state of the deep-sea environment may contribute Fe-Mn metal ions to the formation of Fe–Mn micronodules, which would promote the enrichment of REY in marine sediments [55,56].

Most of the other REY-rich muds reported in the central North Pacific, the western Pacific, and the Indian Ocean appear to be unrelated to hydrothermal fluids [7–9]. However, the influence of hydrothermal fluids on the sediments in the Yupanqui Basin from the eastern South Pacific Ocean is significant.

The formation of the REY-rich muds in the eastern South Pacific Ocean appears to be associated strongly with hydrothermal activity, as suggested by the positive correlations between $\sum$REY and both Fe_2O_3 and MnO (Figure 5). The REY-rich sediments from the eastern South Pacific Ocean are enriched in Fe, Mn, Ba, Co, and Mo, which are much higher than the REY-rich sediments from the Indian Ocean, and the western and central North Pacific Ocean [7,8,14,57]. The clear enrichment in these elements is a typical characteristic of hydrothermal fluids and associated precipitates occurring at EPR [37]. The bulk sediments show LREE-enriched patterns and clear Eu positive anomalies in the seawater-normalized REE diagram, and the ratios of Fe/Ti and Al/(Al+Fe+Mn) range from 26.4–39.8 (greater than 20) and 0.24–0.33 (less than 0.35), which all strongly support the involvement of hydrothermal input during the formation of the REY-rich sediments [37,58]. Moreover, the $^{143}Nd/^{144}Nd$ values of the bioapatites are higher than the average values for Pacific seawater, which confirms the overprint of hydrothermal fluids (Figure 10).

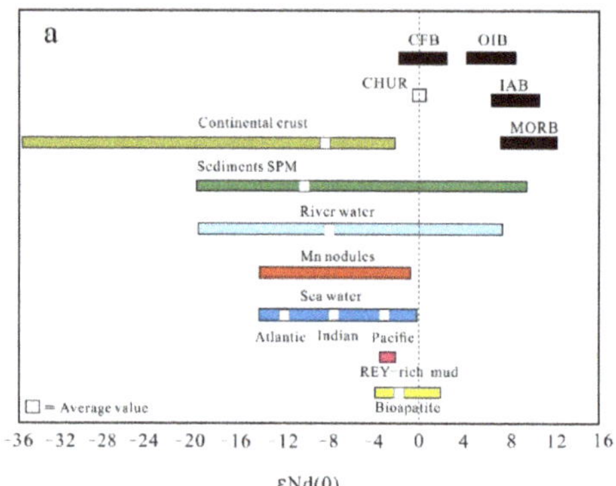

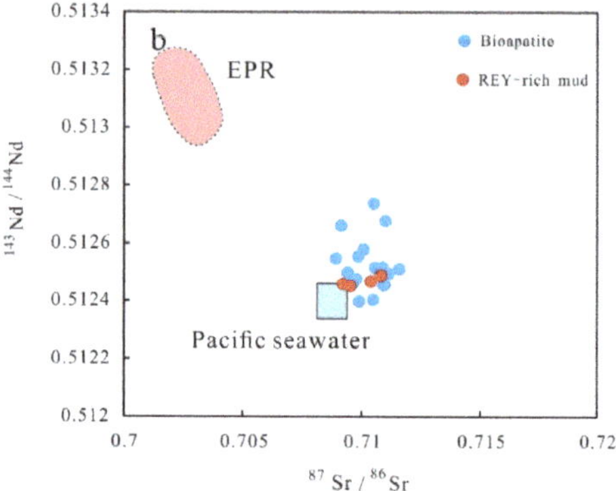

Figure 10. (**a**) Comparison of εNd(0) values in the REY-rich muds and bioapatites with various sources. (**b**) Plot of ($^{87}Sr/^{86}Sr$) vs ($^{143}Nd/^{144}Nd$) for the REY-rich muds and bioapatites. Natural variability of the Nd isotope composition in different rocks, sediments, and waters, given εNd(0) values relative to CHUR, modified after [59]. Mid-oceanic ridge basalts proposed by [60]. River water and river sediment εNd(0) values after [61,62]; seawater εNd(0)values for the different ocean basins after [63] and [64]. CHUR = chondritic uniform reservoir; CFB = continental flood basalt; OIB = ocean island basalt; IAB = island arc basalt; MORB = mid-oceanic ridge basalt. EPR/Chile Ridge field proposed by [65].

Thus, we can draw a conclusion that hydrothermal activity plays an important role in the formation of REY-rich muds in the eastern South Pacific. Hydrothermal fluids from the EPR containing considerable amounts of Fe–Mn metal ions have spread thousands of kilometers, precipitating Fe-Mn micronodules in the Yupanqui Basin, where significant quantities of REY are absorbed by the micronodules.

6. Conclusions

1. Hydrothermal fluids from the EPR may contribute a large amount of Fe-Mn metal ions, leading to the formation of Fe-Mn micronodules that enable the enrichment of REY in the REY-rich deep-sea sediments.

2. Bioapatite fossils are the main carrier for REYs in the REY-rich deep-sea sediments. REYs enter through the roots of the bioapatites and diffuse to the tips. It can be estimated roughly that the 5% bioapatites present in the samples account for ~45% of the REY budget in the analyzed sediments.

3. Ferromanganese micronodules contribute as much as 22.5% of the REY contents of the sediments, thus also making them an important REY-hosting phase in the eastern South Pacific. During early diagenesis, hydrogenetic Fe-Mn micronodules released REYs into the pore water, which are ultimately absorbed by bioapatites.

4. Phillipsite is not a main host of REY in REY-rich muds from the eastern South Pacific Ocean. This indicates a low sedimentation rate.

Supplementary Materials: The following are available online at http://www.mdpi.com/2075-163X/10/12/1141/s1, Table S1: The major elements content (%) and trace elements abundance (ppm) of bioapatite from the REY-rich muds in the S021GC17; Table S2: The major elements content (%) and trace elements abundance (ppm) of micronodules from the REY-rich muds in the S021GC17; Table S3: The major elements content (%) and trace elements abundance (ppm) of phillipsite from the REY-rich muds in the S021GC17

Author Contributions: Conceptualization, T.Z. and X.S.; investigation, X.S., M.H., X.R., G.Y.; data curation, M.Y., D.B., A.Z.; writing—original draft preparation, T.Z.; writing—review and editing, X.S.; visualization, T.Z.; supervision, X.S.; funding acquisition, X.S. All authors have read and agreed to the published version of the manuscript.

Funding: This research was supported by the National Key R&D Program of China (No. 2017YFC0602305), the National "13th Five-Year" Plan Project (DY135-R2-1-01, DY135-R2-1-02), the China Postdoctoral Science Foundation (Grant 2019M660172), Qingdao National Laboratory for Marine Science and Technology (Grant MGQNLM-KF201819), the National Natural Science Foundations of China (No. 41706061, 91858209).

Conflicts of Interest: The authors declare no conflict of interest.

References

1. Kato, Y.; Fujinaga, K.; Nakamura, K.; Takaya, Y.; Kitamura, K.; Ohta, J.; Toda, R.; Nakashima, T.; Iwamori, H. Deep-sea mud in the Pacific Ocean as a potential resource for rare-earth elements. *Nat. Geosci.* **2011**, *4*, 535–539. [CrossRef]
2. Balaram, V.; Banakar, V.K.; Subramanyam, K.S.V.; Roy, P.; Satyanarayanan, M.; Mohan, M.R.; Sawant, S.S. Yttrium and rare earth element contents in seamount cobalt crusts in the Indian Ocean. *Curr. Sci.* **2012**, *103*, 1334–1338.
3. Szamałek, K.; Konopka, G.; Zglinicki, K.; Marciniak-Maliszewska, B. New potential source of rare earth elements. *Gospod. Surowcami Miner. Miner. Resour. Manag.* **2013**, *29*, 59–76. [CrossRef]
4. Shi, X.F.; Huang, M.; Yu, M. *How Much Is Known About Deep-Sea Rare Earths. 10000 Scientific Puzzle*; Beijing Science Press: Beijing, China, 2018; pp. 604–605.
5. Wu, C.; Yuan, Z.; Bai, G. Rare earth deposits in China. In *Rare Earth Minerals: Chemistry, Origin and Ore Deposits. (The Mineralogical Society Series)*; Jones, A.P., Wall, F., Williams, C.T., Eds.; Chapman & Hall: London, UK, 1996; Volume 7, pp. 281–310.
6. Bao, Z.; Zhao, Z. Geochemistry of mineralization with exchangeable REY in the weathering crusts of granitic rocks in South China. *Ore Geol. Rev.* **2008**, *33*, 519–535. [CrossRef]
7. Yasukawa, K.; Liu, H.; Fujinaga, K.; Machida, S.; Haraguchi, S.; Ishii, T.; Nakamura, K.; Kato, Y. Geochemistry and mineralogy of REY-rich mud in the eastern Indian Ocean. *J. Asian Earth Sci.* **2014**, *93*, 25–36. [CrossRef]
8. Yasukawa, K.; Nakamura, K.; Fujinaga, K.; Machida, S.; Ohta, J.; Takaya, Y.; Kato, Y. Rare-earth, major, and trace element geochemistry of deep-sea sediments in the Indian Ocean: Implications for the potential distribution of REY-rich mud in the Indian Ocean. *Geochem. J.* **2015**, *49*, 621–635. [CrossRef]
9. Fujinaga, K.; Yasukawa, K.; Nakamura, K.; Machida, S.; Takaya, Y.; Ohta, J.; Araki, S.; Liu, H.; Usami, R.; Maki, R.; et al. Geochemistry of REY-rich mud in the Japanese exclusive economic zone around Minamitorishima Island. *Geochem. J.* **2016**, *50*, 575–590. [CrossRef]
10. Nakamura, K.; Machida, S.; Okino, K.; Masaki, Y.; Iijima, K.M.; Suzuki, K.; Kato, Y. Acoustic characterization of pelagic sediments using sub-bottom profiler data: Implications for the distribution of REY-rich mud in the Minamitorishima EEZ, western Pacific. *Geochem. J.* **2016**, *50*, 605–619. [CrossRef]
11. Lupton, J.E. Hydrothermal plumes: Near and far field. In *Seafloor Hydrothermal Systems: Physical, Chemical, Biological, and Geological Interactions*; Humphris, S.E., Zierenberg, R.A., Mullineaux, L.S., Thomson, R.E., Eds.; American Geophysical Union: Washington, DC, USA, 1995; Volume 91, pp. 317–346.
12. German, C.; Hergt, J.; Palmer, M.; Edmond, J. Geochemistry of a hydrothermal sediment core from the OBS vent-field, 21°N East Pacific Rise. *Chem. Geol.* **1999**, *155*, 65–75. [CrossRef]
13. Wu, J.; Wells, M.L.; Rember, R. Dissolved iron anomaly in the deep tropical–subtropical Pacific: Evidence for long-range transport of hydrothermal iron. *Geochim. Cosmochim. Acta* **2011**, *75*, 460–468. [CrossRef]
14. Kon, Y.; Hoshino, M.; Sanematsu, K.; Morita, S.; Tsunematsu, M.; Okamoto, N.; Yano, N.; Tanaka, M.; Takagi, T. Geochemical characteristics of apatite in heavy REE-rich deep-sea mud from Minami-Torishima area, southeastern Japan. *Resour. Geol.* **2014**, *64*, 47–57. [CrossRef]
15. Ohta, J.; Yasukawa, K.; Machida, S.; Fujinaga, K.; Nakamura, K.; Takaya, Y.; Iijima, K.M.; Suzuki, K.; Kato, Y. Geological factors responsible for REY-rich mud in the western North Pacific Ocean: Implications from mineralogy and grain size distributions. *Geochem. J.* **2016**, *50*, 591–603. [CrossRef]
16. Liao, J.; Sun, X.; Li, D.; Sa, R.; Lu, Y.; Lin, Z.; Xu, L.; Zhan, R.; Pan, Y.; Xu, H. New insights into nanostructure and geochemistry of bioapatite in REE-rich deep-sea sediments: LA-ICP-MS, TEM, and Z-contrast imaging studies. *Chem. Geol.* **2019**, *512*, 58–68. [CrossRef]

17. Kashiwabara, T.; Toda, R.; Fujinaga, K.; Honma, T.; Takahashi, Y.; Kato, Y. Determination of host phase of lanthanum in deepsea REY-rich mud by XAFS and μ-XRF using high-energy synchrotron radiation. *Chem. Lett.* **2014**, *43*, 199–200. [CrossRef]
18. Holler, G.; Marchig, V. Hydrothermal activity on the East Pacific Rise: Stages of development. In *Geologisches Jahrbuch*; U.S. Department of Energy, Office of Scientific and Technical Information: Oak Ridge, TN, USA, 1990; Volume 75, pp. 3–22.
19. Dick, H.J.; Lin, J.; Schouten, H. An ultraslow-spreading class of ocean ridge. *Nature* **2003**, *426*, 405–412. [CrossRef]
20. Faure, V.; Speer, K. Deep circulation in the eastern South Pacific Ocean. *J. Mar. Res.* **2012**, *70*, 748–778. [CrossRef]
21. Lupton, J.E.; Craig, H. Excess 3He in oceanic basalts: Evidence for terrestrial primordial helium. *Earth Planet. Sci. Lett.* **1975**, *26*, 133–139. [CrossRef]
22. Resing, J.A.; Sedwick, P.N.; German, C.R.; Jenkins, W.J.; Moffett, J.W.; Sohst, B.M.; Tagliabue, A. Basin-scale transport of hydrothermal dissolvedmetals across the South Pacific Ocean. *Nature* **2015**, *523*, 200–203. [CrossRef]
23. Fitzsimmons, J.N.; John, S.G.; Marsay, C.M.; Hoffman, C.L.; Nicholas, S.L.; Toner, B.M.; German, C.R.; Sherrell, R.M. Iron persistence in a distal hydrothermal plume supported by dissolved–particulate exchange. *Nat. Geosci.* **2017**, *10*, 195–201. [CrossRef]
24. Palma, S.; Silva, N. Epipelagic siphonophore assemblages associated with water masses along a transect between Chile and Easter Island (eastern South Pacific Ocean). *J. Plankton Res.* **2006**, *28*, 1143–1151. [CrossRef]
25. Cahill, T.; Isacks, B.L. Seismicity and shape of the subducted Nazca Plate. *J. Geophys. Res. Space Phys.* **1992**, *97*, 17503–17529. [CrossRef]
26. Strub, P.T.; Mesias, J.M.; Montecino, V.; Rutllant, J.; Salinas, S. Coastal Ocean Circulation off Western South America. *Coast. Segm. (6,E)* **1998**, *11*, 273–313.
27. Emery, W.J.; Meincke, J. Global water masses: Summary and review. *Oceanol. Acta* **1986**, *9*, 383–391.
28. Reid, J.L. The shallow salinity minima of the Pacific Ocean. *Deep Sea Res. Oceanogr. Abst.* **1973**, *20*, 51–68. [CrossRef]
29. Tanaka, T.; Togashi, S.; Kamioka, H.; Amakawa, H.; Kagami, H.; Hamamoto, T.; Yuhara, M.; Orihasi, Y.; Yoneda, S.; Shimizu, H.; et al. JNdi-1: A neodymium isotopic reference in consistency with LaJolla neodymium. *Chem. Geol.* **2000**, *168*, 279–281. [CrossRef]
30. Foster, G.L.; Vance, D. In situ Nd isotopic analysis of geological materials by laser ablation MC-ICP-MS. *J. Anal. At. Spectrom.* **2006**, *21*, 288–296. [CrossRef]
31. Taylor, S.R.; McLennan, S.M. *The Continental Crust: Its Composition and Evolution*; Blackwell Science: Oxford, UK, 1985.
32. Alibo, D.S.; Nozaki, Y. Rare earth elements in seawater: Particle association, shale-normalization, and Ce oxidation. *Geochim. Cosmochim. Acta* **1999**, *63*, 363–372. [CrossRef]
33. Bernat, M. Les isotopes de l'uranium et du thorium et les terres rares dans l'environnement marin. *Cah. Ostom. Ser. Geol.* **1975**, *7*, 65–83. (In French)
34. Toyoda, K.; Tokonami, M. Diffusion of rare-earth elements in fish teeth from deep-sea sediments. *Nature* **1990**, *345*, 607–609. [CrossRef]
35. German, C.R.; Klinkhammer, G.P.; Edmond, J.M.; Mura, A.; Elderfield, H. Hydrothermal scavenging of rare-earth elements in the ocean. *Nature* **1990**, *345*, 516–518. [CrossRef]
36. Ren, J.B.; Yao, H.Q.; Zhu, K.C.; He, G.W.; Deng, X.G.; Wang, H.F.; Liu, J.Y.; Fu, P.E.; Yang, S.X. Enrichment mechanism of rare earth elements and yttrium in deep-sea mud of Clarion-Clipperton region. *Earth Sci. Front.* **2015**, *22*, 200–211.
37. Bau, M.; Dulski, P. Comparing yttrium and rare earths in hydrothermal fluids from the Mid-Atlantic Ridge: Implications for Y and REE behavior during near-vent mixing and for the Y/Ho ratio of Proteozoic seawater. *Chem. Geol.* **1999**, *155*, 77–90.
38. Ohta, A.; Ishii, S.; Sakakibara, M.; Mizuno, A.; Kawabe, I. Systematic correlation of the Ce anomaly with the Co/(Ni+Cu) ratio and Y fractionation from Ho in distinct types of Pacific deep-sea nodules. *Geochem. J.* **1999**, *33*, 399–417. [CrossRef]
39. Kanazawa, T.; Sager, W.W.; Escutia, D. Carbonates and bulk sediment geochemistry of ODP Hole 191-1179C. *Pangaea* **2005**. [CrossRef]

40. Shen, H. Rare earth elements in deep-sea sediments. *Geochimica* **1990**, *19*, 340–348, (in Chinese with English abstract).
41. Bonatti, E.; Kraemer, T.; Rydell, H. Classification and genesis of submarine iron-manganese deposits. Papers from a conference on Ferromanganese Deposits on the Ocean Floor. *Natl. Sci. Found.* **1972**, 149–166.
42. Josso, P.; Pelleter, E.; Pourret, O.; Fouquet, Y.; Etoubleau, J.; Cheron, S.; Bollinger, C. A new discrimination scheme for oceanic ferromanganese deposits using high field strength and rare earth elements. *Ore Geol. Rev.* **2017**, *87*, 3–15. [CrossRef]
43. Slomp, C.P.; Epping, E.H.G.; Helder, W.; Van Raaphorst, W. A key role for iron-bound phosphorus in authigenic apatite formation in North Atlantic continental platform sediments. *J. Mar. Res.* **1996**, *54*, 1179–1205. [CrossRef]
44. Tsandev, I.; Reed, D.C.; Slomp, C.P. Phosphorus diagenesis in deep-sea sediments: Sensitivity to water column conditions and global scale implications. *Chem. Geol.* **2012**, *330*, 127–139. [CrossRef]
45. Takahashi, Y.; Hayasaka, Y.; Morita, K.; Kashiwabara, T.; Nakada, R.; Marcus, M.A.; Kato, K.; Tanaka, K.; Shimizu, H. Transfer of rare earth elements (REE) from manganese oxides to phosphates during early diagenesis in pelagic sediments inferred from REE patterns, X-ray absorption spectroscopy, and chemical leaching method. *Geochem. J.* **2015**, *49*, 653–674. [CrossRef]
46. Dubinin, A.V. Geochemistry of rare earth elements in oceanic phillipsites. *Lithol. Miner. Resour.* **2000**, *35*, 101–108. [CrossRef]
47. Piper, D.Z. Rare earth elements in ferromanganese nodules and other marine phases. *Geochim. Cosmochim. Acta* **1974**, *38*, 1007–1022. [CrossRef]
48. Bernat, M.; Church, T.M. Deep-sea phillipsites: Trace geochemistry and modes of formation. *ACM Lisp Bulletin* **1978**, 259–267.
49. Zhang, X.Y.; Deng, H.; Zhang, F.Y.; Zhang, W.Y.; Du, Y.; Jiang, B.B. Enrichment and geochemical characteristics of rare earth elements in deep-sea mud from seamount area of Western Pacific. *J. Chin. Soc. Rare Earths* **2013**, *31*, 729–737, (in Chinese with English abstract).
50. Müller, R.D.; Sdrolias, M.; Gaina, C.; Roest, W.R. Age, spreading rates, and spreading asymmetry of the world's ocean crust. *Geochem. Geophys. Geosystems* **2008**, *9*. [CrossRef]
51. German, C.R.; Lin, J.; Parson, L.M. Mid-ocean ridges: Hydrothermal interactions between the lithosphere and oceans. *Geophys. Monogr. Ser.* **2004**, *148*, 318.
52. Kashiwabara, T.; Toda, R.; Nakamura, K.; Yasukawa, K.; Fujinaga, K.; Kubo, S.; Nozaki, T.; Takahashi, Y.; Suzuki, K.; Kato, Y. Synchrotron X-ray spectroscopic perspective on the formation mechanism of REY-rich muds in the Pacific Ocean. *Geochim. Cosmochim. Acta* **2018**, *240*, 274–292. [CrossRef]
53. Bennett, S.A.; Achterberg, E.P.; Connelly, D.P.; Statham, P.J.; Fones, G.R.; German, C.R. The distribution and stabilization of dissolved Fe in deep-sea hydrothermal plumes. *Earth Planet. Sci. Lett.* **2008**, *270*, 157–167. [CrossRef]
54. Hawkes, J.A.; Connelly, D.P.; Gledhill, M.; Achterberg, E.P. The stabilisation and transportation of dissolved iron from high temperature hydrothermal vent systems. *Earth Planet. Sci. Lett.* **2013**, *375*, 280–290. [CrossRef]
55. Ruhlin, D.E.; Owen, R.M. The rare earth element geochemistry of hydrothermal sediments from the East Pacific Rise: Examination of a seawater scavenging mechanism. *Geochim. Cosmochim. Acta* **1986**, *50*, 393–400. [CrossRef]
56. Olivarez, A.M.; Owen, R.M. REE/Fe variations in hydrothermal sediments: Implications for the REE content of seawater. *Geochim. Cosmochim. Acta* **1989**, *53*, 757–762. [CrossRef]
57. Sa, R.; Sun, X.; He, G.; Xu, L.; Pan, Q.; Liao, J.; Zhu, K.; Deng, X. Enrichment of rare earth elements in siliceous sediments under slow deposition: A case study of the central North Pacific. *Ore Geol. Rev.* **2018**, *94*, 12–23. [CrossRef]
58. Boström, K. Genesis of ferromanganese deposits. In *Hydrothermal Processes at Seafloor Spreading Centers*; Rona, P.A., Boström, K., Laubier, L., Smith, K.L., Eds.; NATO Conference Series (IV Marine Sciences); Springer: Boston, MA, USA, 1983; Volume 12, pp. 473–489.
59. Piepgras, D.J.; Wasserburg, G.J. Isotopic composition of neodymiuminwaters of the Drake Passage. *Science* **1982**, *217*, 207–214. [CrossRef] [PubMed]
60. Goldstein, S.; Hemming, S.R. Long lived isotopic tracers in oceanography. Paleoceanography and ice-sheet dynamics. In *Treatise on Geochemistry*; Elderfield, H., Turekian, K.K., Eds.; Elsevier: Amsterdam, The Netherlands, 2003; Volume 6, pp. 453–489.

61. Goldstein, S.J.; Jacobsen, S.B. The Nd and Sr isotopic systematics of river-water dissolved material: Implications for the sources of Nd and Sr in seawater. *Chem. Geol. Isot. Geosci. Sect.* **1987**, *66*, 245–272. [CrossRef]
62. Goldstein, S.J.; Jacobsen, S.B. Nd and Sr isotopic systematics of river water suspended material: Implications for crustal evolution. *Earth Planet. Sci. Lett.* **1988**, *87*, 249–265. [CrossRef]
63. Piepgras, D.J.; Wasserburg, G.J.; Dasch, E.J. The isotopic composition of Nd in different ocean water masses. *Earth Planet. Sci. Lett.* **1979**, *45*, 223–226. [CrossRef]
64. Piepgras, D.J.; Wasserburg, G. Neodymium isotopic variations in seawater. *Earth Planet. Sci. Lett.* **1980**, *50*, 128–138. [CrossRef]
65. Bach, W.; Erzinger, J.; Dosso, L.; Bollinger, C.; Bougault, H.; Etoubleau, J.; Sauerwein, J. Unusually large Nb–Ta depletions in North Chile ridge basalts at 36°50′ to 38°56′ S: Major element, trace element, and isotopic data. *Earth Planet. Sci. Lett.* **1996**, *142*, 223–240. [CrossRef]

Publisher's Note: MDPI stays neutral with regard to jurisdictional claims in published maps and institutional affiliations.

Article

Garnet Geochemistry of Reduced Skarn System: Implications for Fluid Evolution and Skarn Formation of the Zhuxiling W (Mo) Deposit, China

Xiao-Xia Duan [1,2,*], Ying-Fu Ju [1,2], Bin Chen [3] and Zhi-Qiang Wang [1,2]

1 School of Resources and Environmental Engineering, Hefei University of Technology, Hefei 230009, China; 2019110708@mail.hfut.edu.cn (Y.-F.J.); wangzq@hfut.edu.cn (Z.-Q.W.)
2 Ore Deposit and Exploration Center (ODEC), Hefei University of Technology, Hefei 230009, China
3 Department of Earth and Space Sciences, Southern University of Science and Technology, Shenzhen 518055, China; chenb6@sustech.edu.cn
* Correspondence: duanxiaoxia@hfut.edu.cn

Received: 21 October 2020; Accepted: 12 November 2020; Published: 17 November 2020

Abstract: A newly discovered tungsten ore district containing more than 300,000 tons of WO_3 in southern Anhui Province has attracted great attention. The Zhuxiling W (Mo) deposit in the district is dominated by skarn tungsten mineralization. This paper conducted in suit EPMA and LA-ICPMS spot and mapping analysis of the skarn mineral garnet to reveal the evolution of fluids, metasomatic dynamics, and formation conditions of skarn. Two generations of garnet have been identified for Zhuxiling W (Mo) skarn: 1) Gt-I generation garnet is isotropic, Al-rich grossular without zoning. As a further subdivision, Gt-IB garnet ($Adr_{19\text{-}46}Grs_{49\text{-}77}$ $(Sps+Pyr+Alm)_{4\text{-}5}$) contains significantly high content of Ti and Mn compared with Gt-IA garnet ($Adr_{3\text{-}42}Grs_{53\text{-}96}$ $(Sps+Pyr+Alm)_{1\text{-}5}$). 2) Gt-II generation garnet is anisotropic, Fe-rich andradite with oscillatory zoning. Gt-II garnet displays compositional changes with a decrease of Fe and an increase of Mn from proximal skarn (Gt-IIA) to distal skarn (Gt-IIB) with the presence of subcalcic garnet for Gt-IIB type (Sps+Pyr+Alm = 56–68). The presence of pyrrhotite associated with subcalcic garnet indicates a relatively reduced skarn system. Gt-I grossular is overall enriched in Cr, Zr, Y, Nb, and Ta compared with the Gt-II andradite, and both W and Sn strongly favor Fe-rich garnet compared with Al-rich garnet. Gt-IA grossular garnet presents a REE trend with an upward-facing parabola peaking at Pr and Nd in contrast to low and flat HREE, and Gt-IB grossular garnet has a distinct REE pattern with enriched HREE. Gt-IIA andradite garnet displays a right-dipping REE pattern (enriched LREE and depleted HREE) with a prominent positive Eu anomaly (Eu/Eu* = 3.6–15.3). In contrast, Gt-IIB andradite garnet shows depleted LREE and enriched HREE with a weak positive Eu anomaly (Eu/Eu* = 0–6.0). The incorporation and fractionation of REE in garnet are collectively controlled by crystal chemistry and extrinsic factors, such as P–T–X conditions of fluids, fluid/rock ratios, and mineral growth kinetics. Major and trace elements of two generations of garnet combined with optical and textural characteristics suggest that Gt-I Al-rich grossular garnets grow slowly through diffusive metasomatism under a closed system, whereas Gt-II Fe-rich andradite represent rapid growth garnet formed by the infiltration metasomatism of magmatic fluids in an open system. The Mn-rich garnet implies active fluid–rock interaction with Mn-rich dolomitic limestone of the Lantian Group in the district.

Keywords: LA-ICPMS; Jiangnan tungsten ore belt; Zhuxiling W (Mo) deposit; reduced skarn; Mn-rich garnet

1. Introduction

Mineralogy is fundamental in the definition and study of skarn and skarn deposits [1], as it can provide insight into skarn formations and evaluate their potential for economic importance. As an indicator mineral, garnet composition and texture can effectively reveal the physical and chemical properties of ore-forming fluids and record hydrothermal fluid evolution and water–rock reactions [1–4]. Moreover, garnet chemistry is widely used to reveal skarn formation processes such as the metasomatism mechanism (diffusive metasomatism or advective metasomatism, [3,5]) and the kinetics of mineral growth [6]. Garnets from different skarn deposits (i.e., W, Sn, Cu, and Mo) show enrichment of respective metal elements and demonstrate potential as an indicator for mineralization exploration [7–10].

Recent exploration revealed a NE trending tungsten ore belt, namely the Jiangnan porphyry–skarn tungsten ore belt [11], with many large tungsten deposits located within the Jiangnan orogenic belt, including the giant Zhuxi and Dahutang tungsten deposits in northern Jiangxi Province and numerous W (Mo) deposits in southern Anhui Province. However, few studies have focused on the ore district in Anhui Province [12–15], leading to a lack of detailed mineralogical and geochemical work on skarn minerals. This paper reports new major and trace element data of different types of garnet occurrences in the Zhuxiling deposit, a typical skarn W (Mo) deposit in southern Anhui Province. Based on compositional variation and optical and textural characteristics, we discuss the mechanism and controlling factors for incorporation of trace element (REE in particular) into garnet and constrain the evolution of metasomatic fluids and physicochemical conditions during skarn mineralization from spatial and temporal perspectives.

2. Regional Geological Settings and Ore Deposit Geology

The Jiangnan orogenic belt (JNB) is a Neoproterozoic subduction-collisional zone between the Yangtze and the Cathaysia blocks [16–18]. The JNB is composed of a Precambrian basement and a Phanerozoic sequence. The basement consists mainly of Neoproterozoic volcanoclastic and sedimentary rocks. The overlying Phanerozoic sequence includes the Silurian to Early Jurassic marine clastic and carbonate rocks, Middle Jurassic sedimentary and volcanic rocks, and Cretaceous red-bed sandstone. There are mainly two periods of magmatic events, represented by Jinningian and Yanshanian granitic rocks. The Jinningian intrusions (~821 Ma) are mainly biotite granodiorite and fine-grained granite, including the Jiuling, Xucun, Shexian, and Xiuning plutons. The Yanshanian intrusions in the JNB are further divided as the 149–136 Ma W-related granitoids and the 129–102 Ma W- or Sn-related monzonitic granite [19].

Recent exploration revealed many large-scale tungsten deposits in the Jiangnan orogenic belt (Figure 1), including the giant Zhuxi and Dahutang tungsten deposits in northern Jiangxi Province and numerous W (Mo) deposits in southern Anhui Province. All together those tungsten deposits constitute a NE trending tungsten ore belt, namely the Jiangnan porphyry–skarn tungsten ore belt [11], adjacent to the Middle-Lower Yangtze metallogenic belt (Figure 1). The tungsten ore district in southern Anhui Province hosts more than 50 W(Mo) deposits or occurrences and contains more than 300,000 tons of WO_3 [20,21] which include the Dongyuan W (Mo), Zhuxiling W (Mo), Baizhangyan W (Mo), Xiaoyao W–Mo, and Gaojiabang W–Mo deposits [12–15]. The ore-formation ages are confined within 152–143 Ma and are associated with the Jurassic magmatic activity between 152–139 Ma [22].

The Zhuxiling deposit is a large-scale skarn-type tungsten deposit, situated in Ningguo, a city located in southern Anhui province (Figure 2). The exposed strata in the district are the Neoproterozoic Nantuo and Lantian groups, which consist of tuffaceous gravel-bearing sandstone and pelitic dolomitic limestone, respectively. The dolomitic limestone of the Liantian Group is locally Mn-rich [24]. Magmatic rocks in the district are mainly the Late Mesozoic granodiorite, granodiorite porphyry, and granite porphyry veins. The ore-related granodiorite porphyry has an outcrop area of 1.5 km^2, with zircon U–Pb ages of 139–142 Ma [24,25]. The Zhuxiling W(Mo) deposit, which also contains appreciable Ag, comprises mainly two types of mineralization, the skarn/porphyritic W-Mo mineralization and quartz vein type Ag mineralization. The skarn type W-Mo mineralization occurs within the granodiorite porphyry body as well as along the contact between the granodiorite porphyry and the dolomitic limestone of the Lantian Group. Ore minerals include scheelite, molybdenite, pyrite, sphalerite, pyrrhotite, chalcopyrite, galena, tetrahedrite, argentite, and silver. Gangue minerals mainly include garnet, diopside, quartz, sericite, and calcite. The associated alterations are predominantly skarn, with subordinate silicificated and hornfelsic alterations. For tungsten mineralization, scheelite is the major ore mineral and occurs predominantly as disseminated and in veins intergrown with retrograde minerals. Associated sulfides include large amounts of pyrrhotite, minor sphalerite, chalcopyrite, and pyrite and traces of magnetite (Figure 3a). Limited studies, such as geochronological, petrological, and geochemical analyses of magmatic rocks and ore-forming fluid, have been conducted [24–27].

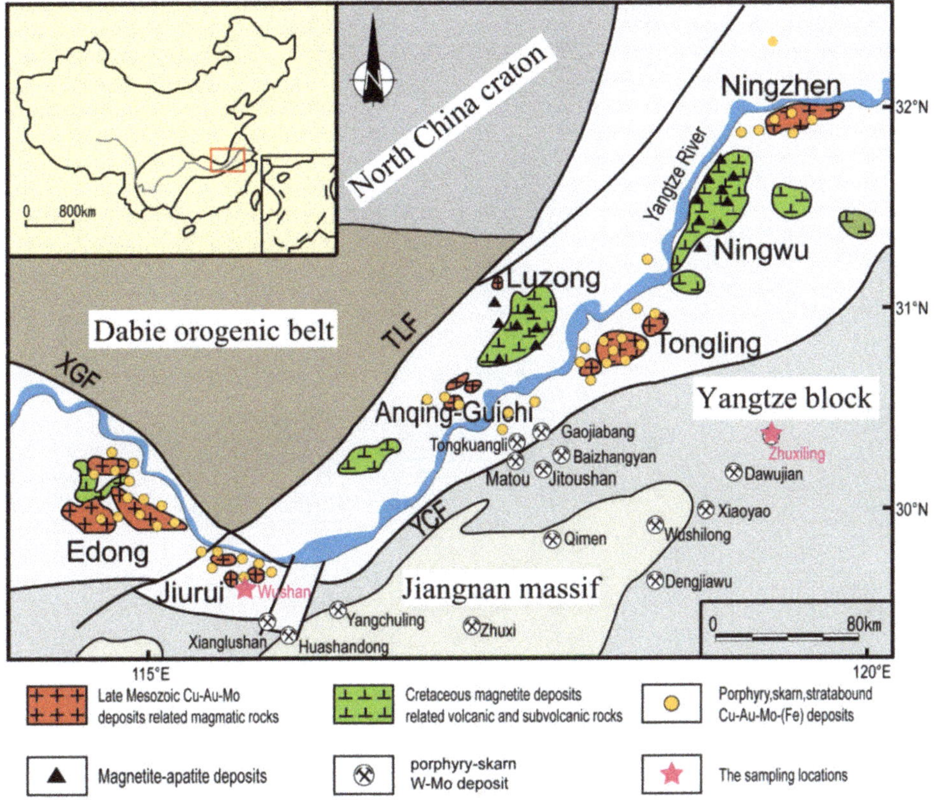

Figure 1. Geological sketch map for the Jiangnan tungsten ore belt showing the location of the Zhuxiling W (Mo) deposit (after [11,23]).

The garnet samples analysed in this study are selected from drill hole ZK502 of the Zhuxiling deposit (Figure 3b–d). Sample ZXL209 is composed predominantly of garnet, quartz, calcite, and diopside, with small amounts of chlorite, titanite, and sphalerite. Sample ZXL211 is diopside-garnet skarn, which is composed mainly of diopside and garnet, as well as minor amounts of sphalerite, pyrite, and scheelite. Mineral assemblages for sample ZXL216 include garnet, diopside, epidote, chlorite, magnetite, pyrite, calcite, and scheelite. The early-stage skarn mineral diopside and garnet are intensively replaced by chlorite, epidote, magnetite, and calcite. Sample ZXL221 is banded skarn, containing mainly garnet and minor amount of chlorite, calcite, and pyrite. Additionally, garnet is widely replaced by calcite and chlorite.

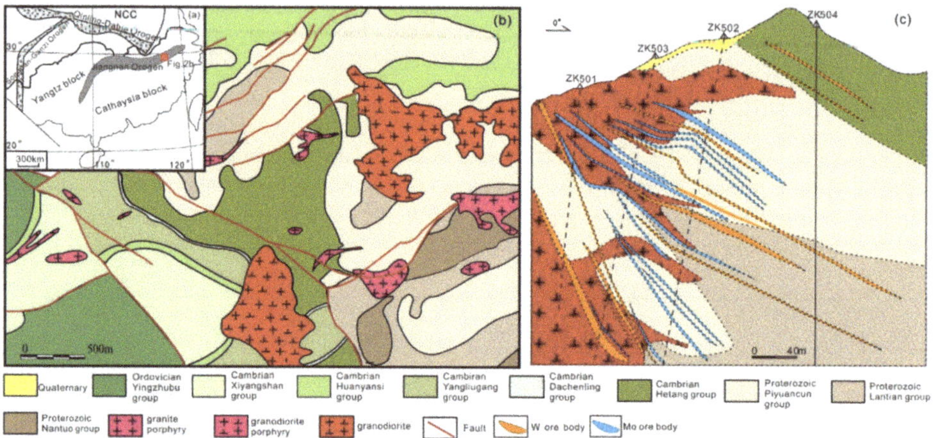

Figure 2. (**a**) Tectonic location of the Jiangnan orogeny; (**b**) geological map of Zhuxiling W(Mo) deposit; (**c**) geological cross section map of the Zhuxiling W (Mo) deposit, illustrating sampling drill hole (after [24]).

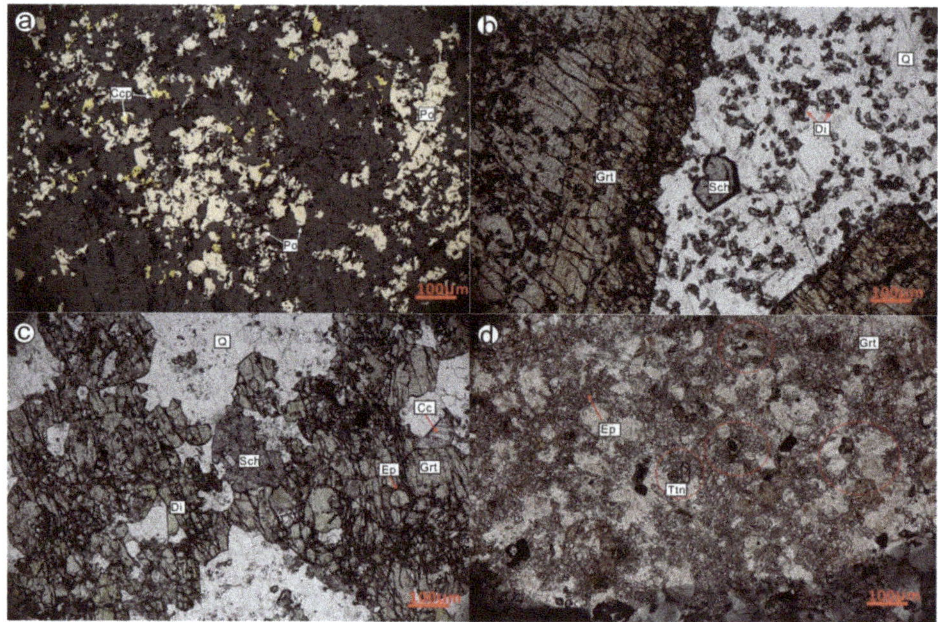

Figure 3. Photomicrographs of skarns from the Zhuxiling W (Mo) deposit. (**a**) Diopside skarn consists of diopside, quartz, and large amounts of pyrrhotite, as well as minor amounts of chalcopyrite and titanite; (**b**) garnet skarn consists of large euhedral garnet crystal, small diopside as inclusions inside quartz and garnet, as well as disseminated scheelite. (**c**) Garnet-diopside skarn where garnet is replaced by epidote, calcite, and scheelite. (**d**) Retrograde skarn with massive epidote, quartz, and disseminated titanite. Abbreviations: Ccp: chalcopyrite; Po: pyrrhotite; Grt: garnet; Sch: scheelite; Di: diopside; Q: quartz; Ep: epidote; Cc: calcite; and Ttn: titanite.

3. Analytical Methods

The major elements of garnet were determined by electron microprobe at the School of Resources and Environmental Engineering at the Hefei University of Technology (HFUT). Backscattered electron (BSE) images were acquired using a JEOL JXA−8900R electron microprobe, and quantitative point analysis of the garnets was conducted using a wavelength-dispersive (WDS) method that employed TAP, PET, and LIF crystals with a 2-μm spatial resolution. Each 120-second analysis was conducted on 2–5 grains for each sample, and the measurement was conducted using an excitation voltage of 15 kV, a beam current of 10 nA, and a spot diameter of 5 μm. The respective peak and background count times were 120 s and 60 s for La, Ce, Sr, and Ba and 60 s and 30 s for all other elements. Garnet molecular formulae were calculated on the basis of 12 oxygen atoms and garnet end-members normative calculations presented proportions of garnet end-members (i.e., grossular (Grs), andradite (Adr), Spessartine (Sps), pyrope (Pyr), and almandine (Alm)) for each analysis.

An in-situ trace element spot and mapping analysis of garnet was performed at HFUT using the Agilent 7900 (quadrupole) ICP−MS coupled with a laser ablation system (PhotonMachines Analyte HE with a 193-nm ArF Excimer). The trace element contents of garnet were measured directly on thin sections. The diameter of the laser beam was 30 μm (with a 10 Hz repetition rate, an output energy of 0.01−0.1 mJ per pulse, and a fluence of ~4 J cm^{-2}). Each analysis consisted of an approximate 30 s background acquisition and 60 s sample acquisition. The dwell time was 3 ms for all REEs and 5 ms for all other elements. Standard reference materials GSE-1g, GSC-1g, BCR-2G, and NIST 612 were used as external standards. The standard reference materials were run after each 10−15 unknowns. Data reduction was performed using ICP−MS DataCal software.

4. Results

Two generations of garnets are identified within the Zhuxiling deposit according to the textural and optical characteristics, which are further divided into four types based on compositional variations. Gt-I generation garnet represents early prograde unzoned garnet, is dark in BSE images, and is replaced by later oscillatory zoned Gt-II generation garnet, which is brighter in BSE images. More specifically, Gt-IA type garnet is colorless under plane polarized light and shows isotropic features with an absence of zoning. Gt-IA garnet is commonly anhedral to subhedral and occurs as irregular-shaped cores encircled by later beige Gt-II oscillatory rims (Figure 4a,b,h). Gt-IA garnet is pervasively modified by retrograde chlorite, scheelite, and calcite. Gt-IB type garnet is also colorless, but it's anisotropic and euhedral compared with Gt-IA garnet. The spotted garnet displays sharp contacts to the later overgrowth of beige garnet. There is no obvious zoning, but it has many cracks filled by massive calcite and quartz inclusions (Figure 4c,e). Gt-IB garnet is relatively scarce and is only observed within the banded skarn (sample ZXL221). Gt-II garnet is subdivided into Gt-IIA and Gt-IIB, representing proximal and distal exoskarn samples, respectively. Gt-II garnet is overall beige under plane polarized light in the form of large euhedral and subhedral crystals. Both the Gt-IIA and Gt-IIB garnet are anisotropic and often exhibit dodecahedral twinning and characteristic oscillatory zoning (Figure 4b,h). In addition to oscillatory zoning, notable irregular zoning was also present for some Gt-IIA garnet shown by BSE imaging (Figure 4f). The trapezohedron {211} faces developing on pre-existing dodecahedral {110} garnets have been observed (Figure 4d). Gt-II garnet contains fine diopside inclusions and is replaced by retrograde minerals such as chlorite, epidote, titanite, tremolite, and calcite.

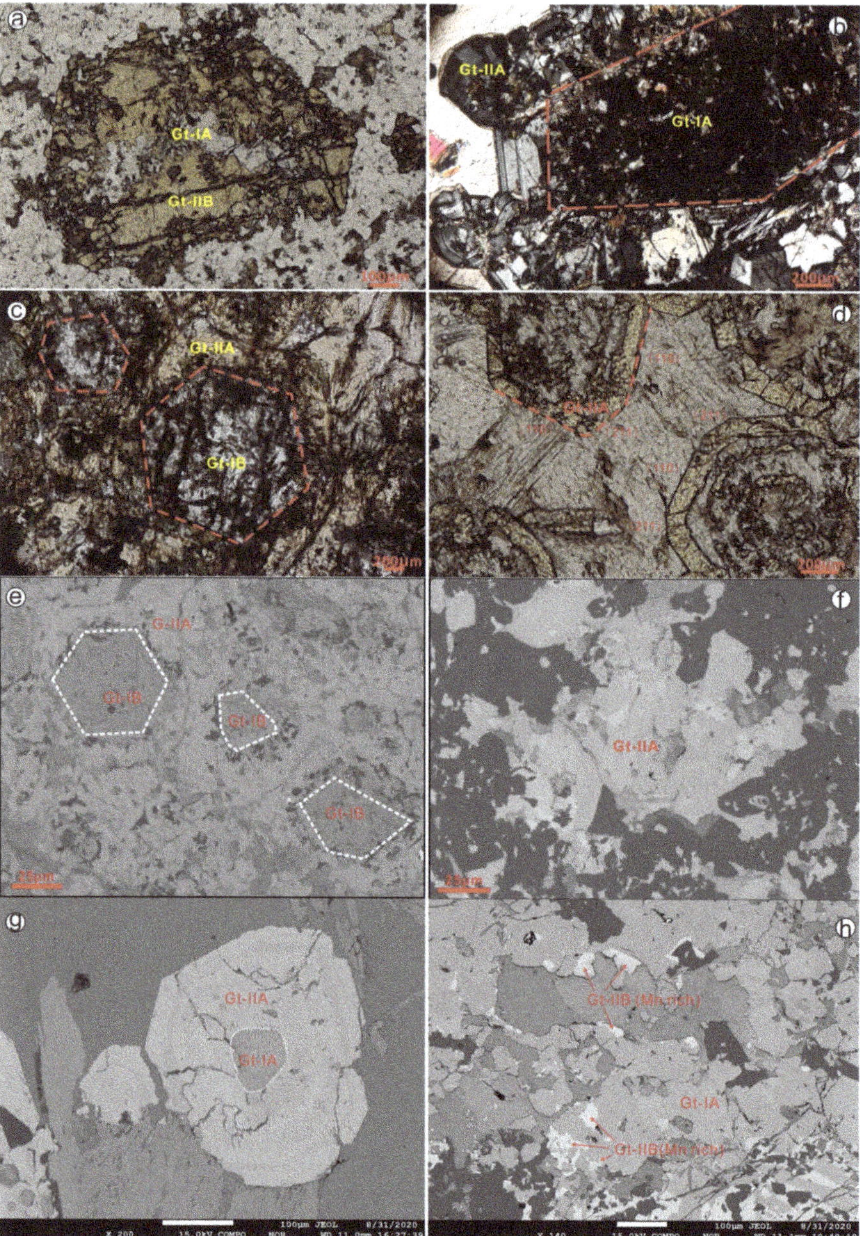

Figure 4. Microscopic and BSE (back scattered electron) characteristics of four types of garnet from the Zhuxiling W (Mo) deposit. (**a**) Pale white Gt-IA garnet is irregularly replaced by beige Gt-IIB garnet; (**b**) Gt-IA is isotropic and shows complete extinction, whereas Gt-IIA garnet is anisotropic with oscillatory zoning; (**c**) spotted euhedral Gt-IB garnet crystal encircled by beige Gt-IIA garnet; (**d**) development of trapezohedron {211} faces in Gt-IIA garnets on pre-existing dodecahedral {110} garnets; (**e**) BSE imaging of euhedral Gt-IB garnet, which is darker than surrounding Gt-IIA garnets; (**f**) patched BSE imaging for typical Gt-IIA garnet; (**g**) BSE image illustrating one euhedral garnet crystal with dark and irregular Gt-IA garnet in the core and oscillatory zoned Gt-IIA garnet in the rim. (**h**) Presence of Mn-rich (spessartine dominate) garnet at the margin of the Gt-IA garnet.

4.1. Major Element Chemistry

The EPMA results for four types of garnets are listed in supplementary Table S1. Garnets have the general chemical formula $X_3Y_2Z_3O_{12}$ (eight formula units per basic cell) and site X corresponds to eight-fold coordination filled by bivalent metal (X = Ca^{2+}, Fe^{2+}, Mg^{2+}, or Mn^{2+}); Y corresponds to six-fold coordination filled by trivalent metal (Y = Al^{3+}, Cr^{3+}, or Fe^{3+}); and Z corresponds to four-fold (tetrahedral) coordination (largely Si). Garnets from the Zhuxiling deposit belong to a grandite solid solution. They have compositions that range from almost pure grossular (Adr_3Grs_{96}) to andradite ($Adr_{85}Grs_{11}$) with variable amounts of spessartine and almandine and traces of pyrope. Different types of garnet show distinct major element compositions (Figure 5), corresponding to the different textural and optical characteristics described before. Gt-IA garnet is characterized by high Al and Mg content and low Fe and Mn content and belongs to grossular ($Adr_{3-42}Grs_{53-96}$ $(Sps+Pyr+Alm)_{1-5}$). Gt-IB garnet is also enriched in Al and depleted in Fe ($Adr_{19-46}Grs_{49-77}$ $(Sps+Pyr+Alm)_{4-5}$) with a significantly high content of Ti (0.08–0.11 apfu) and Mn (0.13–0.15 apfu) compared with Gt-IA garnet (0–0.19 apfu Ti and 0.01–0.05 apfu Mn). Gt-II garnet contains more Fe compared with Gt-I type and is ascribed to andradite. Moreover, the proximal exoskarn Gt-IIA type is Fe-rich andradite ($Adr_{37-85}Grs_{4-58}$ $(Sps+Pyr+Alm)_{4-19}$), whereas the distal exoskarn Gt-IIB is Al-rich andradite ($Adr_{8-60}Grs_{18-47}$ $(Sps+Pyr+Alm)_{11-68}$). Gt-IIB garnet contains elevated Mn (0.28–1.60 apfu) content compared with Gt-IIB garnet (0.09-0.48 apfu). Note that some of Gt-IIB garnet in sample ZXL209 are extremely enriched in Mn (15.834–23.163 wt.%) and those Mn-rich garnet ($Adr_{9-15}Grs_{18-35}$ $(Sps+Pyr+Alm)_{56-68}$) occur as small patches along the margin of Gt-IA grossular (Figure 4g).

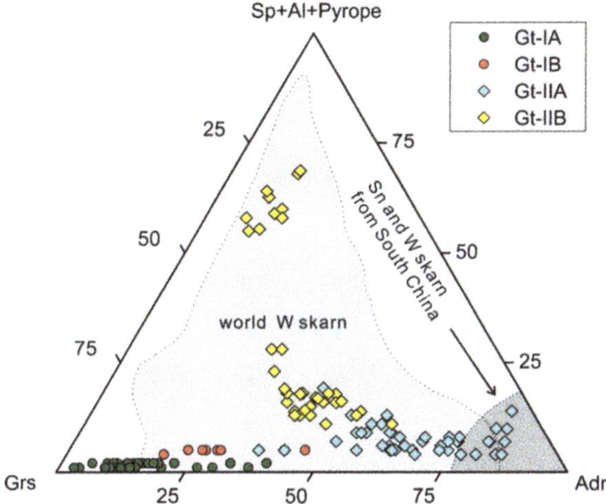

Figure 5. Ternary plots of the garnet compositions from the Zhuxiling W (Mo) deposit. The ranges of garnets in W/Sn skarn around the world are from [1] and the data for Sn and W skarn in South China are from [28].

4.2. Trace Element Chemistry

The trace element data of garnets analyzed by LA-ICPMS are presented in supplementary Table S2. Since the Mn-rich garnet is very small, only two valid spots were obtained. There is no difference between the different types of garnet regarding the large ion lithophile elements (LILE) since all garnets show negligible amounts of LILEs, such as Rb, Cs, Sr, and Ba (Table S2). In contrast, Cr, V, Ga, Zn, Sn, and high field-strength elements (HFSE), such as Zr, Hf, Y, and Sc, are more abundant and show variations corresponding to major element and petrographic features. Gt-I generation grossular is enriched in Cr, Zr, Y, Nb, and Ta and depleted in V, W, and Sn compared with Gt-II generation andradite (Figure 6). Most trace elements (e.g., Zr, Y, W, and Cr) of Gt-II andradite are more scattered compared with those of Gt-I grossular. More specifically, most of the trace element contents in Gt-IA are comparable to those of Gt-IB, with the exception of Cr. The Gt-IB garnet hosts an extremely high content of Cr (1516–9426 ppm, average of 3639 ppm), whereas the Cr content of Gt-IA garnet is almost 2 orders of magnitude lower (4.6–157 ppm, average of 37 ppm). Gt-II generation andradite has significantly higher V, W, and Sn content compared with Gt-I grossular, and Gt-IIB type andradite contains elevated Zr, Y, Nb, Ta, and Ge in contrast to Gt-IIA type andradite, which shows higher U content (Figure 6). The new trace element analysis of the Zhuxiling W (Mo) deposit confirms that the Fe-rich end-members of grandite are more Sn-enriched, consistent with previous studies [8,9,29].

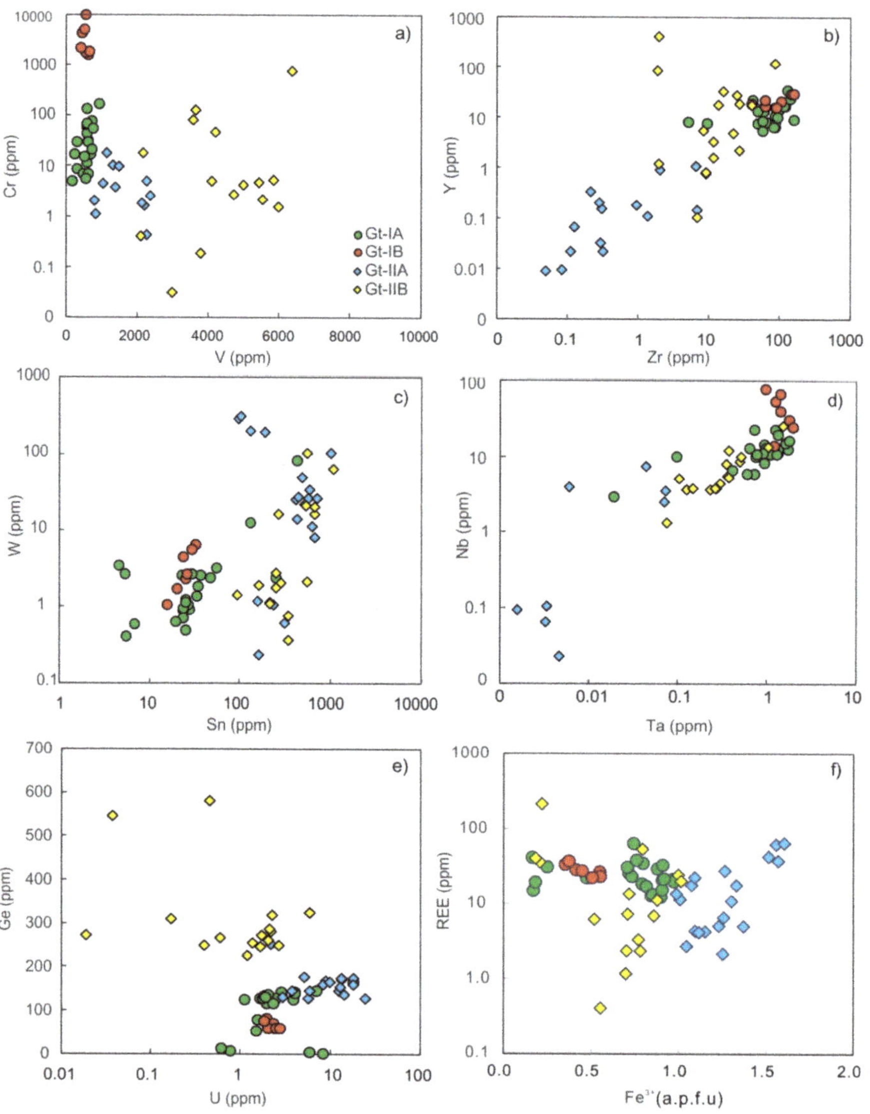

Figure 6. Trace element diagrams of different types of garnets from the Zhuxiling W (Mo) deposit. (**a**) Cr (ppm) versus V(ppm) plot; (**b**) Zr (ppm) versus Y (ppm) plot; (**c**) Sn (ppm) versus W (ppm) plot;

(**d**) Ta (ppm) versus Nb (ppm) plot; (**e**) U (ppm) versus Ge (ppm) plot; (**f**) Fe^{3+} (a.p.f.u) versus REE (ppm) plot.

Despite having the similar average REE content, the REE contents for Gt-II andradite is more dispersed (0.4–214 ppm), whereas the Gt-I grossular are constrained within a narrow range (12.3–64.3 ppm). Moreover, the chondrite-normalized REE pattern of Gt-IA grossular garnet presents a trend with an upward-facing parabola peaking at Pr and Nd, in contrast to low and flat HREE (Figure 7a), due to the closer ionic radii of Pr^{3+} and Nd^{3+} to Ca^{2+}, which is favorable for substitution [30]. Gt-IB grossular has a distinct REE pattern with enriched HREE and depleted LREE (Figure 7b). Both the Gt-IA and Gt-IB grossulars exhibit identical weak negative Eu anomalies (Eu/Eu* = 0.3–1.5 and 0.4–0.9, respectively). On the other hand, Gt-IIA andradite displays a right-dipping REE pattern (enriched LREE and depleted HREE) with a prominent positive Eu anomaly (Eu/Eu* = 3.6–15.3) (Figure 7c). The HREE contents of Gt-IIA andradite are very low with many analyses below detection limits. In comparison, Gt-IIB andradite shows an upward-slopping REE trend manifested by depleted LREE and enriched HREE with weak to no positive Eu anomalies (Eu/Eu* = 0–6.0). Limited analyses of Mn-rich garnets (Sps = 36–54) display prominent fractionation with extremely high HREE and very low LREE (Figure 7d).

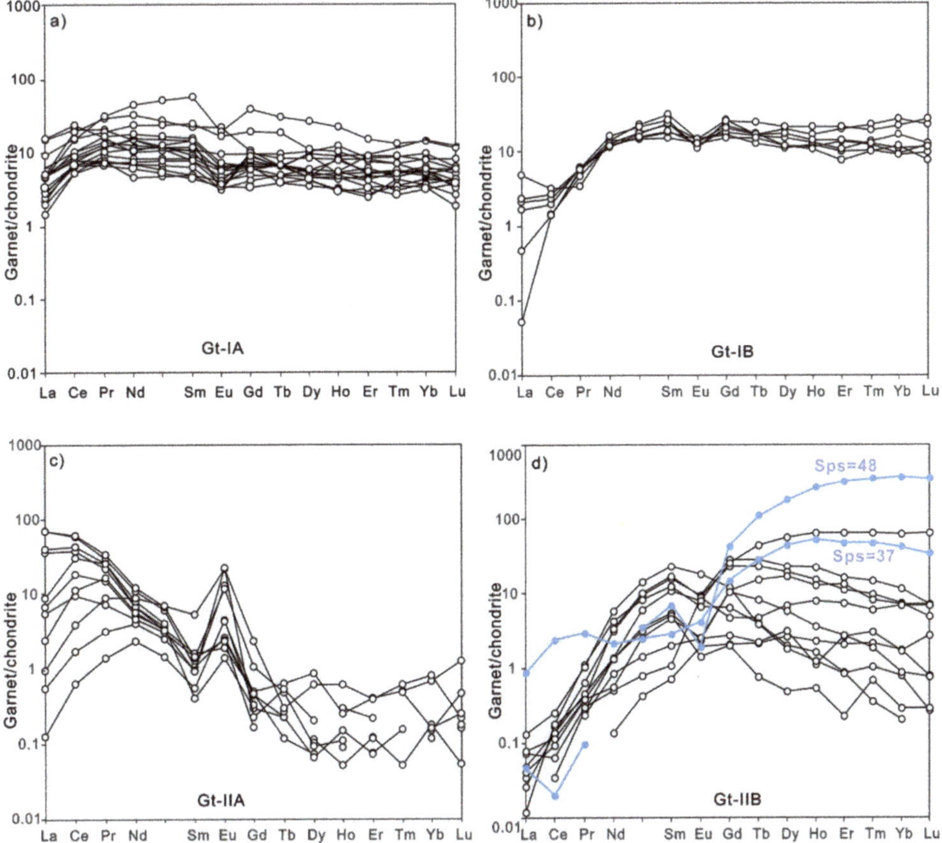

Figure 7. Chondrite-normalized REE patterns of the garnets from the Zhuxiling W (Mo) deposit, chondrite values from [31]. (**a**–**d**) shows REE patterns of Gt-IA, Gt-IB, Gt-IIA and Gt-IIB respectively

and the Mn-rich garnet with high spessartine molecules is marked in blue in Figure 7d. The limit of detection (LOD) is used for the normalized REE patterns when one element value is below the LOD.

LA-ICPMS mapping images of one garnet crystal with a Gt-IA grossular core and Gt-IIA andradite oscillatory rim is presented in Figure 8. The core has relatively homogeneous major and trace element composition, whereas the rim shows rhythmically banded compositional changes corresponding to oscillatory zoning. The mapping results are basically consistent with spot analyses, which shows the Gt-IA garnet of the core is clearly enriched in Al, Ti, and Mg and depleted in Fe and Mn compared with the Gt-II garnet of the rim. The core presents higher contents of Nb and Zr and depleted Sn, In, V, and U compared with the rim. The REE content is characterized by elevated LREE and low HREE for the rim, lacking zoning patterns for HREE.

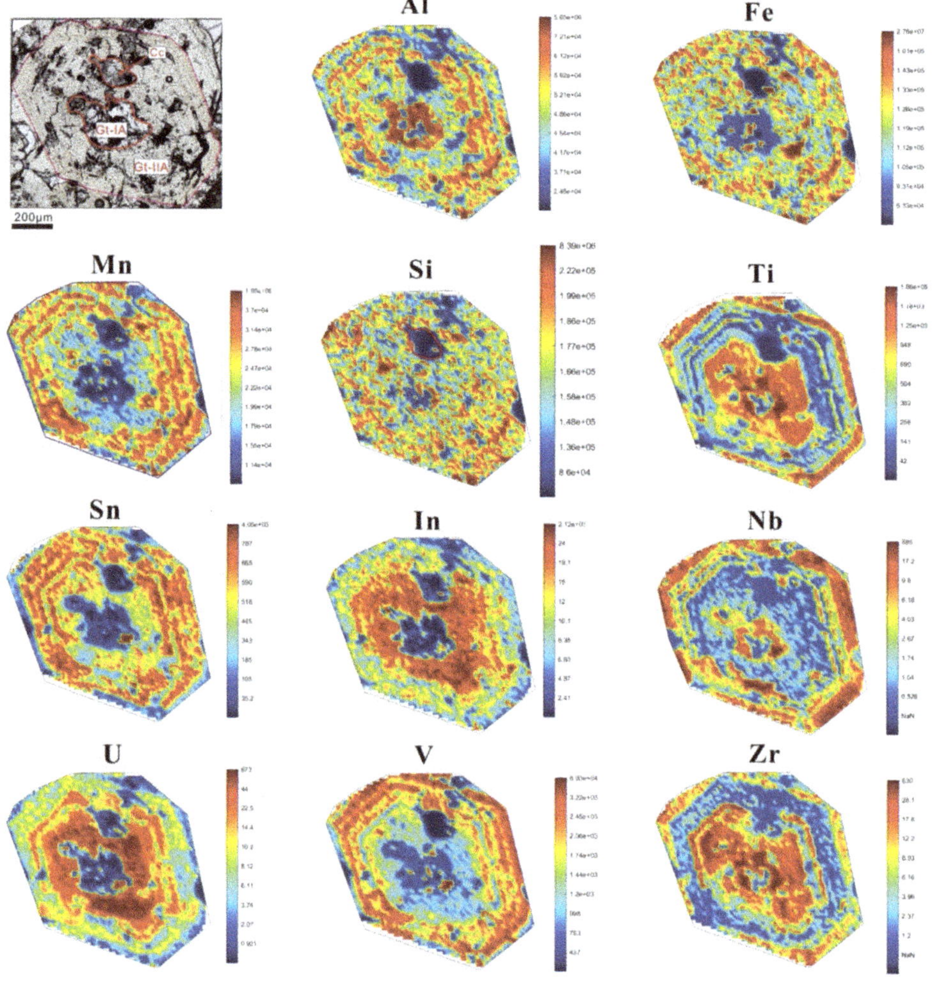

Figure 8. *Cont.*

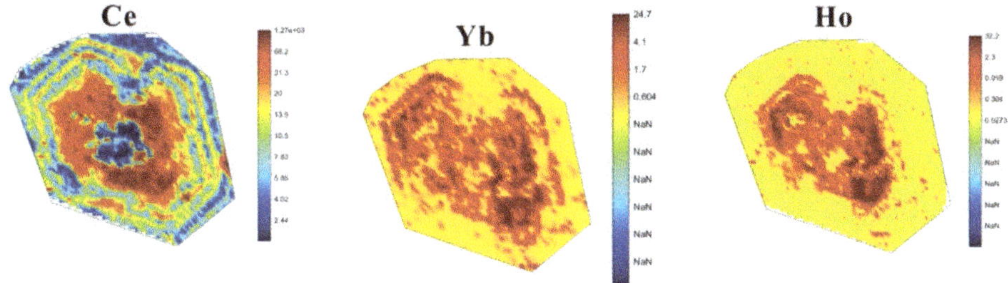

Figure 8. LA-ICPMS probability density maps illustrating element distribution within one garnet crystal with oscillatory-zoned anisotropic rim and isotropic cores, and a calcite inclusion is identified (with very low Si content) and excluded for discussion.

5. Discussion

5.1. Identification of Reduced Skarn System

The tungsten skarn could be divided into reduced and oxidized subtypes based on host rock (carbonaceous versus hematitic), skarn mineralogy (ferrous versus ferric iron ratio), and relative mineralization depth [32]. The reduced W skarn is characterized by lesser grandite garnet in early skarn assemblage and later subcalcic garnet (spessartine and almandine dominant) [1]. The garnets from the Zhuxiling skarn contain significant amount of spessartine and almandine, and the (Sps+Alm) molecule content can reach as high as 56%–68%. The development of subcalcic garnet in Zhuxiling skarn is indicative of a reduced W skarn system. Moreover, associated with subcalcic garnets, the deposit has abundant pyrrhotite and minor sulfides, such as sphalerite and pyrite (Figure 3a). Collectively, the presence of pyrrhotite associated with subcalcic garnet indicate the Zhuxiling skarn belongs to reduced tungsten skarn in comparison to oxidized skarn deposits, such as Kara and King Island deposit [2,3]. Nevertheless, the oxidation state changes during skarn formation indicated by the garnet chemistry. Firstly, the Fe^{2+}/Fe^{3+} ratio is most widely used to reflect oxidation state, which is reflected by the type of garnet and clinopyroxene. For example, Fe-rich andradite with Fe^{3+} as its major composition is considered to be formed under more oxidizing conditions compared with Al-rich grossular [1]. Secondly, the Sn content of garnet could be used as an oxidation indicator. Previous studies have demonstrated that Sn^{4+} substitutes Fe^{3+} at the octathedral site of the garnet lattice [28,33,34], mainly by 1) $Sn^{4+}+Fe^{2+} = 2Fe^{3+}$; 2) $Sn^{4+}+Mg^{2+} = 2Fe^{3+}$; or 3) $3Sn^{4+}+\text{vacancy} = 4Fe^{3+}$. Positive correlation between Sn and andradite molecules of garnets suggests high Sn content is associated with high oxygen fugacity [9]. The Gt-II garnets contain more Fe^{3+} as shown by higher andradite molecules (average Adr = 52) and Sn (average 408 ppm), in contrast to Gt-I garnets (average Adr = 19; Sn = 42 ppm), which implies the former is formed under higher fO_2. Moreover, the Gt-IIA andradite from proximal exoskarn is more Fe-rich and Sn-rich than those of Gt-IIB andradite from distal exoskarn, suggesting a decrease of fO_2 from proximal to distal. The compositional variations of garnet suggested that the oxidization state shifted from a reduced environment of early Gt-I grossular to a relatively oxidized environment of later Gt-II andradite, with the distal exoskarn more reduced than that of proximal skarn.

5.2. Trace Element Incorporation Mechanism

Trace elements could be incorporated into garnet through several ways, such as surface adsorption, occlusion, substitution, and interstitial solid solution [3,35]. Additionally, all those incorporation processes would be influenced by a combination of internal and external factors. The external factors include the fluid composition, P–T condition, fluid/rock ratios, and mineral growth kinetics [4,36,37]. The internal factors mainly refer to crystal-chemical parameters, which play important roles, especially when substitution is the main mechanism. Incorporation mechanisms vary between different elements for the Zhuxiling garnet. For example, some trace elements of garnets, such as V and Cr, are largely controlled by crystal chemical effects, because V and Cr contents of all garnets show linear correlation with MnO and TiO_2, respectively (Figure 9a,b). However, for REE, U, Zr, and Y, the incorporation mechanism and controlling factors are more complicated.

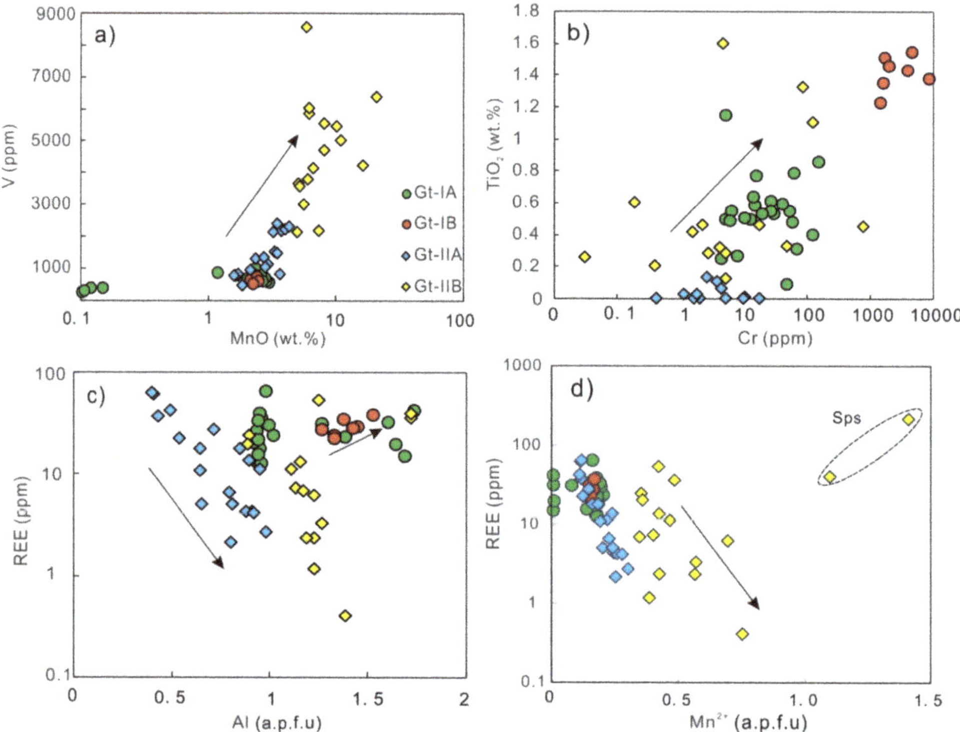

Figure 9. Correlation diagrams of V-Mn, Cr-Ti, Al-REE and Mn-REE of different types of garnets from the Zhuxiling W (Mo) deposit. And (**a**–**d**) represents MnO (%) versus V (ppm), Cr (ppm) versus TiO_2 (%), Al (a.p.f.u) versus REE (ppm), and Mn^{2+} (a.p.f.u) versus REE (ppm) plot respectively.

The incorporation of REE and Y into garnet is basically through the replacement of X^{2+} cations, such as Ca^{2+}, in the dodecahedral position; and the main substitution mechanisms include coupled substitutions, as follows:

$$\left[Ca^{2+}\right]_{-1}^{VIII}\left[REE^{3+}\right]_{+1}^{VIII}\left[Si^{4+}\right]_{-1}^{IV}\left[Al^{3+}\right]_{+1}^{IV}, \left(REE3^{+} + Al3^{+} = Ca2 + Si4^{+}, \text{YAG type}\right); \quad (1)$$

$$\left[Ca^{2+}\right]_{-2}^{VIII}\left[Na^{+}\right]_{+1}^{VIII}\left[REE^{3+}\right]_{+1}^{VIII}, \left(REE3^{+} + Na^{+} = 2Ca2^{+}\right); \quad (2)$$

$$\left[Ca^{2+}\right]_{-3}^{VIII}[\square]_{+1}^{VIII}\left[REE^{3+}\right]_{+2}^{VIII}, \left(2REE3^{+} + \square = 3Ca2^{+}\right); \quad (3)$$

There is no clear correlation between REE content and Al for Gt-IA grossular garnets, and they contain low Na content, which indicates the incorporation of REE might be controlled by $\left[Ca^{2+}\right]_{-3}^{VIII}[\square]_{-1}^{VIII}\left[REE^{3+}\right]_{+2}^{VIII}$ substitution or the REE incorporation is strongly governed by external factors instead of crystal chemistry control. In contrast, Gt-IB grossular garnets show positive correlation between REE content and Al, which indicates a YAG-type substitution. The REE contents of Gt-II andradite garnets display clear linear relationships with Al, Mn, and Fe^{2+}, which suggest a possible substitution control. The clearly negative correlation between REE and Al (Figure 9c) disagrees with a YAG-type substitution mechanism. Negative correlation between REE and Fe^{2+} precludes the $\left[Ca^{2+}\right]_{-1}^{VIII}\left[REE^{3+}\right]_{+1}^{VIII}\left[Fe^{2+}\right]_{+1}^{IV}\left[Al^{3+}\right]_{-1}^{IV}$ substitution mechanism suggested by [4]. The presence of an excellent negative correlation with Mn (Figure 9d) leads us to consider REE incorporation of Gt-II garnet might be partially controlled by $\left[Mn^{2+}\right]_{-1}^{VIII}\left[REE^{3+}\right]_{+1}^{VIII}\left[Ca^{2+}\right]_{+1}^{VIII}\left[Al^{3+}\right]_{-1}^{VIII}$ substitution where REE mainly substitute Mn (and Fe^{2+}) in the octathedral site. On the other hand, the wide range of REE and Y content and distinct REE patterns between Gt-IIA and Gt-IIB suggest the incorporation and fractionation of REE is not only controlled by its crystal chemistry, but more importantly by surface adsorption and external factors, such as fluid composition, fluid/rock ratios, and metasomatism dynamics.

5.3. Fluid Evolution and Metasomatic Dynamics

Garnet composition, especially trace elements and REE fractionation, could reflect the external changes and reveal the evolution of fluid chemistry and physical-chemical conditions and trace the fluid–rock interaction and metasomatism dynamics of the skarn system [6,8,30,38,39]. The Gt-I garnet is Al-rich grossular and represents prograde skarn mineral. The concentrated distribution of trace elements and correlation between REE and Y content of Gt-I grossular (Figure 10) indicate Gt-I grossular is formed by diffusive metasomatism under equilibrium conditions. Under such conditions, the growth rate of garnet is constant and low and surface adsorption and diffusivity are negligible according to the kinetics of mineral growth [3]; and variation of trace element composition and fractionation of REE are basically controlled by crystal chemistry and extrinsic P–T–X conditions of fluid. The REE patterns of Gt-IB grossular garnets are typical HREE-enriched and LREE-depleted, consistent with Al-rich garnets reported in other skarn deposits such as Isle of Skye, Ocna de Fier, and Xinqiao skarn. [5,6,8,40]. As mentioned above, REE incorporation of Gt-IB grossular garnet follows a YAG-type substitution, and the corresponding ideal theoretical radius for REE^{3+} is 0.990 Å, with consideration of the radii of Ca^{2+}, Al^{3+}, and Si^{4+} [41]. HREEs are more favorable for substitution than LREEs, since their radii, Tm (0.994 Å), Yb (0.985 Å), and Lu (0.977 Å) in particular, are closer to the ideal theoretical radius. Therefore, the HREE-enriched and LREE-depleted trend of REE for Gt-IB garnet is largely controlled by ionic radius and charges during substitution. On the other hand, Gt-IA grossular garnet exhibits little fractionation between LREE and HREE, and this flat REE pattern may reflect fluid buffered by composition of host rocks due to long residence time of pore fluid under closed-system conditions.

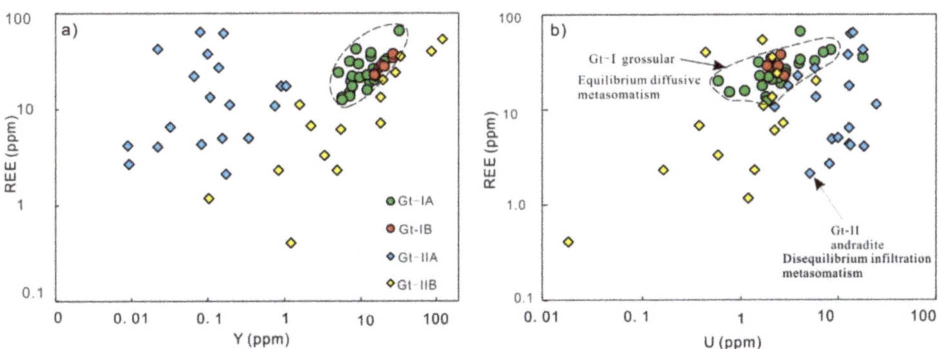

Figure 10. REE and U, Y diagrams of different types of garnets from the Zhuxiling W (Mo) deposit. (**a**) shows Y(ppm) versus REE (ppm) diagram and (**b**) shows U(ppm) versus REE (ppm) diagram.

The Gt-II generation garnet is Fe-rich andradite garnet, which shows common oscillatory zoning. The high Fe^{3+} and Sn content suggests it was formed under oxidized conditions. The development of trapezohedron {211} faces on preexisting dodecahedral {110} garnets (Figure 4D) indicate a fast crystal growth rate under a high fluid/rock ratio [42,43]. Additionally, Gt-II garnet displays dramatic compositional variation, as shown by patched BSE imaging within a single sample (Figure 4F) and scattered REE, Y, and U concentrations (Figure 10), which reflect infiltration metasomatism under disequilibrium conditions in an open system. Fluid fluctuation variations in the open system would facilitate the formation of oscillatory zoning. The trace elements and REE of Gt-II andradite garnet are predominantly controlled by external factors combined with partially crystal chemistry impact. The enriched-LREE and significantly positive Eu anomaly of Gt-IIA garnet are attributed to the following factors: (1) The calculation of enthalpy of mixing between REEAl and REEFe garnet components suggested the HREE are more soluble than the LREE in both grossular and andradite, whereas static interactions taking place in grandite mixtures can lead to opposite trends. High andradite-rich mixtures host more LREE relative to HREE due to a static interaction effect [3]. The Gt-IIA andradite garnets have high andradite molecules (37–85) and the LREE-enriched and HREE-depleted trend of REE, consistent with this crystal chemistry effect. (2) Since Gt-IIA andradite garnet is formed under high fluid/rock ratios in an open system and its REE pattern mimics the hydrothermal fluid composition due to the partition coefficient of unity between garnet and fluid [3]. The REE signature of andradite reflects characteristics of magmatic fluid, which is reported to have enriched LREE and depleted HREE with a variable Eu anomaly [44–46]. (3) As mentioned above, the skarn of the Zhuxiling W (Mo) deposit was formed under a reduced environment and Eu mainly occur in the form of Eu^{2+} in grandite garnet under such conditions. Recent studies verify Cl^- and SO_4^{2-} complexes are major agents for transporting REE^{3+} in hydrothermal fluid [47–49] and chloride complex dominant under low PH conditions [38]. There is no chemical or mineralogical evidence for SO_4^{2-}, and thus chlorine might play a major role. Moreover, chlorine enhances the stability of soluble Eu^{2+} ($EuCl_4^{2-}$ the dominant species) in reduced and slightly acidic fluids at high temperatures (400 °C, [39]), which would lead to positive Eu anomalies. Besides, the Cl-rich fluid bonds with LREE more strongly compared with HREE [50], which fits with LREE-enriched pattern of Gt-IIA andradite garnet.

The Gt-IIB andradite garnet displays a reversed REE pattern with enriched HREE and depleted LREE. The following reasons are considered responsible for the change in REE fractionation from proximal to distal skarn: (1) Continuous crystallization of Fe-rich andradite in proximal exoskarn (Adr 37-85 and enriched in LREE) consumes large amounts of Fe as well as LREE of the fluid and leads to the depletion of Fe and LREE of a subsequent fluid, which formed the distal exoskarn. (2) Lower andradite end-member contents (Adr 8-60 for Gt-IIB) favor the HREE incorporation relative to LREE based on the calculation of enthalpy of mixing between the REEAl and REEFe garnet components, and HREE is preferred during garnet–fluid partitioning [3,51]. This crystal chemistry effect further enhanced LREE depletion and HREE enrichment.

5.4. Formation of Garnets in the Dynamic Skarn System

The first generation garnet is Al-rich grossular formed by diffusive metasomatism with equilibrium with fluids under relatively reduced environment. The elevated contents of Ti and Mn combined with the euhedral to subhedral form and scarce occurrence reveal that Gt-IB grossular garnet represents endoskarn mineral formed very early in prograde-stage, with more incorporation of Ti, Mn, V, and Cr from the intrusion. In contrast, Gt-IA grossular garnet is anhedral and is replaced by latter Gt-II garnet. It represents prograde skarn mineral formed from fluid buffered by composition of host rocks by diffusive metasomatism. The variation of trace element composition and fractionation of REE of the first generation garnet are basically controlled by crystal chemistry and extrinsic P–T–X conditions of fluid. The diffusive metasomatism is responsible for the concentrated REE and Y contents of Gt-I grossular garnet. REE fractionation of Gt-IA garnet is impacted by fluid–water interaction in equilibrium with pore water. In addition, the REE pattern of Gt-IB garnet is largely controlled by crystal chemistry through substitution.

The second generation garnet is Fe-rich andradite, which represents rapid growth garnet formed by infiltration metasomatism with a high water/rock ratio in an open system. Gt-IIA andradite garnets from proximal exoskarn are Fe-rich and LREE-enriched andradite and continuous crystallization of proximal exoskarn consumes large amounts of Fe and consequently leads to the depletion of andradite in the distal exoskarn. Furthermore, garnet shows compositional changes with a decrease of Fe and an increase of Mn, from proximal skarn (Gt-IIA garnet) to distal skarn (Gt-IIB garnet) with enormously high spessartine molecules of 36%–54% for some Gt-IIB garnets. A previous study suggests that Mn-rich garnet forms early crystallization on the contact zone with marble [52]. In this study, the Mn-rich dolomitic limestone of the Lantian Group [24] is most likely the major Mn source for spessartine in Gt-IIB garnets. The presence of Mn-rich garnet implies the fluid must have actively interacted with Mn-rich dolomitic limestone in the distal skarn formation. The trace element and REE signatures of Gt-II andradite garnet are predominantly controlled by external factors and partly by crystallographic properties. The REE patterns of Gt-IIA garnet mainly reflects the characteristics of magmatic fluid, and the positive Eu anomaly suggests chlorine might play a major role as a complex agent ($EuCl_4^{2-}$ the dominant species), which also bonds more strongly with LREE compared with HREE [46,50]. On the contrary, Gt-IIB garnet shows reversed REE patterns compared with Gt-IIA garnet. The change in REE characteristics is possibly caused by the massive consumption of LREE by the crystallization of Fe-rich andradite garnets in the proximal skarn and assisted by a crystal chemical effect. Garnet compositional variations of the two generations and four different types suggest the skarn system evolved from reduced closed conditions gradually to more oxidized, open conditions with fracturing.

6. Conclusions

1) Two generations of garnet have been identified for the Zhuxiling W (Mo) skarn: Gt-I generation garnet is Al-rich grossular ($Adr_{3\text{-}46}Grs_{49\text{-}96}$ $(Sps+Pyr+Alm)_{1\text{-}5}$) and Gt-II generation garnet is Fe-rich andradite. The proximal exoskarn Gt-IIA type garnet is more Fe-rich andradite ($Adr_{37\text{-}85}Grs_{4\text{-}58}(Sps+Pyr+Alm)_{4\text{-}19}$) than Gt-IIB andradite ($Adr_{8\text{-}60}Grs_{18\text{-}47}$ $(Sps+Pyr+Alm)_{11\text{-}68}$), whereas Gt-IIB garnet is Mn-rich, with spessartine and almandine molecule contents as high as 56%–68%. The presence of pyrrhotite associated with subcalcic garnet indicates a relatively reduced tungsten skarn system.
2) Both W and Sn strongly favor Fe-rich garnet compared with Al-rich garnet. Gt-IA grossular garnet presents a flat REE trend and Gt-IB grossular has a distinct REE pattern with enriched HREE. Gt-IIA andradite garnet displays a right-dipping REE trend with a prominent positive Eu anomaly. In comparison, Gt-IIB andradite garnet features depleted LREE and enriched HREE with weak positive Eu anomaly. The REE fractionation of Gt-IB grossular garnet is predominantly controlled by YAG-type substitution, whereas Gt-IA and Gt-II garnets are collectively controlled by crystal chemistry and the external P–T–X condition of fluids.
3) The first generation Al-rich grossular garnets grow slowly under a closed system, whereas the latter formed Fe-rich andradite garnets represent rapid growth garnet associated with magmatic fluids in an open system. Garnet shows compositional changes with a decrease of Fe and an increase of Mn from proximal skarn (Gt-IIA garnet) to distal skarn (Gt-IIB garnet) due to more active fluid–rock interaction with Mn-rich dolomitic limestone of the Lantian Group in the district.

Supplementary Materials: The following are available online at http://www.mdpi.com/2075-163X/10/11/1024/s1. Table S1: EPMA major elements data of garnets from the Zhuxiling W (Mo) deposit; Table S2: LA-ICPMS trace element data of garnets from the Zhuxiling W (Mo) deposit.

Author Contributions: Field work, X.-X.D., Z.-Q.W., and B.C.; experiment, X.-X.D. and Y.-F.J.; data analysis, X.-X.D. and Y.-F.J.; original draft preparation, X.-X.D., Y.-F.J. and B.C.; revision and editing, X.-X.D. and Z.-Q.W. All authors have read and agreed to the published version of the manuscript.

Funding: This research was funded by the National Key Research and Development Project [No. 2016YFC0600108-03-2] and National Nature Science Foundation of China [No.41530206].

Acknowledgments: We thank Dr. Wang Juan and Wang Fangyue of HFUT for their assistance during the EPMA and LA-ICPMS analysis. We would like to express our gratitude to Sun Keke and Professor Wang De'en for their support in the field work. Appreciation is given to Professor Thomas Ulrich and three anonymous reviewers for their careful review and advice. We thank Dr. Zhou Lingli for the organization of this special issue.

Conflicts of Interest: The authors declare no conflict of interest.

References

1. Meinert, L.D.; Dipple, G.M.; Nicolescu, S. World skarn deposits. In *Economic Geology 100th Anniversary Volume*; Hedenquist, J.W., Thompson, J.F.H., Goldfarb, R.J., Richards, J.P., Eds.; Society of Economic Geologists: Littleton, CO, USA, 2005; pp. 299–336.
2. Zaw, K.; Singoyi, B. Formation of magnetite-scheelite skarn mineralization at Kara, Northwestern Tasmania: Evidence from mineral chemistry and stable isotopes. *Econ. Geol.* **2000**, *95*, 1215–1230. [CrossRef]
3. Gaspar, M.; Knaack, C.; Meinert, L.D.; Moretti, R. REE in skarn systems: A LA–ICP–MS study of garnets from the crown jewel gold deposit. *Geochim. Cosmochim. Acta* **2008**, *72*, 185–205. [CrossRef]
4. Ding, T.; Ma, D.S.; Lu, J.J.; Zhang, R.Q. Garnet and scheelite as indicators of multi-stage tungsten mineralization in the Huangshaping deposit, southern Hunan province, China. *Ore Geol. Rev.* **2018**, *94*, 193–211. [CrossRef]
5. Xiao, X.; Zhou, T.F.; Noel, C.W.; Zhang, L.J.; Fan, Y.; Wang, F.Y.; Chen, X.F. The formation and trace elements of garnet in the skarn zone from the Xinqiao Cu-S-Fe-Au deposit, Tongling ore district, Anhui Province, Eastern China. *Lithos* **2018**, *302–303*, 467–479. [CrossRef]
6. Smith, M.P.; Henderson, P.; Jeffries, T.E.R.; Long, J.; Williams, C.T. The rare earth elements and uranium in garnets from the Beinnan Dubhaich Aureole, Skye, Scotland, UK: Constraints on processes in a dynamic hydrothermal system. *J. Petrol.* **2004**, *45*, 457–484. [CrossRef]
7. Somarin, A.K. Garnet composition as an indicator of Cu mineralization: Evidence from skarn deposits of NW Iran. *J. Geochem. Explor.* **2004**, *81*, 47–57. [CrossRef]
8. Park, C.P.; Song, Y.G.; Kang, I.M.; Shim, J.; Chung, D.H.; Park, C.S. Metasomatic changes during periodic fluid flux recorded in grandite garnet from the Weondong Wskarn deposit, South Korea. *Chem. Geol.* **2017**, *451*, 135–153. [CrossRef]
9. Zhou, J.H.; Feng, C.Y.; Li, D.X. Geochemistry of the garnets in the Baiganhu W-Sn orefield, NW China. *Ore Geol. Rev.* **2017**, *82*, 70–92. [CrossRef]
10. Yang, Y.L.; Ni, P.; Wang, Q.; Wang, J.Y.; Zhang, X.L. In situ LA-ICP-MS study of garnets in the Makeng Fe skarn deposit, eastern China: Fluctuating fluid flow, ore-forming conditions and implication for mineral exploration. *Ore Geol. Rev.* **2020**, *126*, 103725. [CrossRef]
11. Mao, J.W.; Xiong, B.K.; Liu, J.; Pirajno, F.; Cheng, Y.B.; Ye, H.S.; Song, S.W.; Dai, P. Molybdenite Re/Os dating, zircon U-Pb age and geochemistry of granitoids in the Yangchuling porphyry W-Mo deposit Jiangnan tungsten ore belt, China: Implications for petrogenesis, mineralization and geodynamic setting. *Lithos* **2017**, *296–297*, 35–52. [CrossRef]
12. Song, G.X.; Qin, K.Z.; Li, G.M.; Eavans, N.J.; Chen, L. Scheelite elemental and isotopic signatures: Implications for the genesis of skarn-type W(Mo) deposits in the Chizhou Area, Anhui Province, Eastern China. *Am. Mineral.* **2014**, *99*, 303–317. [CrossRef]
13. Song, G.X.; Qin, K.Z.; Li, G.M.; Li, X.; Qu, W. Geochronology and ore-forming fluids of the Baizhangyan W-Mo Deposit in the Chizhou Area, Middle-Lower Yangtze Valley, SE China. *Res. Geol.* **2013**, *63*, 57–71. [CrossRef]
14. Zhang, D.Y.; Zhou, T.F.; Yuan, F.; Fan, Y.; Chen, X.F.; White, N.C.; Ding, N.; Jiang, Q.S. Petrogenesis and W(Mo) fertility indicators of the Gaojiabang "satellite" granodiorite porphyry in southern Anhui Province, South China. *Ore Geol. Rev.* **2017**, *88*, 550–564. [CrossRef]
15. Su, Q.W.; Mao, J.W.; Wu, S.H.; Zhang, Z.C.; Xu, S.F. Geochronology and geochemistry of the granitoids and ore-forming age in the Xiaoyao tungsten polymetallic skarn deposit in the Jiangnan Massif tungsten belt, China: Implications for their petrogenesis, geodynamic setting, and mineralization. *Lithos* **2018**, *296–299*, 365–381. [CrossRef]

16. Wang, X.L.; Zhou, J.C.; Griffin, W.L.; Wang, R.C.; Qiu, J.S.; O'Reilly, S.Y.; Xu, X.S.; Liu, X.M.; Zhang, G.L. Detrital zircon geochronology of Precambrian basement sequences in the Jiangnan orogen: Dating the assembly of the Yangtze Cathaysia blocks. *Precambrian Res.* **2007**, *159*, 117–131. [CrossRef]
17. Zhang, C.L.; Santosh, M.; Zou, H.B.; Li, H.K.; Huang, W.C. The Fuchuan ophiolite in Jiangnan Orogen: Geochemistry, zircon U-Pb geochronology, Hf isotope and implications for the Neoproterozoic assembly of South China. *Lithos* **2013**, *179*, 263–274. [CrossRef]
18. Yao, J.; Shu, L.; Santosh, M.; Li, J. Neoproterozoic arc–related andesite and orogeny–related unconformity in the eastern Jiangnan orogenic belt: Constraints on the assembly of the Yangtze and Cathaysia blocks in South China. *Precambrian Res.* **2015**, *262*, 80–100. [CrossRef]
19. Mao, Z.H.; Liu, J.J.; Mao, J.W.; Deng, J.; Zhang, F.; Meng, X.Y.; Xiong, B.K.; Xiang, X.K.; Luo, X.H. Geochronology and geochemistry of granitoids related to the giant Dahutang tungsten deposit, middle Yangtze River region, China: Implications for petrogenesis, geodynamic setting, and mineralization. *Gondwana Res.* **2015**, *29*, 816–836. [CrossRef]
20. Geological Survey of Anhui Province (GSAP). *Geological Report of the Anhui Tungsten (Tin and Molybdenum) Mineral Resources Potential Evaluation*; Geological Survey of Anhui Province (GSAP): Anhui, China, 2011; p. 263.
21. Ding, N. Metallogenic Regularity of Tungsten Deposits in Anhui Province. Master's Thesis, Hefei University of Technology, Hefei, China, 2012; p. 154. (In Chinese).
22. Yan, J.; Hou, T.J.; Wang, A.G.; Wang, D.E.; Zhang, D.Y.; Weng, W.F.; Liu, J.M.; Liu, X.Q.; Li, Q.Z. Petrogenetic contrastive studies on the Mesozoic early stage ore-bearing and late stage ore-barren granites from the southern Anhui Province. *Sci. China Earth Sci.* **2017**, *60*, 1920–1941. [CrossRef]
23. Zhou, T.F.; Wang, S.W.; Fan, Y.; Yuan, F.; Zhang, D.Y.; White, N.C. A review of the intracontinental porphyry deposits in the Middle-Lower Yangtze River Valley metallogenic belt, Eastern China. *Ore Geol. Rev.* **2015**, *65*, 433–456. [CrossRef]
24. Huang, M. Metallogenetic Ages and Geochemical Characteristics of Zhuxiling W-Ag Polymetallic Deposit in Ningguo City, Southern Anhui Province. Master's Thesis, Hefei University of Technology, Hefei, China, 2017; p. 64. (In Chinese).
25. Chen, X.F.; Wang, Y.G.; Sun, W.D.; Yang, X.Y. Zircon U-Pb chronology, geochemistry and genesis of the Zhuxiling granite in Ningguo, Southern Anhui. *Acta Geol. Sin.* **2013**, *87*, 1662–1678. (In Chinese)
26. Wang, Y.; van den Kerkhof, A.; Xiao, Y.; Sun, H.; Yang, X.; Lai, J.; Wang, Y. Geochemistry and fluid inclusions of scheelite-mineralized granodiorite porphyries from southern Anhui Province, China. *Ore Geol. Rev.* **2017**, *89*, 988–1005. [CrossRef]
27. Duan, X.X.; Chen, B.; Sun, K.K.; Wang, Z.Q.; Yan, X.; Zhang, Z. Accessory mineral chemistry as a monitor of petrogenetic and metallogenetic processes: A comparative study of zircon and apatite from Wushan Cu- and Zhuxiling W(Mo)-mineralization-related granitoids. *Ore Geol. Rev.* **2019**, *111*, 102940. [CrossRef]
28. Chen, J.; Halls, C.; Stanley, C.J. Tin-bearing skarns of South China: Geological setting and mineralogy. *Ore Geol. Rev.* **1992**, *7*, 225–248. [CrossRef]
29. Yao, Y.; Chen, J.; Lu, J.J.; Zhang, R.Q.; Zhao, L.H. Composition, trace element and infrared spectrum of garnet from three types of W-Sn bearing skarns in the South of China. *Acta Mineral. Sin.* **2013**, *33*, 315–329. (In Chinese)
30. Van Westrenen, W.; Allan, N.L.; Blundy, J.D.; Purton, J.A.; Wood, B.J. Atomistic simulation of trace element incorporation into garnets-comparison with experimental garnet-melt partitioning data. *Geochim. Cosmochim. Acta* **2000**, *64*, 1629–1639. [CrossRef]
31. Sun, S.S.; McDonough, W.F. Chemical and isotopic systematics of oceanic basalts: Implications for mantle composition and processes. *Geol. Soc. Spec. Publ.* **1989**, *42*, 313–345. [CrossRef]
32. Newberry, R.J.; Einaudi, M. T Tectonic and geochemical setting of tungsten skarn mineralization in the Cordillera. *Ariz. Geol. Soc. Dig.* **1981**, *14*, 99–111.
33. Butler, B.C.M. Tin-rich garnet, pyroxene, and spinel from a slag. *Mineral. Mag.* **1978**, *42*, 487–492. [CrossRef]
34. Amthauer, G.; McIver, J.R.; Viljoen, E.A. ^{57}Fe and ^{119}Sn Mössbauer studies of natural tin-bearing garnets. *Phys. Chem. Miner.* **1979**, *4*, 235–244. [CrossRef]
35. McIntire, W.L. Trace element partition coefficients—A review of theory and applications to geology. *Geochim. Cosmochim. Acta* **1963**, *27*, 1209–1264. [CrossRef]
36. Jamtveit, B.; Hervig, R.L. Constraints on transport and kinetics in hydrothermal systems from zoned garnet crystals. *Science* **1994**, *263*, 505–508. [CrossRef] [PubMed]

37. Chernoff, C.B.; Carlson, W.D. Trace element zoning as a record of chemical disequilibrium during garnet growth. *Geology* **1999**, *27*, 555–558. [CrossRef]
38. Bau, M. Rare-earth element mobility during hydrothermal and metamorphic fluid-rock interaction and the significance of the oxidation state of europium. *Chem. Geol.* **1991**, *93*, 219–230. [CrossRef]
39. Allen, D.E.; Seyfried, W.E. REE controls in ultramafic hosted MOR hydrothermal systems: An experimental study at elevated temperature and pressure. *Geochim. Cosmochim. Acta* **2005**, *69*, 675–683. [CrossRef]
40. Nicolescu, S.; Cornell, D.H.; Sodervall, U.; Odelius, H. Secondary ion mass spectrometry analysis of rare earth elements in grandite garnet and other skarn related silicates. *Eur. J. Mineral.* **1998**, *10*, 251–259. [CrossRef]
41. Shannon, R.D. Revised effective ionic radii and systematic studies of interatomic distances in halides and calcogenides. *Acta Crystallogr.* **1976**, *32*, 751–767. [CrossRef]
42. Jamtveit, B.; Wogelius, R.A.; Fraser, D.G. Zonation patterns of skarn garnets: Records of hydrothermal system evolution. *Geology* **1993**, *21*, 113–116. [CrossRef]
43. Jamtveit, B. Oscillatory zonation patterns in hydrothermal grossular-andradite garnet. Nonlinear dynamics in regions of immiscibility. *Am. Mineral.* **1991**, *76*, 1319–1327.
44. Kravchuk, I.F.; Ivanova, G.F.; Varezhkina, N.S.; Malinin, S.D. REE fractionation in acid fluid-magma systems. *Geochem. Intern.* **1995**, *32*, 60–68.
45. Reed, M.J.; Candela, P.A.; Piccoli, P.M. The distribution of rare earth elements between monzogranitic melt and the aqueous volatile phase in experimental investigations at 800 C and 200 MPa. *Contrib. Mineral. Petrol.* **2000**, *140*, 251–262. [CrossRef]
46. Yang, X.M. Using Rare Earth Elements (REE) to decipher the origin of ore fluids associated with granite intrusions. *Minerals* **2019**, *9*, 426. [CrossRef]
47. Mayanovic, R.A.; Anderson, A.J.; Bassett, W.A.; Chou, I.M. On the formation and structure of rare-earth element complexes in aqueous solutions under hydrothermal conditions with new data on gadolinium aquo and chloro complexes. *Chem. Geol.* **2007**, *239*, 266–293. [CrossRef]
48. Migdisov, A.A.; Williams-Jones, A.E. Hydrothermal transport and deposition of the rare earth elements by fluorine-bearing aqueous liquids. *Miner. Depos.* **2014**, *49*, 987–997. [CrossRef]
49. Xing, Y.; Etschmann, B.; Liu, W.; Mei, Y.; Testemale, D.; Shvarov, Y.; Tomkins, A.; Brugger, J. The role of fluorine in hydrothermal mobilization and transportation of Fe, U and REE and the formation of IOCG deposits. *Chem. Geol.* **2019**, *504*, 158–176. [CrossRef]
50. Rolland, Y.; Cox, S.; Boullier, A.-M.; Pennacchioni, G.; Mancktelow, N. Rare earth and trace element mobility in mid-crustal shear zones: Insights from the Mont Blanc Massif (Western Alps). *Earth Planet Sci. Lett.* **2003**, *214*, 203–219. [CrossRef]
51. Moretti, R.; Ottonello, G. An appraisal of endmember energy and mixing properties of the rare earth garnets. *Geochim. Cosmochim. Acta* **1998**, *62*, 1147–1173. [CrossRef]
52. Einaudi, M.T.; Meinert, L.D.; Newberry, R.J. Skarn deposits. *Econ. Geol.* **1981**, *75*, 317–391.

Publisher's Note: MDPI stays neutral with regard to jurisdictional claims in published maps and institutional affiliations.

Article

Fluid Evolution, H-O Isotope and Re-Os Age of Molybdenite from the Baiyinhan Tungsten Deposit in the Eastern Central Asian Orogenic Belt, NE China, and Its Geological Significance

Ruiliang Wang [1], Qingdong Zeng [2,3,4,*], Zhaochong Zhang [1], Yunpeng Guo [2,3] and Jinhang Lu [5]

1 Faculty of Geosciences and Resources, China University of Geosciences (Beijing), Beijing 100083, China; rlwang@cugb.edu.cn (R.W.); zczhang@cugb.edu.cn (Z.Z.)

2 Key Laboratory of Mineral Resources, Institute of Geology and Geophysics, Chinese Academy of Sciences, Beijing 100029, China; guoyunpeng@mail.iggcas.ac.cn

3 Innovation Academy of Earth Sciences, Chinese Academy of Sciences, Beijing 100029, China

4 University of Chinese Academy of Sciences, Beijing 100049, China

5 264 Brigade of the Jiangxi Nuclear Industry Geological Bureau, Ganzhou 341000, China; wrl3001190049@163.com

* Correspondence: zengqingdong@mail.iggcas.ac.cn

Received: 28 June 2020; Accepted: 23 July 2020; Published: 26 July 2020

Abstract: The quartz-vein-type Baiyinhan tungsten deposit is located at the eastern part of the Central Asian Orogenic Belt, NE China. Analyses of fluid inclusions, H-O isotope of quartz and Re-Os isotope of molybdenite were carried out. Three stages of mineralization were identified: The early quartz + wolframite + bismuth stage, the middle quartz + molybdenite stage and the late calcite + fluorite stage. Quartz veins formed in the three stages were selected for the fluid inclusion analysis. The petrographic observation and fluid inclusion microthermometry results revealed three types of fluid inclusions: CO_2-H_2O (C-type), liquid-rich (L-type) and vapor-rich (V-type). The homogenization temperatures of C-type, V-type and L-type inclusions were 233–374 °C, 210–312 °C, and 196–311 °C, respectively. The salinity of the three types of inclusions was identical, varying in the range of 5–12 wt%. The H-O isotope analyses results showed that quartz had $\delta^{18}O_{H2O}$ and δD_{SMOW} compositions of −2.6‰ to 4.3‰ and −97‰ to −82‰, respectively, indicating that the ore-forming fluids were mainly derived from magmatic water with a minor contribution of meteoric water. The addition of meteoric water reduces the temperature and salinity of the ore-forming fluids, which leads to a decrease of the solubility of tungsten and molybdenum in the fluids and eventually the precipitation of minerals. Re-Os isotopic analysis of five molybdenite samples yielded an isochron age of 139.6 ± 7.6 Ma (2σ) with an initial ^{187}Os of −0.05 ± 0.57 (MSWD = 3.5). Rhenium concentrations of the molybdenite samples were between 3.1 ug/g and 8.5 ug/g. The results suggest that the metals of the Baiyinhan deposit have a crust origin, and the mineralization is one episode of the Early Cretaceous tungsten mineralization epoch which occurred at the eastern part of the Central Asian Orogenic Belt.

Keywords: Re-Os dating; fluid inclusions; H-O isotope; quartz-vein tungsten deposit; eastern Central Asian Orogenic Belt

1. Introduction

The eastern Central Asian Orogenic Belt (CAOB) in NE China is characterized by orogenic events of the Paleo-Asian domain superimposed by the magmatism of the Paleo-Pacific and Ohotsk Ocean tectonic domains, which resulted in extensive distribution of Mesozoic volcanic and granitic rocks together with widespread Cu, Mo, Au, Pb-Zn-Ag and Sn polymetal deposits [1–7]. In recent years, an increasing number of tungsten deposits have been discovered in this region, revealing a great potential of large-scale tungsten mineralization.

The previous geochronology studies of the tungsten mineralization in eastern CAOB have shown Triassic, Jurassic and Early Cretaceous ages [8–29] and are in close association with the magmatic activities. The mineralization includes porphyry, skarn and quartz-type deposits. The quartz-vein-type tungsten deposits can be divided into two subtypes: Quartz-wolframite and quartz-scheelite vein types. The Baiyinhan quartz-vein-type tungsten deposit belongs to the quartz-wolframite subtype.

At present, there is no unified understanding of the precipitation mechanism of tungsten in ore-forming fluids, especially whether CO_2 plays an important role in the migration and precipitation of tungsten. In this study, we selected the newly discovered Baiyinhan deposit, which has not been reported in any article to date. The Re-Os isotope test of molybdenite in the Baiyinhan quartz vein tungsten deposit was carried out to determine the ore-forming age and source of ore-forming materials. Fluid inclusions and quartz H-O isotopes were used to discuss the fluid evolution and fluid source of the deposit, especially the role of CO_2 in the migration and precipitation of tungsten.

2. Geological Setting

The eastern CAOB is divided into four tectonic blocks: Erguna, Xing'an, Songliao and Jiamusi massifs, which are separated by the regional-scale Tayuan-Xiguitu, Hegenshan-Heihe and Mudanjiang faults, respectively [1] (Figure 1). The Baiyinhan tungsten deposit is located in the Liaoyuan massif and belongs to the southern part of the Da Hinggan Mountains polymetal metallogenic belt (SDMB) (Figure 2). The SDMB is bounded by the western Hegenshan Fault, the eastern Nenjiang Fault and the southern Xilamulun Fault (Figure 2). The SDMB is composed of the Permian strata and is overlain by the Mesozoic-Cenozoic volcanic and sedimentary rocks and intruded by Mesozoic intrusive rocks (Figure 2). The Permian strata are composed of four units [30]: (1) The Lower Permian Qingfengshan Formation, which consists of graywacke and siltstone with tuffaceous intercalation; (2) the Lower Permian Dashizhai Formation, which is mainly composed of submarine lava and tuff (andesitic, felsic and basaltic) with arenite; (3) the Lower Permian Huanggangliang Formation, which consists of mix-bedded sandstone and shale with limestone and tuffite; and (4) the Upper Permian Linxi Formation, which is mainly composed of terrestrial sandstone, siltstone and mudstone.

The Mesozoic formations include Jurassic and Cretaceous strata. The Jurassic strata are composed of conglomerate, sandstone, siltstone, mudstone, coal bed, volcaniclastic rock, tuff, rhyolite and dacite. The Cretaceous strata consist of andesite, volcaniclastic rock, tuff, tuff sandstone, siltstone and basal conglomerate. The Cenozoic corresponds to continental red strata. The Major structures are the Erlian-Hegenshan Fault, the Xilamulun River Fault and the Nenjiang Fault [31,32].

A scatter of a late Paleozoic granite is exposed at the Haolibao mine area and at the northern part of the Xilamulun River [33]. Mesozoic granites include Triassic, Late Jurassic and Early Cretaceous granites [3]. The Mesozoic granites are characterized by low $(^{87}Sr/^{86}Sr)_i$ and high $\varepsilon Nd(t)$ values, showing that juvenile crust materials are the major source of the Mesozoic granites [1].

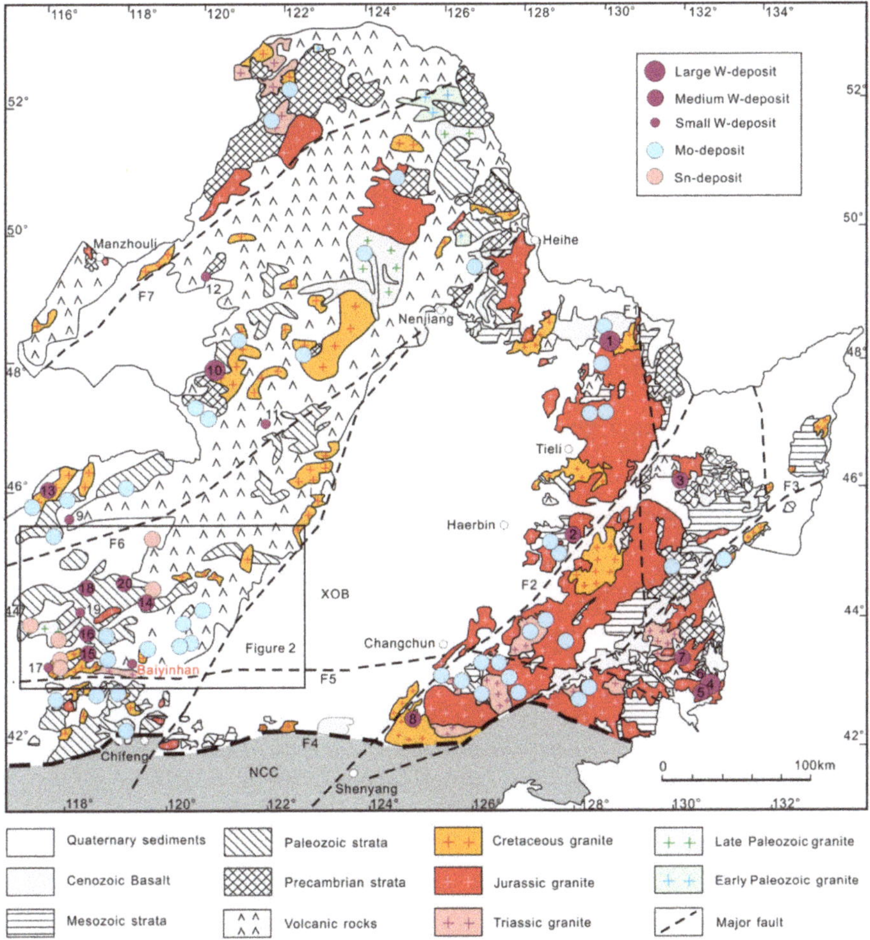

Figure 1. Geological map of the eastern part of the Central Asian Orogenic Belt and distribution of tungsten deposits in NE China (modified from [3]). Fault: F1—Nengjiang Fault; F2—Yitong-Yilan Fault; F3—Dunhua-Mishan Fault; F4—Chifeng-Kaiyuan Fault; F5—Xilamulun River Fault; F6—Heganshan-Heihe Fault; F7—Xiguitu-Tayuan Fault. Locations: 1—Cuihongshan; 2—Gongpengzi; 3—Yangbishan; 4—Yangjingou; 5—Wudaogou; 6—Liudaogou; 7—Baishilazi; 8—Sanjiazi; 9—Dayana; 10—Honghuaerji; 11—Weilianhe; 12—Zishi; 13—Shamai; 14—Dongshanwan; 15—Xiaolaogualinzi; 16—Weilasituo; 17—Chamuhan; 18—Daolundaba; 19—Wulegerejidaban; 20—Xiaohaiqing.

The SDMB is an important polymetal metallogenic belt in China, and contains abundant porphyry Mo deposits, skarn Pb-Zn deposits, skarn Fe-Sn deposits, hydrothermal-vein-type Ag-polymetal, Sn-polymetal, Pb-Zn polymetal and Cu-polymetal deposits (Figure 2).

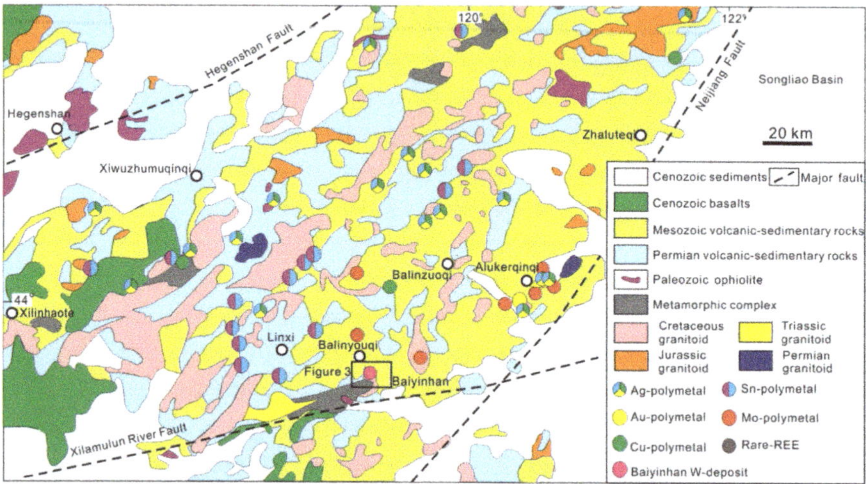

Figure 2. Regional geological map of south DaHinggan mountains showing the distribution of the metal ore deposits (modified from [34,35]).

3. The Baiyinhan Tungsten Deposit

The Baiyinhan tungsten deposit occurs within the Yanshanian granite and is mainly composed of wolframite-bearing quartz veins (Figure 3). The Yanshanian granite (biotite granite, Figure 4A) is constituted of quartz (30–40%), K-feldspar (30–40%), plagioclase (~15%), biotite (~8%), and minor accessory minerals. The xenomorphic granular or granular quartz occurs as clusters, and the grain size varies from 0.2 mm to 2.0 mm. The grain size of K-feldspar varies from 0.5 mm to 2.0 mm and shows a xenomorphic granular texture. The grain size of plagioclase ranges from 0.5 mm to 1.5 mm. The tungsten mineralization at Baiyinhan is spatially confined to gently dipping faults, with a striking length of 10–150 m. The tungsten ore bodies occur mainly as quartz veins, and five ore bodies have been defined (Figure 3). The features of these ore bodies are similar.

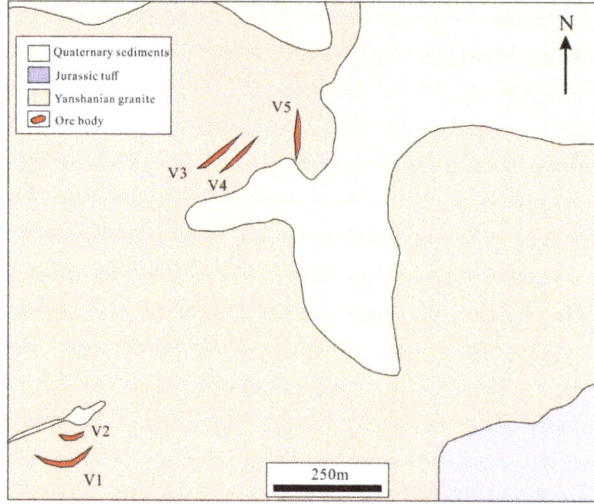

Figure 3. Geological map of the Baiyinhan tungsten deposit.

Figure 4. Photographs of the biotite granite (**A**) and the mineralization (**B**–**I**) in the Baiyinhan deposit showing: (**A**) Biotite granodiorite; (**B**) Wolframite-bearing quartz vein within the granite (dipping to north); (**C**) Quartz + pyrite + wolframite vein; (**D**) Quartz + molybdenite + wolframite vein with drusy cavity; (**E**) Euhedral wolframite in greisen; (**F**) massive ore; (**G**) Bismuthinite in massive ore; (**H**) Veined wolframite in greisen; (**I**) Veined wolframite in quartz vein. Abbreviations: Bi—biotite, Pl—plagioclase, Kfs—K-feldspar, Mb—molybdenite, Py—pyrite, Gs—greisen, Qv—quartz vein, Qz—quartz, Wf—wolframite, Bm—bismuthinite.

The V1 ore body is located at the southwest part of the ore deposit. It strikes EW and dips north at angles of 10–15°, with a length of 150 m and thickness of 0.1–1.5 m (Figure 3). The V2 ore body is smaller and located at the north of the V1 ore body (Figure 3). It also strikes EW and dips north at an angle of 15°, but the scale of V2 ore body is smaller and thinner than the V1 ore body. The V3, V4 and V5 ore bodies are located at the northern part of ore deposit (Figure 3). The V3 and V4 ore bodies strike NE, and the V5 ore body strike NS. These three ore bodies dip northwest at angles of 12–20°, with a length of several tens to 150 m and thickness of 0.2–1.8 m. The orebody has a WO_3 grade of 0.5–2.5%. These orebodies are mainly composed of tungsten-bearing quartz veins and minor greisen alteration rock.

The vein-type ore includes wolframite + pyrite + molybdenite (Figure 4C,D) and minor amounts of bismuthinite (Figure 4G). Gangue minerals consist of quartz and muscovite. The ore minerals occur in major veins and massive structures, along with minor drusy cavity structures (Figure 4F,H,I). The ore minerals are mainly euhedral and subhedral (Figure 4C–E,G). Alteration at Baiyinhan is well-developed and characterized by a fracture-controlled greisen alteration. A small proportion of tungsten mineralization occurs within the greisen (Figure 4E,H). Mineralogically, the greisen alteration comprises quartz and muscovite (Figure 4E).

Based on the structure, texture and mineralogy of the ore veins, three hydrothermal stages can be distinguished: (1) Wolframite + quartz vein stage; (2) molybdenite + quartz + pyrite stage and

(3) calcite + fluorite stage. The first stage of quartz-wolframite veins contains significant amounts of disseminated wolframite and minor bismuthinite, accompanying intense greisen alteration. The quartz is typically milky white, coarse-grained, anhedral and fractured. The wolframite is also commonly fractured and is locally cemented together with fractured quartz that was formed during the second and third stages of activity. The second hydrothermal stage is characterized by an enrichment in pyrite and molybdenite. The third stage is characterized by coarse translucent euhedral calcite and fluorite crystals occurring in fractures and veinlets, which crosscut earlier formed veins.

4. Sampling and Analytical Methods

4.1. Samples

Thirty-six samples of quartz veins were collected from the open pit, all of which were collected from the V3, V4 and V5 ore bodies in the north of the ore deposit for the study of fluid inclusions shape, type, abundance and spatial distribution. Twenty-one of these samples were used for the analysis of microthermometry of fluid inclusions, and 10 samples were used for H-O isotope analysis. Five molybdenite samples were also collected from the ore pile for Re-Os geochronology analysis.

4.2. Analytical Methods

The microthermometric measurements on the fluid inclusions were carried out using a Linkam THMS 600 programmable heating-freezing stage combined with a Zeiss microscope at the China University of Geosciences, Beijing (CUGB), China. The process has been described in detail by Yu et al. [36]. Salinities of H_2O-NaCl [37] and CO_2-bearing [38] fluids were calculated using the final melting temperatures of ice (Tm-ice) and clathrate (Tm-cla). The pressure of the fluid inclusions was calculated according to Steele et al. [39].

The H-O isotope analyses were accomplished with a MAT253 mass spectrometer at the Institute of Geology and Geophysics, Chinese Academy of Sciences. Oxygen was liberated from quartz through reaction with BrF_5 [40] and converted to CO_2 on a platinum-coated carbon rod. In accordance with the method described by Simon [41], H isotopic ratios of bulk fluid inclusions in quartz were measured by mechanically crushing ~5 g of quartz grains to a grain size of 1–5 mm. Samples were first degassed of labile volatiles and secondary fluid inclusions by heating under vacuum to 120 °C for 3 h. The released water was trapped and reduced to H_2 by reaction with Zn. The analytical precision was better than ±0.2‰ (1σ) for $\delta^{18}O$ and ±2‰ (1σ) for δD. H-O isotopic data were reported relatively to the Standard Mean Ocean Water standard. $\delta^{18}O_{H2O}$ of fluids were calculated using the equation for quartz-water isotopic equilibrium: $1000 \times \ln\alpha_{quartz\text{-}water} = (3.38 \times 10^6)T^{-2} - 3.40$, where T is the temperature in Kelvins [42], and the average fluid inclusion temperature of each stage was used to calculate the $\delta^{18}O_{H2O}$ value.

The Re-Os isotopic analyses were performed at the National Research Center of Geoanalysis, Chinese Academy of Geosciences. The details of the chemical procedure have been described by Du et al. [43,44], Shirey et al. [45] and Stein et al. [46] Five molybdenite samples from the Baiyinhan deposit were collected in the footrill. Microprobe screening of these samples revealed euhedral molybdenite crystals.

5. Results

5.1. Fluid Inclusion Petrography

Fluid inclusions were widely developed in the quartz veins (Figure 5). Most fluid inclusions were isolated or randomly clustered, had relatively regular shapes (polygonal, oval and semicircular) and were interpreted as primary in origin. However, secondary and pseudosecondary inclusions have also been observed as intergranular arrays or aligned along microfractures in transgranular crystal trails [47]. During this study, we focused on the primary inclusions. According to the composition of inclusions and the phase state at room temperature [47,48], fluid inclusions in the Baiyinhan deposit can be divided into three types: C-, L- and V-type (Figure 5). On the basis of the microthermometry results of fluid inclusions, the discussion of fluid evolution corresponded to the aforementioned three hydrothermal stages.

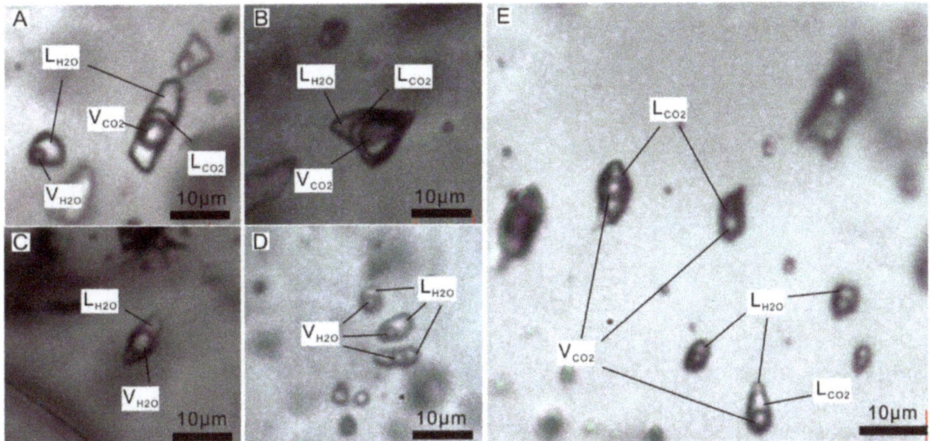

Figure 5. Photomicrographs of typical fluid inclusions in quartz from the Baiyinhan tungsten deposit. (**A**) Liquid-rich C-type inclusion. (**B**) Vapor-rich C-type inclusion. (**C**) V-type inclusions. (**D**) Clustered L-type inclusions. (**E**) Clustered C-type inclusions. Abbreviation: L_{CO2} = liquid CO_2; L_{H2O} = liquid H_2O; V_{CO2} = vapor CO_2; V_{H2O} = vapor H_2O.

5.1.1. C-Type (CO_2-H_2O Inclusions)

C-type inclusions were irregular-shaped to rounded and were ~4–20 μm in size. They usually consisted of three phases at room temperature. In the first phase, the proportion of CO_2 was less than 50% (Figure 5A). In the second phrase, the proportion of CO_2 was more than 50% and the proportion of V_{CO2} was larger (Figure 5B). In the third phase, the proportion of CO_2 was greater than 50% and the proportion of L_{CO2} was larger (Figure 5E). Although some inclusions showed only two phases at room temperature, they turned into three phases during cooling. C-type inclusions occurred mainly in veins of early stages. Based on the ratio of liquid to vapor, C-type inclusions can be divided into liquid-rich and vapor-rich inclusions. The liquid-rich inclusions were homogenized to the liquid phase, and the most vapor-rich inclusions were homogenized to the vapor phase.

5.1.2. L-Type (Liquid-Rich Inclusions)

The L-type fluid inclusions were found in quartz veins from all stages. Most L-type inclusions contained liquid and a vapor bubble that occupied ~5–45 vol% of the inclusion (Figure 5D,E). They occurred as irregular to rounded forms of ~4–19 μm in size (mainly 5–9 μm; Figure 5D). In the early stage, different from C-type inclusions, only three L-type inclusions were observed. The massive development of L-type fluid inclusions occurred in the middle and late stages.

5.1.3. V-Type (Vapor-Rich Inclusions)

The V-type fluid inclusions were rarely developed in all quartz veins. They contained two distinct phases, namely liquid H_2O and vapor H_2O. They were rounded rectangles, and the size ranged from ~3–12 μm (Figure 5C). The bubble usually accounted for >55 vol% of the inclusion. V-type fluid inclusions appeared mainly in the veins of middle and late stages.

5.2. Microthermometry

A summary of the microthermometric data obtained from fluid inclusions from the three mineralization stages is listed in Table 1, and the results are plotted in Figure 6.

Table 1. Microthermometric data of fluid inclusions from the Baiyinhan tungsten deposit.

Stage	Type	N	Tm_{-CO2} (°C)	Tm_{-ice} (°C)	Tm_{-cla} (°C)	Th_{-CO2} (°C)	Th_{-tot} (°C)	Salinity (wt.%)	Pressure (bars)	Density (g/cm^3)
I	L-type	3		−8.1 to −4.4			299–311	7.02–11.81	80–92	0.79–0.82
	C-type	31	−63.9 to −60.8		3.9–7.9	28.1–31.1	294–374	4.14–10.72		
II	L-type	72		−9.5 to −0.5			228–298	0.88–13.40	25–82	0.78–0.93
	V-type	5		−6.6 to −0.6			248–297	1.05–9.98	35–82	0.72–0.88
	C-type	16	−62.2 to −60.9		6.4–8.1	27.9–31.1	233–305	3.19–6.81		
III	L-type	103		−9.4 to −0.3			171–244	0.53–13.29	8–34	0.84–0.96
	V-type	4		−5.0 to −0.6			210–245	1.05–7.86	14–36	0.81–0.91

Abbreviations: N = numbers of measured fluid inclusion, Tm_{-CO2} = final melting temperature of solid CO_2, Tm_{-cla} = final melting temperature of CO_2-H_2O clathrate, Th_{-CO2} = homogenization temperature of CO_2 phases, Tm_{-ice} = final melting temperature of ice, Th_{-tot} = total homogenization temperature, wt.% NaCl equiv. = weight percent NaCl equivalent.

In the quartz veins of the early stage, there were mainly C-type inclusions and only a few L-type inclusions. C-type inclusions yielded final melting temperatures of solid CO_2 ranging from −63.9 to −60.8 °C, lower than the pure CO_2 critical point of −56.6 °C. The melting temperatures of the clathrate were 3.9–7.9 °C, and the corresponding salinity were 4.14–10.72 wt% NaCl equiv. The homogenization temperatures of the CO_2 phase were 28.1–31.1 °C, and the final homogenization temperatures ranging from 294 °C to 374 °C. Only three L-type inclusions were found in quartz from this stage, yielding ice-melting temperatures ranging from −8.1 °C to −4.4 °C, corresponding to salinities of 7.02–11.82 wt% NaCl equiv. The final homogenization temperatures ranged ~299–311 °C, and the bulk densities of the inclusions were 0.79–0.82 g/cm^3. The pressure of L-type inclusions was ~80–92 bars.

The middle stage was dominated by L-type and C-type inclusions with minor V-type inclusions. C-type inclusions yielded final melting temperatures of solid CO_2 ranging from −62.2 °C to −60.9 °C, indicating the presence of minor concentrations of other volatiles. The melting temperatures of the clathrate were 6.4–8.1 °C and the corresponding salinity were 3.19–6.81 wt% NaCl equiv. Partial homogenization of CO_2 phases occurred between 27.9 °C and 31.1 °C, and the final homogenization temperatures ranged ~233–305 °C. L-type inclusions yielded ice-melting temperatures ranging from −9.5 °C to −0.5 °C, corresponding to salinities of 0.88–13.40 wt% NaCl equiv. The final homogenization temperatures ranged ~228–298 °C, and the bulk densities of the inclusions ranged 0.78–0.93 g/cm^3. The pressure of L-type inclusions was ~25–82 bars. V-type inclusions were the least abundant, yielding ice-melting temperatures ranging from −6.6 °C to −0.6 °C, corresponding to salinities of 1.05–9.98 wt% NaCl equiv. The final homogenization temperatures ranged ~248–297 °C, and the bulk densities of the inclusions ranged 0.72–0.88 g/cm^3. The pressure of V-type inclusions was ~35–82 bars.

The late stage was dominated by L-type inclusions with minor V-type inclusions. L-type inclusions yielded ice-melting temperatures ranging from −9.4 °C to −0.3 °C, corresponding to salinities of 0.53–13.29 wt% NaCl equiv. The final homogenization temperatures ranged ~171–244 °C, and the bulk densities of the inclusions ranged 0.84–0.96 g/cm^3. The pressure of L-type inclusions was ~8–34 bars. Only four L-type inclusions were found in the late stage, yielding ice-melting temperatures ranging from −5.0 °C to −0.6 °C, corresponding to salinities of 1.05–7.86 wt% NaCl equiv. The final homogenization temperatures ranged ~210–245 °C, and the bulk densities of the inclusions ranged 0.81–0.91 g/cm^3. The pressure of V-type inclusions was ~14–36 bars.

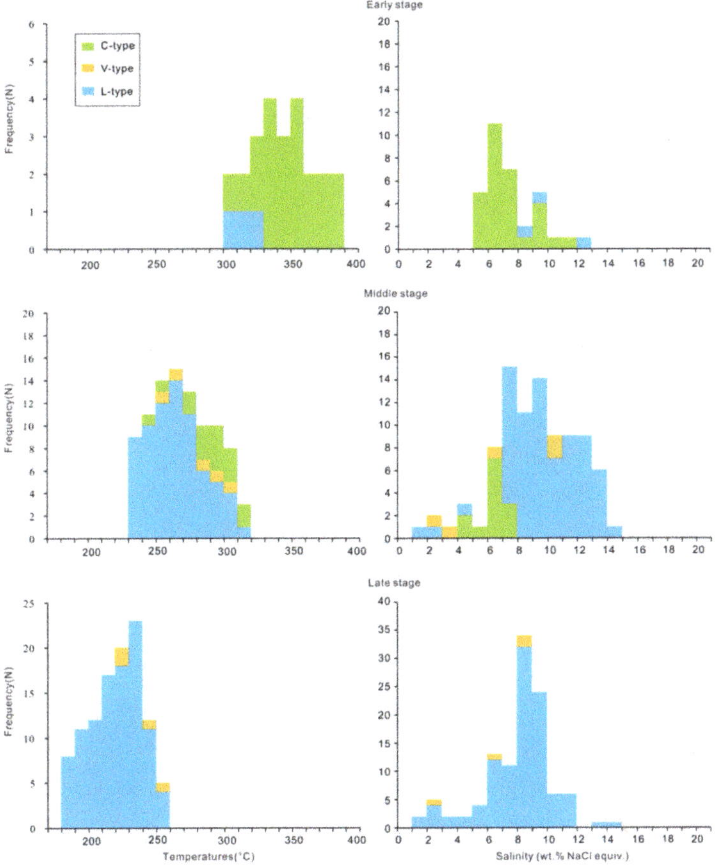

Figure 6. Histograms of total homogenization temperatures and salinities of different fluid inclusions from the three stages in the Baiyinhan tungsten deposit.

5.3. H-O Isotopes

Among the 21 quartz vein samples that were analysed for microthermometry, 9 of them were selected and analysed for the O (quartz) and H (fluid inclusions) isotopes. The H-O isotope ratios are listed in Table 2. The $\delta^{18}O$ and δD values of hydrothermal quartz samples ranged 8.5–10.2‰ and −97‰ to −92‰, respectively. The calculated $\delta^{18}O_{H2O}$ of fluids varied from −2.6‰ to 4.3‰.

Table 2. Hydrogen and oxygen isotopic compositions of quartz from the Baiyinhan tungsten deposit.

Sample No.	Mineral	Stage	δD (‰)	$\delta^{18}O_{quartz}$ (‰)	Th (°C)	$\delta^{18}O_{water}$ (‰)
17BYH-8	Quartz	Early	−81.65	9.96	338	4.3
17BYH-10	Quartz	Early	−84.23	9.27	338	3.6
17BYH-31	Quartz	Early	−83.68	9.20	338	3.5
17BYH-11	Quartz	Middle	−91.76	9.32	255	0.6
17BYH-13	Quartz	Middle	−82.06	8.83	255	0.1
17BYH-26	Quartz	Middle	−87.37	9.82	255	1.1
17BYH-14	Quartz	Late	−96.54	8.46	210	−2.6
17BYH-16	Quartz	Late	−97.07	10.16	210	−0.9
17BYH-17	Quartz	Late	−93.27	9.68	210	−1.4

5.4. Re-Os Isochron Ages

The abundance of Re and Os and the osmium isotopic compositions of molybdenite from the quartz-vein-type ores of Baiyinhan are shown in Table 3. The concentration range of Re was 3108–8486 ng/g and that of ^{187}Os was 4.52–12.53 ng/g. The Baiyinhan tungsten deposit had low Re and Os contents. The Re content of the Baiyinhan deposit was less than that in the porphyry Mo deposit (where the average value ranged from 14–68 μg/g) [49]. Five samples gave a model age of 137–141 Ma and a weight mean age of 138.8 ± 1.6 Ma (Table 3). Data, processed using the ISOPLOT/Ex program [50], yielded an isochron age of 139.6 ± 7.6 Ma and with MSWD = 3.5 (Figure 7).

Table 3. Re-Os isotope data for molybdenite samples from the Baiyinhan tungsten deposit.

Sample No.	Weight (g)	Re (ng/g)	Uncertainty	^{187}Re (ng/g)	Uncertainty	^{187}Os (ng/g)	Uncertainty	Model Age (Ma)	Uncertainty
BYH-13	0.03025	7464	57	4691	36	10.73	0.09	137.1	2.0
BYH-14	0.03016	8486	82	5333	52	12.53	0.10	140.8	2.3
BYH-15	0.03031	8223	60	5168	38	11.99	0.11	139.1	2.1
BYH-16	0.03092	8169	78	5134	49	11.87	0.08	138.6	2.1
BYH-18	0.0.001	3108	20	1953	13	4.52	0.03	138.7	1.9

Enriched ^{190}Os and ^{185}Re were obtained from the Oak Ridge National Laboratory. Decay constant: λ (^{187}Re) = 1.666 × 10^{-11}/year [51]. The uncertainty in each individual age determination was about 1.02% including the uncertainty of the decay constant of ^{187}Os, uncertainty in isotope ratio measurement, and spike calibration. BYH-13, -14, -15 were sampled form the V3 orebody, BYH-16, -18 were sampled from the V4 orebody.

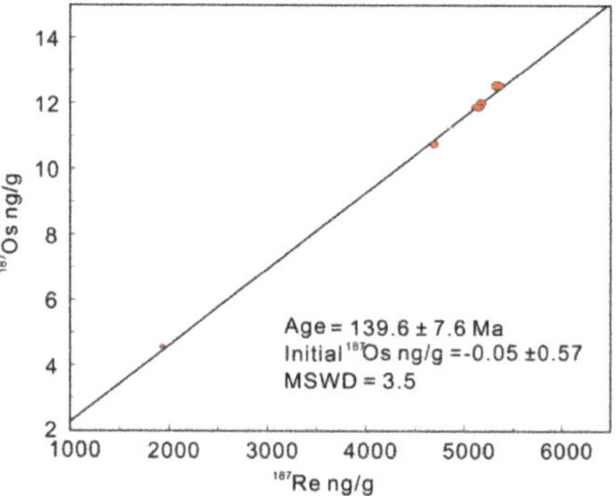

Figure 7. Re-Os isochron plot for molybdenite samples from the Baiyinhan tungsten deposit.

6. Discussion

6.1. Evolution of Ore-Forming Fluids-the Important Role of CO_2

The study of fluid inclusions shows that the ore-forming fluid of the Baiyinhan tungsten deposit is a NaCl-H_2O-CO_2 system with medium temperature and low salinity. The temperature of the fluid gradually decreased from the early stage to the late stage. As indicated from the types and abundance of fluid inclusions in the three stages, the early stage was mainly characterized by high-temperature CO_2-rich inclusions, and the middle stage was mainly composed of gas-liquid two-phase inclusions, while the late stage only contained low-temperature gas-liquid two-phase inclusions. The content of CO_2 in the ore-forming fluid tended to decrease with the evolution of the fluid.

CO_2 is the most abundant gas composition in fluid inclusions in many typical wolframite deposits [52,53]. Rios et al. [54], Li et al. [55] and Chen et al. [48] also found primary fluid inclusions containing CO_2 in wolframite, which shows that CO_2 is closely related to the dissolution and migration of tungsten. Experimental research has shown that the carbonate or CO_2 in the fluid is conducive to the dissolution and migration of tungsten [52,56,57]. Higgins [58] stated that tungsten may migrate in the form of carbonate and bicarbonate in CO_2-rich fluids under the condition of high temperature and high pressure. The experimental study on the crystallization of tungsten-manganese ore in alkali carbonate solution showed that tungsten exists mainly in the form of WO_4^{2-} in the hydrothermal system, and the solubility of wolframite is affected by the pH value of the system. The effect of alkali carbonate or CO_2 on the solubility of wolframite cannot be attributed to complexation because it can affect the pH value of the fluid, resulting in the change of tungsten solubility [55]. At the same time, CO_2 can act as pH buffer [59,60], and the high content of CO_2 in the fluid can maintain the stability and enhance the migration of tungsten [61].

It can be seen from the pressure calculation of the three stages (Table 1) that the pressure of the fluid inclusion gradually decreased. A large number of liquid-rich inclusions were developed in the deposit. There was an obvious negative correlation between density and homogenization temperature (Figure 8). These characteristics may have been caused by the CO_2 dissipation during the process of temperature and pressure reduction, resulting in the gradual decrease of CO_2 content and increase of fluid density, and the evolution of fluids into a NaCl-H_2O system [62–64]. The CO_2-rich fluid formed during the hydrothermal differentiation in the late magmatism can carry sufficient ore-forming element tungsten. During the migration of ore-forming fluid along the fractures developed in the area, the addition of meteoric water precipitation and decrease of temperature and pressure leads to the immiscibility of the fluid. The escape of CO_2 results in the change of pH and the decrease of the solubility of tungsten, which leads to the precipitation and enrichment of WO_4^{2-} in the ore-forming fluid by combining with cations such as Fe^{2+} and Mn^{2+} [65–68]. The dilution and escape of CO_2 can effectively enrich the ore-forming elements and even cause the fluid to supersaturate in an instant [69,70]. The content of CO_2 in the ore-forming fluid of the Baiyinhan deposit tends to decrease with the evolution of the fluid, which is similar to that of the Grey River tungsten deposit and Parrila tungsten-tin deposit [71,72]. It also further explains the important role of CO_2 in tungsten migration, and that the precipitation and enrichment of wolframite were closely related to the dispersion of CO_2 at the Baiyinhan deposit.

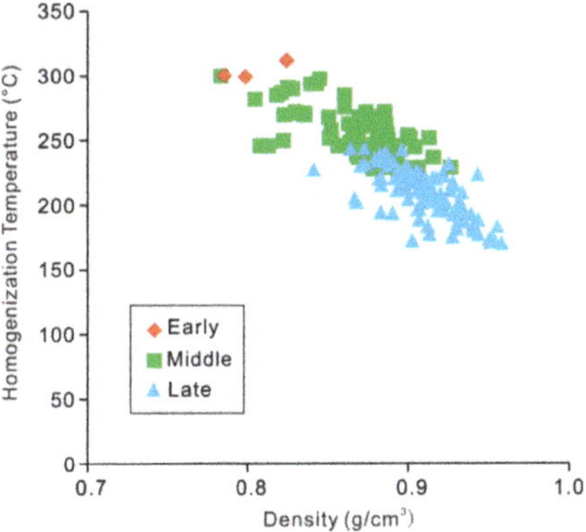

Figure 8. Temperature-density diagram for fluid inclusions from the Baiyinhan tungsten deposit.

6.2. Source of Ore-Forming Fluids and Materials

The H-O isotopes of fluid inclusions show that the samples from three stages have an obvious evolutionary trend, gradually transitioning from early residual magmatic water to late meteoric water (Figure 9). The three samples in the early stage were plotted in the lower part of the residual magma water, and their δD values were lower than the range of the residual magma water. There are two possible explanations for the low fluid δD value for the Baiyinhan deposit: One is the existence of CH_4 observed from the Raman test of fluid inclusions and the hydrogen isotope exchange between CH_4 and H_2 in the fluid, which leads to the fluid depletion of δD [73]. However, in the hydrothermal system containing CH_4, the proportion of water was much larger than that of CH_4 and H_2. Therefore, this process has a trivial impact on hydrogen isotope production [73]. Another possibility is that in actual geological conditions, when magma intrudes into the shallow part of the surface, it is likely to mix with meteoric water directly or indirectly [74]. The mixing of meteoric water and magmatic water has a significant impact on H-O isotopes, and the effect on hydrogen is generally greater than that on oxygen [75]. The second possibility well explains the fact that the δD values of H-O isotopes in quartz in the early stage of the deposit consistently fell in the lower part of the residual magmatic water. Therefore, we conclude that the initial fluid of the Baiyinhan tungsten deposit was a mixture of residual magmatic water and meteoric water. With the migration and evolution of the fluid, the proportion of meteoric water gradually increased.

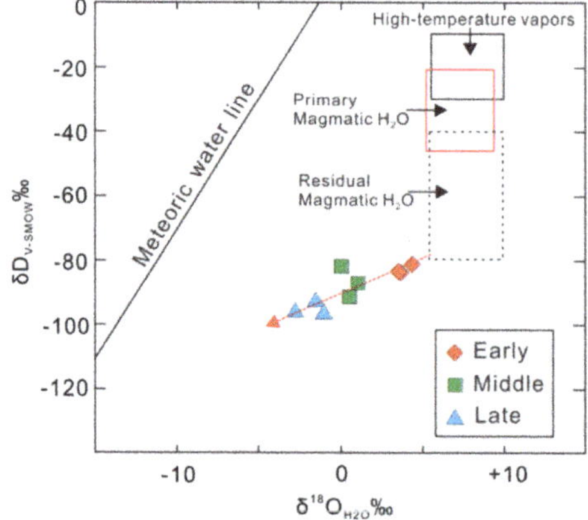

Figure 9. The quartz δD versus $\delta^{18}O_{H2O}$ diagram of the Baiyinhan tungsten deposit (Primary Magmatic H_2O after Taylor, [76]; high-temperature vapours after Giggenbach, [77]; residual magmatic H_2O after Taylor, [73]).

The Re content in molybdenite from the W-Mo-bearing quartz veins in the Baiyinhan tungsten deposit is consistent with many porphyry Mo deposits but is significantly smaller than those from porphyry Cu-Mo deposits [78]. The average Re contents of molybdenites from porphyry Cu–Mo deposits and porphyry Mo deposits range from 118–19,800 ug/g, and 4–68 ug/g, respectively [76]. The Re content in molybdenites decreases gradually from the mantle source to a mixture of mantle and crust and then to the crustal source [79,80]. The Re contents of molybdenite from the Baiyinhan tungsten deposit ranged from 3.1–8.4 ug/g with an average of 7.1 ug/g (Table 3). The relatively lower Re contents of molybdenites form the Baiyinhan tungsten deposit may indicate that the ore-forming materials source of the ore-bearing quartz vein was a crust source.

6.3. The Significance of Re-Os Age of the Baiyinhan Tungsten Deposit

The analysis of five molybdenite samples yielded an isochron age of 139.6 ± 7.6 Ma with an initial ^{187}Os of − 0.05 ± 0.57. It was shown that the initial ^{187}Os values from the molybdenite were close to zero, and the Re-Os isochron ages reflected the time of sulfide deposition [46,79,80]. Some W (-Mo) deposits were discovered in the eastern part of the CAOB in recent years. The ages of W (-Mo) mineralization in the eastern part of the CAOB were reported and recognized in three periods (Figure 10): Triassic (~230 Ma), Early-Middle Jurassic (198–170 Ma) and Early Cretaceous (~140 Ma). The Baiyinhan deposit was formed in the early Cretaceous with an Re-Os age of 139.6 ± 7.6 Ma, which further indicates that the early Cretaceous is an important metallogenic period with great metallogenic prospect in the eastern part of the CAOB.

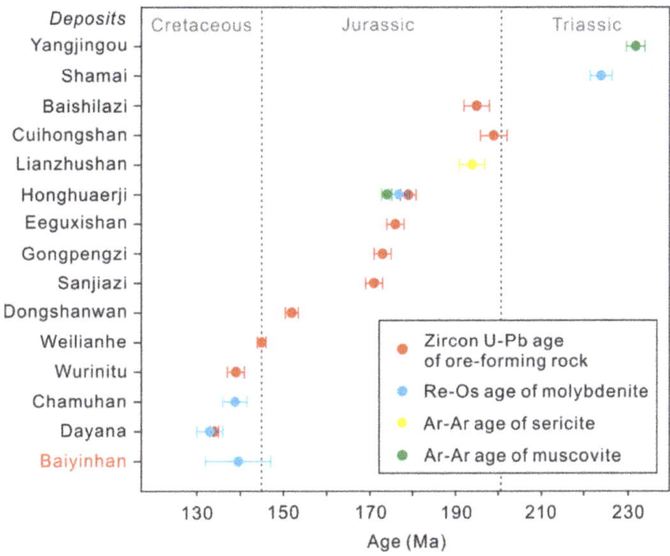

Figure 10. Ages of W-(Mo) deposits in NE China (data from [11,23,81–86] and this paper).

7. Conclusions

(1). The Baiyinhan quartz vein tungsten deposit occurs within the Yanshanian granite. The ore mineral assemblage is wolframite + pyrite + molybdenite ± bismuthinite. The greisen is the main hydrothermal alteration in the deposit.

(2). The thermometric analysis of inclusions showed that the fluid presented a medium-temperature and a low-salinity H_2O-CO_2-NaCl system. The ore-forming temperature in the three stages gradually decreased, and the CO_2 content gradually decreased with the fluid evolution. The mineralization was closely related to CO_2 dispersion. The H-O isotope of quartz showed that the ore-forming fluid was a mixture of residual magmatic water and meteoric water.

(3). Five molybdenite samples yielded an isochron age of 139.6 ± 7.6 Ma (MSWD = 3.5), and model ages for individual analyses ranged from ~138–141Ma. Combined with the regional mineralization age data, the tungsten mineralization of the Baiyinhan deposit indicates that the Early Cretaceous is an important tungsten mineralization epoch in the eastern CAOB.

Author Contributions: Conceptualization, R.W. and Q.Z.; formal analysis, R.W.; investigation, R.W., Q.Z. and Y.G.; resources, Q.Z. and J.L.; writing—original draft preparation, R.W.; writing—review and editing, Q.Z. and Z.Z.; visualization, R.W.; supervision, Q.Z.; funding acquisition, Q.Z. All authors have read and agreed to the published version of the manuscript.

Funding: This research was funded by the National Natural Science Foundation of China (Grant No. 91962104) and the National Key Research Project of China (Grant No. 2017YFC0601306).

Acknowledgments: We thank Fangming Lai and Xiaoping Zhang from Chifeng Jinghong Mining Company for their help in the field work. We are also grateful to anonymous reviewers for constructive suggestions.

Conflicts of Interest: The authors declare no conflict of interest.

References

1. Wu, F.Y.; Sun, D.Y.; Ge, W.C.; Zhang, Y.B.; Grant, M.L.; Wilde, S.A.; Jahn, B.M. Geochronology of the Phanerozoic granitoids in northeastern China. *J. Asian Earth Sci.* **2011**, *41*, 1–30. [CrossRef]
2. Zeng, Q.D.; Liu, J.M.; Yu, C.M.; Ye, J.; Liu, H.T. Metal DEPOSITS in the Da Hinggan Mountains, NE China: Styles, characteristics, and exploration potential. *Int. Geol. Rev.* **2011**, *53*, 846–878. [CrossRef]

3. Zeng, Q.D.; Liu, J.M.; Chu, S.X.; Wang, Y.B.; Sun, Y.; Duan, X.X.; Zhou, L.L. Mesozoic molybdenum deposits in the East Xingmeng orogenic belt, northeast China: Characteristics and tectonic setting. *Int. Geol. Rev.* **2012**, *54*, 1843–1869. [CrossRef]
4. Zeng, Q.D.; Qin, K.Z.; Liu, J.M.; Li, G.M.; Zhai, M.G.; Chu, S.X.; Guo, Y.P. Porphyry molybdenum deposits in the Tianshan–Xingmeng orogenic belt, northern China. *Int. J. Earth Sci.* **2015**, *104*, 991–1023. [CrossRef]
5. Ouyang, H.G.; Mao, J.W.; Santosh, M.; Zhou, J.; Zhou, Z.H.; Wu, Y.; Hou, L. Geodynamic setting of Mesozoic magmatism in NE China and surrounding regions: Perspectives from spatio–temporal distribution patterns of ore deposits. *J. Asian Earth Sci.* **2013**, *78*, 222–236. [CrossRef]
6. Xu, W.L.; Pei, F.P.; Wang, F.; Meng, E.; Ji, W.Q.; Yang, D.B.; Wang, W. Spatial–temporal relationships of Mesozoic volcanic rocks in NE China: Constraints on tectonic overprinting and transformations between multiple tectonic regimes. *J. Asian Earth Sci.* **2013**, *74*, 167–193. [CrossRef]
7. Sun, J.G.; Zhang, Y.; Han, S.J.; Men, L.J.; Li, Y.X.; Chai, P.; Yang, F. Timing of formation and geological setting of low-sulphidation epithermal gold deposits in the continental margin of NE China. *Int. Geol. Rev.* **2013**, *55*, 608–632. [CrossRef]
8. Ren, Y.S.; Lei, E.; Zhao, H.L.; Wang, H.; Ju, N. Characteristics of fluid inclusions and ore genesis of Yangjingou large scheelite deposit in Yanbian area, NE China. *J. Jilin Univ. Earth Sci. Ed.* **2010**, *40*, 764–772. (In Chinese)
9. Ren, Y.S.; Niu, J.P.; Lei, E.; Wang, H.; Wang, X. Geological&geochemical Characteristics and metallogenesis of Sanjiazi scheelite deposit in Siping area. Jilin Province. *J. Jilin Univ. Earth Sci. Ed.* **2010**, *40*, 314–320. (In Chinese)
10. Ren, Y.S.; Ju, N.; Zhao, H.L.; Wang, H.; Lu, X.Q.; Wu, C.Z. Geological characteristics and fluid inclusions of Wudaogou lode scheelite deposit in eastern Yanbian, Jilin Province. *J. Jilin Univ. Earth Sci. Ed.* **2011**, *41*, 1736–1744. (In Chinese)
11. Xiang, A.P.; Wang, Y.J.; Qin, D.J.; She, H.Q.; Han, Z.G.; Guan, J.D.; Kang, Y.J. Metallogenic and diagenetic age of Honghuaerji tungsten polymetallic deposit in Inner Mongolia. *Miner. Depos.* **2014**, *33*, 428–439. (In Chinese)
12. Xiang, A.P.; She, H.Q.; Chen, Y.C.; Qin, D.J.; Wang, Y.J.; Han, Z.G.; Kang, Y.J. Ar-Ar age of muscovite from the greisenization alteration zones of the Honghuaerji tungsten polymetallic deposit, inner Mongolia, and its geological significance. *Rock Miner. Anal.* **2016**, *35*, 108–116. (In Chinese)
13. Xiang, A.P.; Chen, Y.C.; She, H.Q.; Li, G.M.; Li, Y.X. Geochronology and geochemical characteristic of biotite granite of the Dayana W-Mo deposit in Dong Uj imqin banner, inner Mongolia, and its geological significance. *Acta Geol. Sin.* **2018**, *92*, 107–124. (In Chinese)
14. Xiang, A.P.; Chen, Y.C.; She, H.Q.; Li, G.M.; Li, Y.X. Chronology and geochemical characteristics of granite in Weilianhe of Inner Mongolia and its geological significance. *Geol. China* **2018**, *45*, 963–976. (In Chinese)
15. Guo, Z.J.; Li, J.W.; Chang, Y.L.; Han, Z.G.; Dong, X.Z.; Yang, Y.C.; Tian, J.; She, H.Q.; Xiang, A.P.; Kang, Y.J. Genetic types and ore-forming geological significance of granites in the Honghuaerji scheelite deposit, Inner Mongolia. *Acta Petrol. Mineral.* **2015**, *34*, 322–342. (In Chinese)
16. Guo, Z.J.; Li, J.W.; Huang, M.H.; Guo, W.J.; Dong, X.Z.; Tian, J.; Yang, Y.C.; She, H.Q.; Xiang, A.P.; Kang, Y.J. Characteristics of ore-forming fluid in Honghuaerji scheelite deposit, Inner Mongolia. *Miner. Depos.* **2016**, *35*, 1–17. (In Chinese)
17. Li, J.J.; Fu, C.; Tang, W.L.; Li, H.M.; Lin, Y.X.; Zhang, T.; Wang, S.G.; Zhao, Z.L.; Dang, Z.C.; Zhao, L.J. The metallogenic age of the Shamai wolframite deposit in Dong Ujimqin Banner, Inner Mongolia. *Geol. Bull. China* **2016**, *35*, 524–530. (In Chinese)
18. Chen, G.Z.; Wu, G.; Li, T.G.; Liu, R.L.; Wu, L.W.; Zhang, P.C.; Zhang, T.; Chen, Y.C. LA-ICP-MS zircon and cassiterite U-Pb ages of Daolundaba copper-tungstentin deposit in Inner Mongolia and their geological significance. *Miner. Depos.* **2018**, *37*, 225–245. (In Chinese)
19. Ma, Y.P.; Ren, Y.S.; Hao, Y.J.; Lai, K.; Zhao, H.L.; Liu, J. Genesis and Material Source of Scheelite of Yangbishan Iron-Tungsten Deposit in Heilongjiang, NE China. *J. Jilin Univ. Earth Sci. Ed.* **2018**, *48*, 105–117. (In Chinese)
20. Yuan, X.P. Geological characteristics and genesis study of Xiaolaogualinzi W-Mo polymetallic deposit in Hexigten Banner, Inner Mongolia. *Miner. Resour. Geol.* **2018**, *32*, 276–282. (In Chinese)
21. Nie, F.J.; Jiang, S.H. Geological setting and origin of Mo–W–Cu deposits in the honggor–shamai District, inner Mongolia, North China. *Resour. Geol.* **2011**, *61*, 344–355. [CrossRef]

22. Hu, X.L.; Ding, Z.J.; He, M.C.; Yao, S.Z.; Zhu, B.P.; Shen, J.; Chen, B. Two epochs of magmatism and metallogeny in the Cuihongshan Fe-polymetallic deposit, Heilongjiang Province, NE China: Constrains from U-Pb and Re-Os geochronology and Lu-Hf isotopes. *J. Geochem. Explor.* **2014**, *143*, 116–126. [CrossRef]
23. Zeng, Q.D.; Sun, Y.; Chu, S.X.; Duan, X.X.; Liu, J.M. Geochemistry and geochronology of the Dongshanwan porphyry Mo-W deposit, Northeast China: Implications for the Late Jurassic tectonic setting. *J. Asian Earth Sci.* **2015**, *97*, 472–485. [CrossRef]
24. Guo, Z.; Li, J.; Xu, X.; Song, Z.; Dong, X.; Tian, J.; Yang, Y.; She, H.; Xiang, A.; Kang, Y. Sm-Nd dating and REE Composition of scheelite for the Honghuaerji scheelite deposit, Inner Mongolia, Northeast China. *Lithos* **2016**, *261*, 307–321. [CrossRef]
25. Jiang, S.-H.; Bagas, L.; Hu, P.; Han, N.; Chen, C.L.; Liu, Y.; Kang, H. Zircon U-Pb ages and Sr-Nd-Hf isotopes of the highly fractionated granite with tetrad REE patterns in the Shamai tungsten deposit in eastern Inner Mongolia, China: Implications for the timing of mineralization and ore genesis. *Lithos* **2016**, *261*, 322–339. [CrossRef]
26. Xiang, A.; Chen, Y.; Bagas, L.; She, H.; Kang, Y.; Yang, W.; Li, C. Molybdenite Re-Os and U-Pb zircon dating and genesis of the Dayana W-Mo deposit in eastern Ujumchin, Inner Mongolia. *Ore Geol. Rev.* **2016**, *78*, 268–280. [CrossRef]
27. Fei, X.; Zhang, Z.; Cheng, Z.; Santosh, M.; Jin, Z.; Wen, B.; Li, Z.; Xu, L. Highly differentiated magmas linked with polymetallic mineralization: A case study from the Cuihongshan granitic intrusions, Lesser Xing'an Range, NE China. *Lithos* **2018**, *302*, 158–177. [CrossRef]
28. Wang, Y.H.; Zhang, F.F.; Liu, J.J.; Xue, C.J.; Zhang, Z.C. Genesis of the Wurinitu W-Mo deposit, Inner Mongolia, northeast China: Constraints from geology, fluid inclusions and isotope systematics. *Ore Geol. Rev.* **2018**, *94*, 367–382. [CrossRef]
29. Li, Y.; Yu, X.; Mi, K.; Carranza, E.J.M.; Liu, J.; Jia, W.; He, S. Deposit geology, geochronology and geochemistry of the Gongpengzi skarn Cu-Zn-W polymetallic deposit, NE China. *Ore Geol. Rev.* **2019**, *109*, 465–481. [CrossRef]
30. Bureau of Geology and Mineral Resources of Neimongol Autonomous Region (BGMR). *Regional Geology of Neimonggol Autonomous Region*; Geological Publishing House: Beijing, China, 1991; 532p. (In Chinese)
31. Wang, Y. Geological Information Were Discovered and Their Plate Tectonic Significance on the Northern Bank of Xarmoron River. *Geol. Inn. Mong.* **1999**, *90*, 6–28. (In Chinese)
32. Yang, B.J.; Liu, W.S.; Wang, X.C.; Li, Q.X.; Wang, J.M.; Zhao, X.P.; Li, R.L. Geophysical characteristics of Daxinganling gravitational gradient zone in Eastern China and its geodynamic mechanism. *Chin. J. Geophys.* **2005**, *48*, 86–97. (In Chinese) [CrossRef]
33. Zeng, Q.D.; Liu, J.M.; Liu, H.T. Geology and geochemistry of the Bianbianshan Au-Ag-Cu-Pb-Zn deposit, Southern Da Hinggan mountains, Northeastern China. *Acta Geol. Sin. Engl. Ed.* **2012**, *86*, 630–639.
34. Zeng, Q.D.; Liu, J.M.; Chu, S.X.; Guo, Y.P.; Gao, S.; Guo, L.X.; Zhai, Y.Y. Poly-metal mineralization and exploration potential in southern segment of the Da Hinggan mountains. *J. Jilin Univ. Earth Sci. Ed.* **2016**, *46*, 1100–1123. (In Chinese)
35. Xiang, G.G.; Jin, W.L.; De, H.Z.; Fei, X.; Han, B.X.; Shuai, J.W.; Tian, L.J. Petrogenesis and tectonic setting of igneous rocks from the Dongbulage porphyry Mo deposit, Great Hinggan Range, NE China: Constraints from geology, geochronology, and isotope geochemistry. *Ore Geol. Rev.* **2020**, *120*, 103326.
36. Yu, B.; Zeng, Q.; Frimmel, H.E.; Wang, Y.; Guo, W.; Sun, G.; Zhou, T.; Li, J. Genesis of the Wulong gold deposit, northeastern North China Craton: Constraints from fluid inclusions, H-O-S-Pb isotopes, and pyrite trace element concentrations. *Ore Geol. Rev.* **2018**, *102*, 313–337. [CrossRef]
37. Bodnar, R.J. Revised equation and table for determining the freezing point depression of H_2O-NaCl solutions. *Geochim. Cosmochim. Acta* **1993**, *57*, 683–684. [CrossRef]
38. Collins, P.L.F. Gas hydrates in CO_2-bearing fluid inclusions and the use of freezing data for estimation of salinity. *Econ. Geol.* **1979**, *74*, 1435–1444. [CrossRef]
39. Steele, M.M.; Lecumberri, S.P.; Bodnar, R.J. HokieFlincs_H_2O-NaCl: A Microsoft Excel spreadsheet for interpreting microthermometric data from fluid inclusions based on the PVTX properties of H_2O-NaCl. *Comput. Geosci.* **2012**, *49*, 334–337. [CrossRef]
40. Clayton, R.N.; Mayeda, T.K. The use of bromine pentafluoride in the extraction of oxygen from oxides and silicates for isotopic analysis. *Geochim. Cosmochim. Acta* **1963**, *27*, 43–52. [CrossRef]

41. Simon, K. Does δD from fluid inclusion in quartz reflect the original hydrothermal fluid? *Chem. Geol.* **2001**, *177*, 483–495. [CrossRef]
42. Clayton, R.N.; O'Neil, J.L.; Mayeda, T.K. Oxygen isotope exchange between quartz and water. *J. Geophys. Res.* **1972**, *77*, 3057–3067. [CrossRef]
43. Du, A.D.; He, H.L.; Yin, N.W.; Zou, X.Q.; Sun, Y.L.; Sun, D.Z.; Cheng, S.Z.; Qu, W.J. A study of the rhenium-osmium geochronometry of molybdenites. *Acta Geol. Sin.* **1994**, *68*, 339–347. (In Chinese)
44. Du, A.D.; Wu, S.Q.; Sun, D.Z.; Wang, S.X.; Qu, W.J.; Markey, R.; Stein, H.; Morgan, J.; Malinovskiy, D. Preparation and certification of Re-Os dating reference materials: Molybdenite HLP and JDC. *Geostand. Geoanal. Res.* **2004**, *28*, 41–52. [CrossRef]
45. Shirey, S.B.; Walker, R.J. Carius tube digestion for low-blank rhenium-osmium analysis. *Anal. Chem.* **1995**, *67*, 2136–2141. [CrossRef]
46. Stein, H.J.; Markey, R.J.; Morgan, J.W.; Du, A.; Sun, Y. Highly precise and accurate Re-Os ages for molybdenite from the East Qinling molybdenum belt, Shanxi Province, China. *Econ. Geol.* **1997**, *92*, 827–835. [CrossRef]
47. Lu, H.Z.; Fan, H.R.; Ni, P.; Ou, G.X.; Shen, K.; Zhang, W.H. *Fluid Inclusions*; Science Press: Beijing, China, 2004; pp. 1–450. (In Chinese)
48. Chen, L.L.; Ni, P.; Li, W.S.; Ding, J.Y.; Pan, J.Y.; Wang, J.J.; Yang, Y.L. The link between fluid evolution and vertical zonation at the Maoping tungsten deposit, Southern Jiangxi, China: Fluid inclusion and stable isotope evidence. *J. Geochem. Explor.* **2018**, *192*, 18–32. [CrossRef]
49. Berzina, A.N.; Sotnikov, V.I.; Economou, E.M.; Eliopoulos, D.G. Distribution of rhenium in molybdenite from porphyry Cu-Mo and Mo-Cu deposits of Russia (Siberia) and Mongolia. *Ore Geol. Rev.* **2005**, *26*, 91–113. [CrossRef]
50. Ludwig, K.R. *Users Manual for Isoplot/Ex rev. 2.49*; Geochronological Center Special Publication: Berkeley, CA, USA, 2001; Volume 1a, pp. 1–56.
51. Smoliar, M.I.; Walker, R.J.; Morgan, J.W. Re-Os ages of groupIIA, IIIA, IVA, and IVB iron meteorites. *Science* **1996**, *271*, 1099–1102. [CrossRef]
52. Liu, Y.J.; Ma, D.S. *Geochemistry of Tungsten*; Science Press: Beijing, China, 1987; pp. 1–222. (In Chinese)
53. Wood, S.A.; Samson, I.M. The hydrothermal geochemistry of tungsten in granitoid environments: I. relative solubilities of ferberite and scheelite as a function of T, P, pH, and m_{NaCl}. *Econ. Geol.* **2000**, *95*, 143–182. [CrossRef]
54. Rios, F.J.; Villas, R.N.; Fuzikawa, K. Fluid evolution in the Pedra Preta wolframite ore deposit, Paleoproterozoic Musa granite, eastern Amazon craton, Brazil. *J. S. Am. Earth Sci.* **2003**, *15*, 787–802. [CrossRef]
55. Li, J.K.; Liu, Y.C.; Zhao, Z.; Chou, I.M. Roles of carbonate/CO_2 in the formation of quartz-vein wolframite deposits: Insight from the crystallization experiments of huebnerite in alkal-icarbonate aqueous solutions in a hydrothermal diamond-anvil cell. *Ore Geol. Rev.* **2018**, *95*, 40–48. [CrossRef]
56. Xu, Y.S.; Zhang, B.; Han, Y.W. An experimental study on the partitioning of tungsten between aqueous fluid and silicate melts. *Geochimica* **1992**, *3*, 272–281. (In Chinese)
57. Liu, Y.C.; Li, J.K.; Zhao, Z. A preliminary experimental study of the crystallization of wolframite using hydrothermal diamond anvil cell. *Earth Sci. Front.* **2017**, *24*, 159–166. (In Chinese)
58. Higgins, N.C. Fluid inclusion evidence for the transport of tungsten by carbonate complex in hydrothermal solutions. *Can. J. Earth Sci.* **1980**, *17*, 823–830. [CrossRef]
59. Drummond, S.E.; Ohmoto, H. Chemical evolution and mineral deposition in boiling hydrothermal systems. *Econ. Geol.* **1985**, *80*, 126–147. [CrossRef]
60. Phillips, G.N.; Evans, K.A. Role of CO_2 in the formation of gold deposits. *Nature* **2004**, *429*, 860–863. [CrossRef] [PubMed]
61. Yin, X.B.; Zhang, D.H.; Wang, C.S.; Zhao, G.J. Characteristics of fluid inclusion for typical tungsten, stannary vein deposit and porphyry molybdenum deposit in China. *Guilin Ligong Daxue Xuebao* **2011**, *31*, 524–532. (In Chinese)
62. Linnen, R.L.; Williams, J.A.E. The evolution of pegmatite-hosted Sn-W mineralization at Nong Sua, Thailand: Evidence from fluid inclusions and stable isotopes. *Geochim. Cosmochim. Acta* **1994**, *58*, 735–747. [CrossRef]
63. Wang, X.D.; Ni, P.; Yuan, S.D.; Wu, S.H. Fluid inclusion studies of the Huangsha quartz-vein type tungsten deposit, Jiangxi Province. *Acta Petrol. Sin.* **2012**, *28*, 122–132. (In Chinese)

64. Zhang, D.Q.; Feng, C.Y.; Li, D.X.; Chen, Y.C.; Zeng, Z.L. Fluid inclusions characteristics and ore genesis of Taoxikeng tungsten and tin deposit in Chongyi County, Jiangxi Province. *J. Jilin Univ. Earth Sci. Ed.* **2012**, *42*, 374–383. (In Chinese)
65. Spycher, N.F.; Reed, M. Evolution of a broadlands-type epithermal ore fluid along alternative P-T paths: Implications for the transport and deposition of base, precious, and volatile metals. *Econ. Geol.* **1989**, *84*, 328–359. [CrossRef]
66. So, C.S.; Yun, S.T. Origin and evolution of W-Mo-producing fluids in a granitic hydrothermal system: Geochemical studies of quartz vein deposits around the Susan Granite, Hwanggangri District, Republic of Korea. *Econ. Geol.* **1994**, *89*, 246–267. [CrossRef]
67. Graupner, T.; Kempe, U.; Dombon, E.; Pätzold, O.; Leeder, O.; Spooner, E.T.C. Fluid regime and ore formation in the tungsten (-yttrium) deposits of Kyzyltau (Mongolian Altai): Evidence for fluid variability in tungsten-tin ore systems. *Chem. Geol.* **1999**, *154*, 21–58. [CrossRef]
68. Liu, C.; Zhao, Z.; Lu, L.N.; Zeng, Z.L.; Liu, C.H.; Xu, H. Metallogenic fluid study of the Yanqian skarn type tungsten deposit in eastern Nanling region. *Acta Geol. Sin.* **2018**, *92*, 2485–2507. (In Chinese)
69. Zhang, D.H. Some new advances in ore-forming fluid geochemistry on boiling and mixing of fluids during the progress of hydrothermal deposits. *Adv. Earth Sci.* **1997**, *12*, 546–552. (In Chinese)
70. Van Hinsberg, V.J.; Berlo, K.; Migdisov, A.A.; Williams, J.A.E. CO_2-fluxing collapses metal mobility in magmatic vapour. *Geochem. Perspect. Lett.* **2016**, *2*, 169–177. [CrossRef]
71. Higgins, N.C.; Kerrich, R. Progressive ^{18}O depletion during CO_2 separation from a carbon dioxide-rich hydrothermal fluid: Evidence from the Grey River tungsten deposit, Newfoundland. *Can. J. Earth Sci.* **1982**, *19*, 2247–2257. [CrossRef]
72. Mangas, J.; Arribas, A. Hydrothermal fluid evolution of the Sn-W mineralization in the Parrilla ore deposit (Caceres, Spain). *J. Geol. Soc.* **1988**, *145*, 147–155. [CrossRef]
73. Taylor, H.P. The application of oxygen and hydrogen isotope studies to problems of hydrothermal alteration and ore deposition. *Econ. Geol.* **1974**, *69*, 843–883. [CrossRef]
74. Rye, R.O.; Ohmoto, H. Sulfur and carbon isotopes and ore genesis: A review. *Econ. Geol.* **1974**, *69*, 826–842. [CrossRef]
75. Barnes, H.L. Solubilities of ore minerals. In *Geochemistry of Hydrothermal Ore Deposits*; Wiley-Blackwell: Hoboken, NJ, USA, 1979; pp. 404–460.
76. Taylor, B.E. Degassing of H_2O from rhyolite magma during eruption and shallow intrusion, and the isotopic composition of magmatic water in hydrothermal systems. Magmatic contributions to hydrothermal systems. *Geol. Surv. Jpn. Rep.* **1992**, *279*, 190–194.
77. Giggenbach, W.F. Isotopic shifts in waters from geothermal and volcanic systems along convergent plate boundaries and their origin. *Earth Planet. Sci. Lett.* **1992**, *113*, 495–510. [CrossRef]
78. Zeng, Q.D.; Guo, F.; Zhou, L.L.; Duan, X.X. Two periods of mineralization in Xiaoxinancha Au-Cu deposit, NE China: Evidences from the geology and geochronology. *Geol. J.* **2016**, *51*, 51–64. [CrossRef]
79. Mao, J.W.; Zhang, Z.; Zhang, Z.; Du, A.D. Re-Os isotopic dating of molybdenites in the Xiaoliugou W (Mo) deposit in the northern Qian mountains and its geological significance. *Geochim. Cosmochim. Acta.* **1999**, *63*, 1815–1818.
80. Mao, J.W.; Du, A.; Seltmann, R.; Yu, J.J. Re–Os ages for the Shameika porphyry Mo deposit and the Lipovy Log rare metal pegmatite, central Urals, Russia. *Miner. Depos.* **2003**, *38*, 251–257. [CrossRef]
81. Zhao, H.L.; Ren, Y.S.; Ju, N.; Wang, H.; Hou, K.J. Geochronology and geochemistry of metallogenic intrusion in Baishilazi tungsten deposit of eastern Yanbian area, Northeast China. *J. Jilin Univ. Earth Sci. Ed.* **2011**, *41*, 1726–1735. (In Chinese)
82. Hao, Y.J.; Ren, Y.S.; Zhao, H.L.; Zou, X.T.; Chen, C.; Hou, Z.S.; Qu, W.J. Re-Os isotopic dating of the molybdenite from the Cuihongshan W-Mo polymetallic deposit in Heilongjiang province and its geological significanc. *J. Jilin Univ. Earth Sci. Ed.* **2013**, *43*, 1840–1850. (In Chinese)
83. Liu, Y. The Geological Characteristics and Prospecting Direction of Gongpengzi Cu-Zn-W Deposit in Heilongjiang Province. Master's Thesis, Jilin University, Changchun, China, 2013. (In Chinese).
84. Liang, B.S. Geological Characteristics and Genesis of Iron Polymetallic Orefield in Ergu Area, Heilongjiang Province. Master's Thesis, Jilin University, Changchun, China, 2014. (In Chinese).

85. Zhao, H.L. Ore Genesis and Geodynamic Settings of Tungsten Deposits in Eastern Jilin and Heilongjiang Provinces. Ph.D. Thesis, Jilin University, Changchun, China, 2014. (In Chinese).
86. Ren, L.; Sun, J.G.; Tang, C.; Men, L.J.; Li, Y.X.; Cui, P.L. Petrogenic and metallogenic geochronology and its geological implications of Lianzhushan gold deposit, Jiayin, Heilongjiang Province. *Acta Petrol. Sin.* **2015**, *8*, 2435–2449. (In Chinese)

Article

The Effect of Co-Crystallising Sulphides and Precipitation Mechanisms on Sphalerite Geochemistry: A Case Study from the Hilton Zn-Pb (Ag) Deposit, Australia

Bradley Cave [1,*], Richard Lilly [1] and Wei Hong [1,2]

[1] Mawson Centre for Geoscience, Department of Earth Sciences, The University of Adelaide, Adelaide 5005, SA, Australia; richard.lilly@adelaide.edu.au (R.L.); wei.hong@adelaide.edu.au (W.H.)

[2] Mineral Exploration Cooperative Research Centre (MinEx CRC), Adelaide, SA 5005, Australia

* Correspondence: bradley.cave@adelaide.edu.au

Received: 3 August 2020; Accepted: 7 September 2020; Published: 9 September 2020

Abstract: High-tech metals including Ge, Ga and In are often sourced as by-products from a range of ore minerals, including sphalerite from Zn-Pb deposits. The Hilton Zn-Pb (Ag) deposit in the Mount Isa Inlier, Queensland, contains six textural varieties of sphalerite that have formed through a diverse range of processes with variable co-crystallising sulphides. This textural complexity provides a unique opportunity to examine the effects of co-crystallising sulphides and chemical remobilisation on the trace element geochemistry of sphalerite. Early sphalerite (sph-1) is stratabound and coeval with pyrrhotite, pyrite and galena. Disseminated sphalerite (sph-2) occurs as isolated fine-grained laths rarely associated with co-crystallising sulphides and represents an alteration selvage accompanying the precipitation of early stratabound sphalerite (sph-1). Sphalerite (sph-3) occurs in early ferroan-dolomite veins and formed from the chemical remobilisation of stratabound sphalerite (sph-1) during brittle fracturing and interstitial fluid flow. This generation of veins terminate at the interface, and occurs within clasts of the paragenetically later sphalerite-dominated breccias (sph-4). Regions of high-grade Cu (>2%) mineralisation contain a late generation of sphalerite (sph-5), which formed from the recrystallisation of breccia-type sphalerite (sph-4) during the infiltration of a paragenetically late Cu- and Pb-rich fluid. Late ferroan-dolomite veins crosscut all previous stages of mineralisation and also contain chemically remobilised sphalerite (sph-6). Major and trace elements including Fe, Co, In, Sn, Sb, Ag and Tl are depleted in sphalerite associated with abundant co-crystallised neighbouring sulphides (e.g., pyrite, pyrrhotite, galena and chalcopyrite) relative to sphalerite associated with minor to no co-crystallising sulphides. This depletion is attributed to the incorporation of the trace elements into the competing sulphide minerals. Chemically remobilised sphalerite is enriched in Zn, Cd, Ge, Ga and Sn, and depleted in Fe, Tl, Co, Bi and occasionally Ag, Sb and Mn relative to the primary minerals. This is attributed to the higher mobility of Zn, Ge, Ga and Sn relative to Fe and Co during the chemical remobilisation process, coupled with the effect of co-crystallising with galena and ferroan-dolomite. Results from this study indicate that the consideration of co-crystallising sulphides and post-depositional processes are important in understanding the trace element composition of sphalerite on both a microscopic and deposit-scale, and has implications for a range of Zn-Pb deposits worldwide.

Keywords: sphalerite; trace elements; Hilton Zn-Pb (Ag); sulphides; Mount Isa; critical metals

1. Introduction

The growth in technical innovation and an increasing drive for 'green' technology has prompted a growth in the demand for a previously underutilised set of metals and non-metals [1]. The elements Ge, Ga, In, Co and Cd are used in the production of a wide variety of technologies such as solar panels, LCD glasses, batteries, transistors and optical fibres [2]. These elements are often concentrated in polymetallic sulphide deposits, preferentially residing in sphalerite [3–5]. In order to understand the mechanisms responsible for the concentration of trace elements in sphalerite, an increasing amount of research has been dedicated to assessing the geochemistry of sphalerite from a range of Zn-deposits worldwide (e.g., [6–9]).

Previous studies have indicated that the concentrations of Mn, Fe, Ge, Ga and In are proportional to the precipitation temperature of sphalerite [8]. Ge can be upgraded in deposits by the dynamic recrystallisation of sphalerite during metamorphism [10] and high-In sphalerite can be produced by the infiltration of late Cu-bearing fluids [6]. It has also been shown that sphalerite from high-temperature magmatic fluids are enriched in In, while sphalerite derived from low-temperature crustal fluids are relatively enriched in Ga and Ge [11]. George et al. [5] showed the factors governing the partitioning trends of trace elements between sphalerite, chalcopyrite and galena included: (1) the redox state of the elements; (2) the trace element budget for each mineral and; (3) the ionic radius of the substituting element. Multiple textural varieties of sphalerite, or individual generations of sphalerite within a single ore deposit has been the focus of multiple studies [12,13], and show that trace element contents can differ substantially. Despite the continued research, two key aspects that greatly affect the trace element content of sphalerite require further investigation: (1) the effect of co-crystallising sulphides and; (2) the effect of chemical remobilisation.

The Hilton Zn-Pb (Ag) deposit, northern Australia (Figure 1) contains numerous textural varieties of sphalerite that have precipitated via different formation mechanisms and alongside a range of co-crystallising sulphides [14,15]. This study uses detailed petrography to determine the multiple stages of sphalerite and co-genetic minerals present within the deposit. This is accompanied by the trace element analysis of each sphalerite variety to determine the relationship between the trace element composition of sphalerite, the presence/absence of co-crystallising sulphides and the effect of small-scale chemical remobilisation. Understanding and interpreting the cause of variations in the trace element geochemistry of sphalerite can have potential applications to a variety of Zn-Pb deposits worldwide.

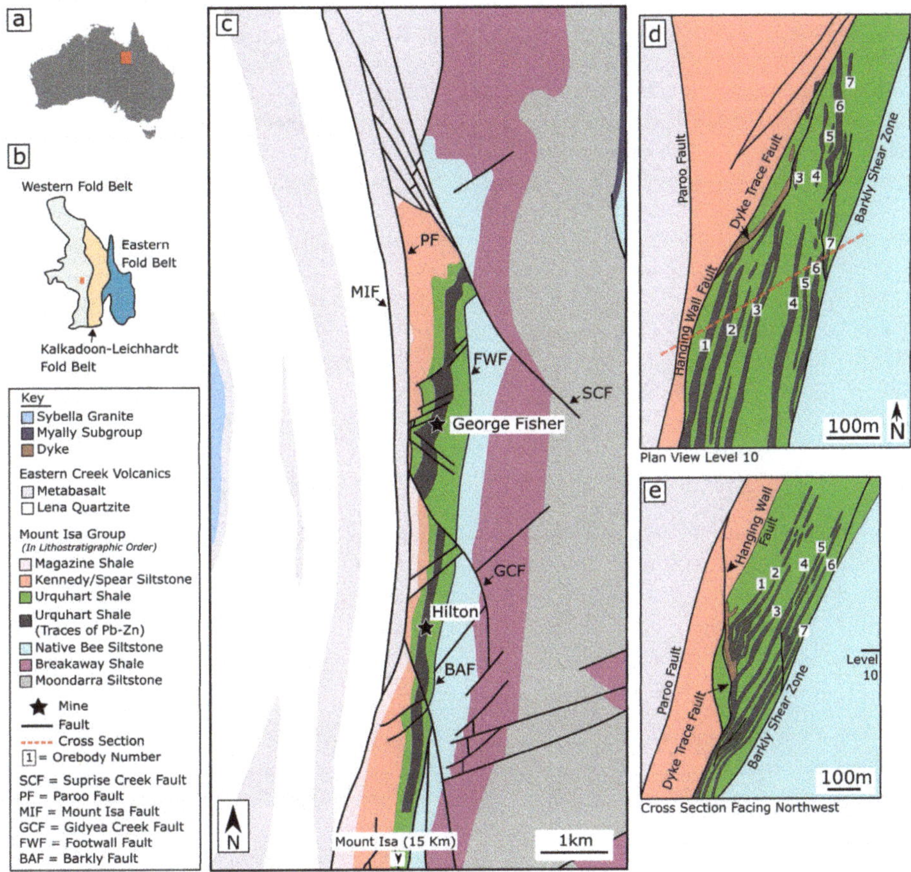

Figure 1. Location of the study area with reference to (**a**) Australia and (**b**) the Mount Isa Inlier. (**c**) A simplified geological map of the study area after Valenta [16] with the location of the Hilton and George Fisher Zn-Pb (Ag) deposits. (**d**) Plan view of the Hilton Zn-Pb (Ag) deposit showing the seven north to south trending stratabound ore zones. (**e**) A schematic cross-section of the Hilton Zn-Pb (Ag) deposit facing northwest. Figure 1d,e have been updated from Valenta [17] using available mine models.

2. Background

2.1. Regional Geology

The late Paleoproterozoic to early Mesoproterozoic McArthur Basin and the Mount Isa Inlier in northern Australia are host to numerous Zn-Pb deposits that comprise the Carpentaria—Mount Isa Zinc Belt [18]. The Hilton Zn-Pb (Ag) deposit is the focus of this study. This deposit is located in the Western Fold Belt of the Paleoproterozoic Mount Isa Inlier, approximately 20 km north of the Mount Isa Cu-Zn-Pb (Ag) deposit (Figure 1a–c). The Western Fold Belt hosts three major Superbasins; (1) Leichhardt Superbasin (1800–1750 Ma); (2) Calvert Superbasin (1750–1690 Ma); and (3) Isa Superbasin (1690–1575 Ma) [19]. The Isa Superbasin is further divided into the Gun, Loretta, River, Term, Lawn, Wilde and Doom supersequences, which are composed of various turbidites, stromatolitic dolostones and carbonaceous shales that were deposited in a shallow to deep marine environment [19,20]. The Gun and Loretta supersequences contain the regionally extensive Mount Isa Group (Figure 1c) and are dominated by deep water turbiditic rhythmites [21]. Basin development was terminated by the onset of the 1610–1500 Ma Isan Orogeny [19,22–24].

The Mount Isa region has undergone four major deformation events (D_1–D_4) during the Isan Orogeny. D_1 deformation consisted of N–S directed compression that produced a bedding-parallel foliation in carbonaceous and phyllosilicate units in the Hilton area [14,25]. D_2 deformation is characterised by E–W shortening that produced district-scale upright folds with N–S trending axes [25,26]. Early studies dated this deformation event at ca 1555–1530 Ma [25,27]. However, more recent geochronology studies favour D_2 deformation at 1575 Ma [28,29]. D_2 deformation was coeval with regional peak metamorphism, locally constrained to temperatures of ~200 °C by bitumen reflectance data [30] at the nearby George Fisher Zn-Pb (Ag) deposit (Figure 1c). D_3 deformation (formally $D_{2.5}$ by Bell and Hickey [31]) is only present in localised regions throughout the Mount Isa Inlier, and produced highly asymmetric deflections in S_2 fabric and bedding with top to the east vergence [31]. D_4 deformation produced NNW–SSE trending inclined to recumbent folding with amplitudes of cm to 10's of meters at the Hilton deposit [14]. Post to late-D_4 faults offset Pb-Zn orebodies at Hilton from 2–100 m [14].

2.2. Hilton Zn-Pb (Ag) Deposit Geology

Economic Zn-Pb mineralisation is hosted by the ca. 1650 Ma Urquhart Shale and occurs as seven (1–7) stratabound orebodies (Figure 1d,e). The genesis of the Hilton Zn-Pb (Ag) deposit is controversial, with both syn-sedimentary [14] and syn-deformational carbonate replacement [32] models proposed. The north of the orebody is bounded by the Dyke Trace Fault Zone, the east by the Barkly Shear Zone, the west by the Hanging Wall Fault and the south by an unnamed northwest-southeast striking sinistral fault (Figure 1d,e; [17]). The orebodies are divided into the hanging wall orebodies (1–3) and the footwall orebodies (4–7), which are separated by an un-mineralised siltstone pillar. The highest Zn and Pb values occur near the Dyke Trace Fault in the hanging wall orebodies and laterally up-dip [15]. Pb grades shows a northerly trend associated with the hanging wall fault zone, with high-grade Pb (>8%) primarily constrained to the lower 2 and 3 orebodies [15]. Mineralisation is constrained to stratabound regions that have undergone multiple stages of pre-mineralisation alteration, and occur as bedding-parallel veins and various polymetallic sphalerite- and galena-dominant breccias. A chalcopyrite and pyrrhotite-rich zone occurs adjacent to the Hanging Wall Fault within an area associated with syn- to post-D_2 movement [17]. Elevated Cu (>1%) is restricted to the 1–3 orebodies, and shows a strong spatial relationship to the major north-south sub-vertical shear/dyke system [15,17]. High-grade Cu (>2%) appears in Pb- and Zn-bearing breccias, and has a close spatial relationship with high-grade Pb (>8%) [15].

3. Methods

3.1. Sampling Selection

Five sections of drill core that intersect multiple orebodies on level 5, 12 and 14 were logged in detail and sampled. Samples were selected on the basis of mineralisation styles and temporal relationships. This resulted in a diverse sample suite of 21 sphalerite-bearing samples. The samples were made into polished resin blocks and polished thin sections following standard procedures. The polished resin blocks were used for subsequent LA-ICP-MS analysis.

3.2. LA-ICP-MS Anslyses

3.2.1. Spot Analyses

LA-ICP-MS spot analyses of sphalerite were carried out at Adelaide Microscopy using a NWR213 (Omaha, NE, USA) solid state 213 nm laser ablation system coupled with an Agilent 7900 (Santa Clara,

CA, USA) quadrupole ICP-MS. Ablation was conducted in an atmosphere of He (0.6 L/min) that was mixed with Ar (0.88 L/min) upon leaving the ablation cell. The material then passes through a pulse-homogenisation device before being introduced to the ICP torch. Ablation spot size ranged from 10–70 μm. The repetition rate of the laser was 5 Hz with a consistent fluence of ~4 J/cm^2. Pre-ablation consisted of 5 pulses, followed by a 20 s delay to allow adequate cell wash out before the proceeding analyses. Data collection consisted of 30 s background collection, followed by 40 s of ablation. The Isotopes monitored include: ^{27}Al, ^{34}S, ^{29}Si, ^{55}Mn, ^{57}Fe, ^{59}Co, ^{60}Ni, ^{65}Cu, ^{66}Zn, ^{69}Ga, ^{71}Ga, ^{72}Ge, ^{73}Ge, ^{75}As, ^{77}Se, ^{95}Mo, ^{109}Ag, ^{111}Cd, ^{113}In, ^{115}In, ^{118}Sn, ^{121}Sb, ^{125}Te, ^{182}W, ^{197}Au, ^{201}Hg, ^{205}Tl, ^{206}Pb, ^{207}Pb, ^{208}Pb, and ^{209}Bi. The dwell time of each isotope was 0.02 s, except for ^{27}Al, ^{29}Si, ^{34}S, and ^{55}Mn with dwell times of 0.005 s and ^{59}Co, ^{60}Ni, ^{65}Cu, ^{66}Zn ^{206}Pb, ^{207}Pb, and ^{208}Pb with dwell times of 0.01 s. The GSD-1G (USGS) and STDGL3 (CODES, University of Tasmania; [33]) standards were analysed using a 70 μm diameter spot size with a 25:2 standard sample bracketing approach.

3.2.2. Trace Element Maps

LA-ICP-MS trace element maps were produced using a spot size of 16 μm with a scan speed of 16 μm/s. The isotopes: ^{27}Al, ^{34}S, ^{29}Si, ^{55}Mn, ^{57}Fe, ^{59}Co, ^{60}Ni, ^{65}Cu, ^{66}Zn, ^{69}Ga, ^{71}Ga, ^{72}Ge, ^{73}Ge, ^{77}Se, ^{109}Ag, ^{111}Cd, ^{115}In, ^{118}Sn, ^{121}Sb, ^{125}Te, ^{205}Tl, ^{208}Pb and ^{209}Bi were analysed. A dwell time of 0.002 s was used for isotopes ^{27}Al, ^{29}Si and ^{34}S, 0.005 s for isotopes ^{55}Mn and ^{57}Fe, and a dwell time of 0.01 s was used for the collection of the remaining isotopes. The repetition rate used was 10 Hz with a fluence of ~4 J/cm^2. Pre-ablation consisted of a single transect across the same path used for analysis, followed by a 20 s delay to allow adequate cell wash out. The scan time for each line was 153.875 and 138.369 s for maps that are located between sph-4 and sph-5, and sph-4 and sph-6 respectively, which followed a 10 s background collection. The GSD-1G (USGS) and STDGL3 (CODES, University of Tasmania; [33]) standards were analysed every 20 transects using a 70 μm spot size.

3.2.3. Data Processing

LA-ICP-MS spot analyses were processed using LADR (Moonah, TAS, Australia) data reduction software [34]. Each analysis was checked for mineral inclusions by inspecting the time-resolved profile of each individual isotope. Data compromised by obvious inclusions, as inferred by large irregular spikes in any element, were rejected, and inclusion-free portions of the time-resolved profile was selected. The STDGL3 standard was used as the primary standard to correct for instrument drift and mass bias, with GSD-1G used as a secondary standard. For the elements Al, Si and Ga, GSD-1G was used as the primary standard. ^{113}In and ^{115}In were corrected for isobaric interferences from ^{111}Cd and ^{118}Sn respectively by using the measured ^{111}Cd and ^{118}Sn isotopes and assuming normal isotopic abundances. To account for solid-solution between Zn, Fe and Cd in the analysed sphalerite, the isotopes listed in Section 3.2.1. (excluding: ^{29}Si, ^{27}Al ^{71}Ga, ^{73}Ge, ^{113}In, ^{206}Pb and ^{207}Pb) were normalised to a total of 100 wt% assuming mono-sulphide stoichiometry. Where multiple isotopes were measured, the concentration of the isotopes ^{115}In, ^{208}Pb, ^{69}Ga and ^{72}Ge were preferentially used. To check the validity of this method, the concentration of Zn, Fe and Cd was measured on multiple samples using a Cameca SXFive (Fitchburg, MA, USA) Electron Microprobe at Adelaide Microscopy (Supplementary Materials Table S1). A comparison between the normalised LA-ICP-MS data and the microprobe show good agreement, and are always within analytical error of each instrument.

LA-ICP-MS trace element maps were produced using Iolite [35]. Instrumental drift was corrected through the Baseline_Subtract data reduction scheme [36]. As the major chemistry of each sphalerite variety differs considerably, counts per second (cps) maps were produced. This allows for a direct comparison between the relative abundance of trace-elements in each variety of sphalerite, as well as all other co-crystallising minerals.

4. Sphalerite Textures and Generations

The paragenesis documented from the samples collected at Hilton (Figure 2) is similar to the paragenesis recognised by Chapman [37] and Rieger et al. [38] at the nearby George Fisher Zn-Pb (Ag) deposit. Broadly, early stratabound Zn mineralisation is followed by breccia-style Zn-Pb mineralisation with intermittent stages of ferroan-dolomite veining (Figure 2). Each sphalerite stage is documented in detail below.

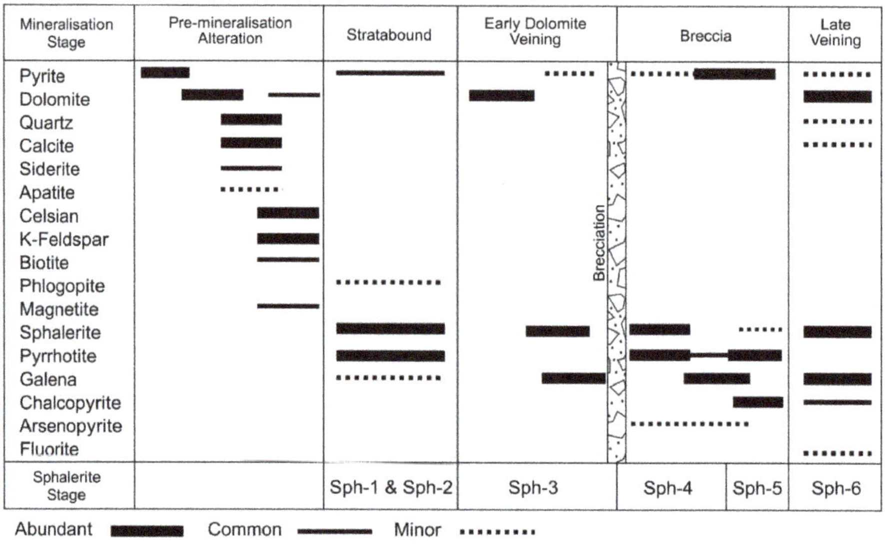

Mineralisation Stage	Pre-mineralisation Alteration	Stratabound	Early Dolomite Veining	Brecciation	Breccia	Late Veining
Pyrite	Abundant	Common	Minor		Minor, Abundant	Minor
Dolomite	Abundant, Common		Abundant			Abundant
Quartz	Abundant					Minor
Calcite	Abundant					Minor
Siderite	Common					
Apatite	Minor					
Celsian	Abundant					
K-Feldspar	Abundant					
Biotite	Common					
Phlogopite		Minor				
Magnetite	Common					
Sphalerite		Abundant	Abundant		Abundant, Minor	Abundant
Pyrrhotite		Abundant			Abundant, Common, Abundant	
Galena		Minor	Abundant		Abundant	Abundant
Chalcopyrite					Abundant	Common
Arsenopyrite					Minor	
Fluorite						Minor
Sphalerite Stage		Sph-1 & Sph-2	Sph-3		Sph-4 / Sph-5	Sph-6

Abundant ▬ Common — Minor ········

Figure 2. Paragenesis of the Hilton Zn-Pb (Ag) deposit based on the samples from this study. Pre-mineralisation alteration is followed by stratabound mineralisation containing sph-1 and sph-2. Early dolomite veins contain sph-3 and are followed by a stage of brecciation and the emplacement of sph-4. This is overprinted by Pb and Cu mineralisation where sph-4 is recrystallised to produce sph-5. Multiple varieties of late veins crosscut all types of mineralisation and contain sph-6.

4.1. Stratabound Sphalerite (Sph-1)

Stratabound sphalerite (sph-1) constitutes the first major Zn mineralisation event, accounting for 6–38% of the total Zn in the drill core used in this study. Sph-1 is paragenetically early and broadly constrained to regions affected by pre-mineralisation dolomite + quartz + calcite + siderite ± apatite (Figure 3a), and often K-feldspar + celsian + biotite ± magnetite alteration. The abundance of sphalerite in this stage varies from a major (>90%) component of the mineralogy to inter-connected sphalerite grains that make up a minor (<10%) component (Figure 3a–d). Co-crystallising pyrrhotite varies from a minor (<20%) to major (>60%) component (Figure 3b,c). Fine-grained (<50 μm) galena, phlogopite and euhedral pyrite form a minor component of this stage (Figure 3d,e). Sph-1 and coeval sulphides envelope and partially replace minerals that comprise the various stages of pre-mineralisation alteration. Individual mineralised horizons vary from <1 to >20 cm in width. Sph-1 is stratabound over a deposit scale, but commonly envelopes and partially replaces discordant pre-mineralisation dolomite + quartz + calcite + siderite veins (Figure 4a).

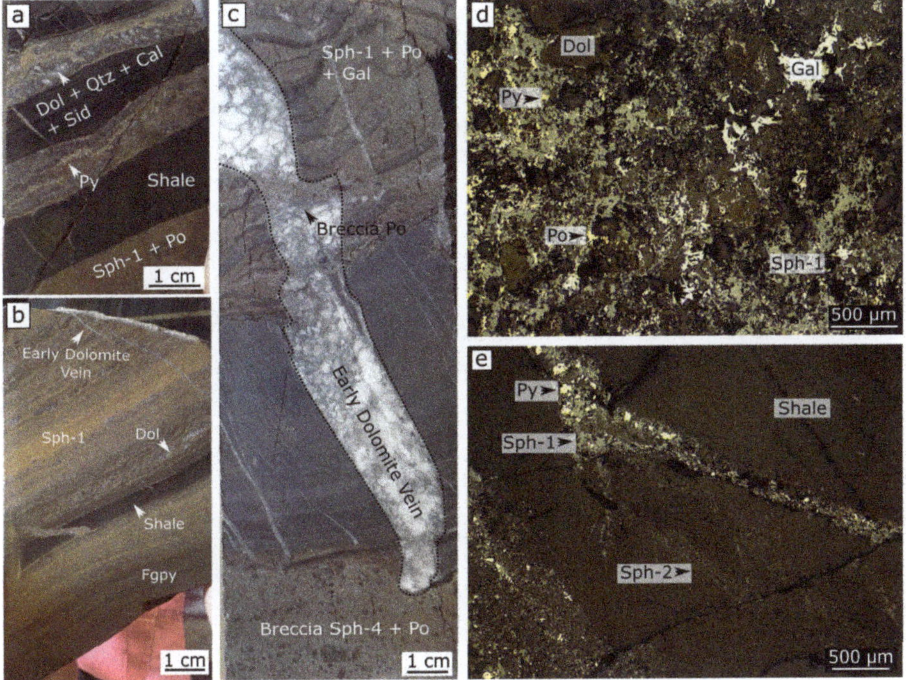

Figure 3. Sphalerite paragenesis at the Hilton Zn-Pb (Ag) deposit. (**a**) Sph-1 replacing a stratabound pre-mineralisation dolomite + quartz + calcite + siderite vein. (**b**) Sph-1 dominated stratabound mineralisation that is crosscut by an early ferroan dolomite vein. (**c**) Stratabound Zn mineralisation crosscut by an early dolomite vein. The early dolomite vein terminates at the interface of a Zn-dominated breccia. (**d**) Reflected light photomicrograph showing sph-1 with co-crystallising pyrite, pyrrhotite and galena. (**e**) Reflected light photomicrograph showing fine-grained disseminated sph-2 splaying from a sph-1 vein. Abbreviations: qtz = quartz, cal = calcite, sid = siderite, sph = sphalerite, dol = dolomite, fgpy = fine-grained pyrite, py = pyrite, gal = galena, po = pyrrhotite.

4.2. Disseminated Sphalerite Alteration (Sph-2)

Disseminated sphalerite (sph-2) occurs as fine to medium grains (<50 μm) that form stratabound layers that are often <1 cm in width (Figure 4a,b). On a deposit scale, disseminated sphalerite is volumetrically insignificant and does not constitute economic Zn mineralisation. Sph-2 occurs in carbonaceous shale that has been variably affected by pre-mineralisation dolomite + quartz + calcite + siderite ± apatite and K-feldspar + celsian + biotite veining/alteration. Co-crystallising sulphides are rare in this stage, with only very minor pyrite and galena documented in the samples from this study. Disseminated sphalerite often splays from pre-mineralisation dolomitic veins (Figures 3e and 4a) and graphitic seams containing sph-1. Consistent with the interpretation of this variety of sphalerite at George Fisher by Chapman [37], sph-2 is interpreted to represent an alteration selvage associated with the emplacement of sph-1.

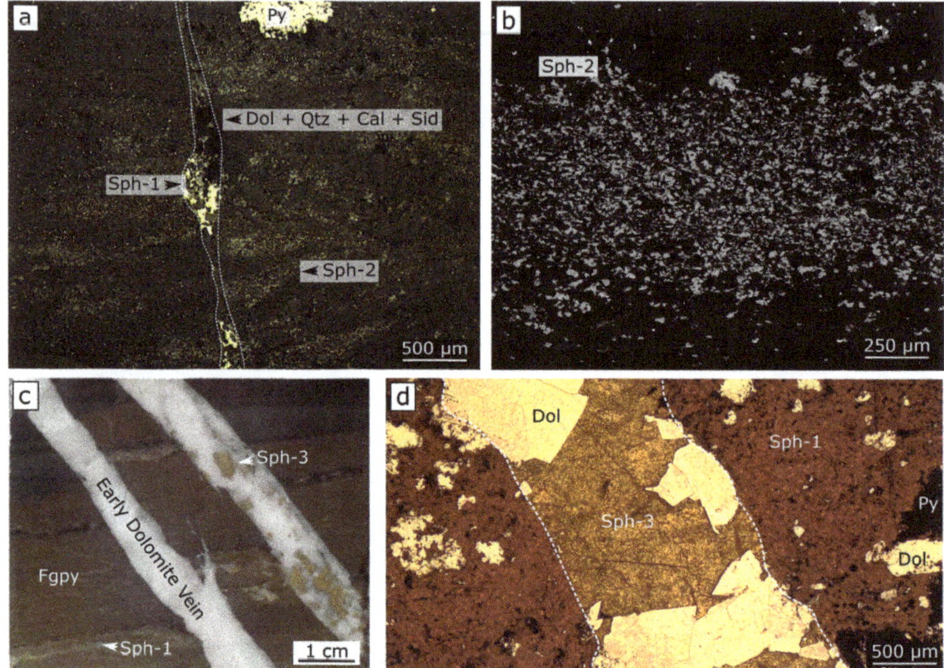

Figure 4. Sphalerite paragenesis at the Hilton Zn-Pb (Ag) deposit (continued). (**a**) Reflected light photomicrograph showing fine-grained disseminated sph-2 splaying out from a discordant sph-1 bearing vein. (**b**) Back-scattered electron image of fine-grained disseminated sph-2. (**c**) An early dolomite vein containing sph-3 cross-cutting a sample containing abundant fine-grained pyrite with minor sph-1. (**d**) Plane polarised light photomicrograph of an early dolomite vein cross-cutting stratabound sph-1 mineralisation. Note the colour difference between sph-1 and sph-3. Abbreviations: sph = sphalerite, dol = dolomite, fgpy = fine-grained pyrite, py = pyrite, qtz = quartz, cal = calcite, sid = siderite.

4.3. Sphalerite in Early Ferroan-Dolomite Veins (Sph-3)

Discordant ferroan-dolomite veins occur throughout the deposit and crosscut early sph-1 and disseminated sph-2 (Figure 3b,c and Figure 4c,d). Ferroan-dolomite veins range from 0.1–10 cm in width. Ferroan-dolomite is typically coarse-grained and highly zoned. Sph-3 and galena are present as medium- to coarse-grained components in these veins (Figure 4c,d), with fine-grained sphalerite occurring along the vein boundaries. Sph-3 is a lighter colour in plane polarised light relative to sph-1 (Figure 4d). Pyrite is rare in these veins and only occurs as isolated grains (<50 µm) along the grain boundaries of coarse-grained ferroan-dolomite. Minor veins of this stage contain abundant chlorite, which splays out from the veins and preferentially replace pre-mineralisation magnetite. Sph-3 and the co-crystallising galena in these veins are interpreted to have formed from the chemical remobilisation of sph-1, with an insignificant contribution from sph-2 during initial fracturing and subsequent interstitial fluid flow.

4.4. Sphalerite Dominated Breccias (Sph-4)

Matrix-supported breccia mineralisation constitutes the second major Zn mineralising event and is associated with the highest grades of Zn at Hilton, accounting for 62–94% of Zn in the drill core used in this study. Shale clasts within the breccia contain fine-grained pyrite, dolomite + quartz + calcite +

siderite ± apatite veins, K-feldspar + celsian + magnetite alteration, sph-1 and sph-2 mineralisation, and early ferroan-dolomite veins (Figure 5a,b). Core samples clearly outline the relationship between stratabound mineralisation, early ferroan-dolomite veins and breccia mineralisation. Figure 3c shows that stratabound sph-1 is crosscut by early ferroan-dolomite veins, which terminate at the interface of sph-4 bearing breccia mineralisation. Similar to stratabound mineralisation, the co-crystallising sulphides present in sphalerite-dominated breccias are pyrite, pyrrhotite (Figure 5a–d) and arsenopyrite.

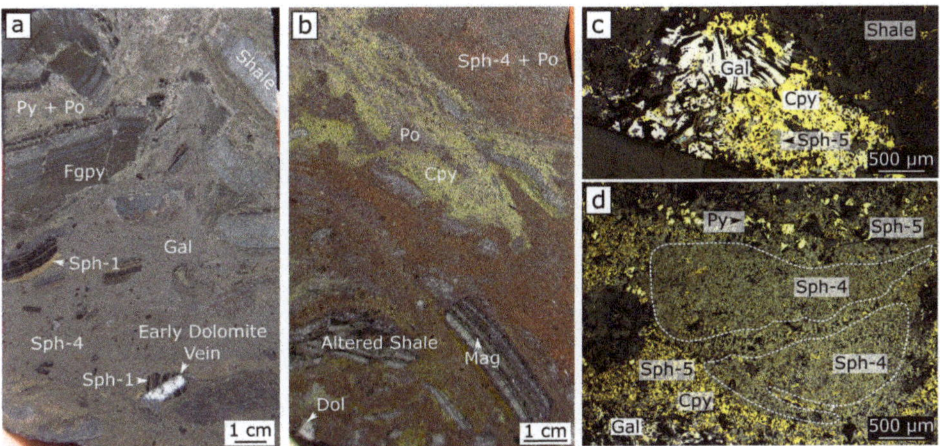

Figure 5. Sphalerite paragenesis at the Hilton Zn-Pb (Ag) deposit (continued). (**a**) Polymetallic sulphide-rich breccia containing sph-4, abundant pyrrhotite and high grade Pb mineralisation. Note the presence of early stratabound Zn mineralisation and early dolomite veins within the clasts of shale. (**b**) Polymetallic breccia containing abundant pyrrhotite and chalcopyrite that overprints sph-4. (**c**,**d**) Reflected light photomicrographs highlighting the relationship between high-grade Cu mineralisation and sph-5. Note the recrystallisation of sph-4 to form sph-5 in Figure 5d. Abbreviations: sph = sphalerite, dol = dolomite, fgpy = fine-grained pyrite, py = pyrite, gal = galena, po = pyrrhotite, cpy = chalcopyrite, mag = magnetite.

4.5. Sphalerite in Chalcopyrite Dominated Breccias (Sph-5)

Chalcopyrite-bearing breccias are the main source of Cu mineralisation at Hilton and are restricted to the 1–3 orebodies. Chalcopyrite and pyrrhotite are the dominant sulphides with variable amounts of pyrite and galena (Figure 5b–d). Chalcopyrite-dominated breccias frequently overprint and replace sph-4 dominant breccias (Figure 5b–d). Chalcopyrite and pyrrhotite often occupy texturally late sites between brecciated clasts of shale and the sphalerite-dominated breccia, are situated in veins that penetrate fragmented breccia clasts, and in discordant veins that splay out from sulphide-dominated breccias into the neighbouring shale. Core from this study records Cu values as high as 2.77%, with Cu values regularly exceeding 0.5%. In regions where Cu mineralisation has overprinted sphalerite-dominated breccias, a relatively uncommon late stage of sphalerite (sph-5) is recognised. Sph-5 penetrates sph-4 and is coeval with abundant neighbouring chalcopyrite, pyrrhotite, pyrite, galena (Figure 5d) and arsenopyrite. Sph-5 is interpreted as sph-4 that has been recrystallised during subsequent Cu mineralisation.

4.6. Sphalerite in Late Ferroan-Dolomite Veins (Sph-6)

Multiple styles of late veins crosscut all previous stages of mineralisation (Figure 6a–d). Ferroan-dolomite + calcite + quartz veins are typically crystalline and contain medium- to coarse-grained sphalerite and galena with minor chalcopyrite, pyrrhotite and pyrite. Galena is more abundant in these late veins relative to the early ferroan-dolomite veins. Other styles of these veins include sphalerite- and galena-dominant (Figure 6c), galena-dominant with minor sphalerite, chalcopyrite and pyrite (Figure 6d) and pyrite-, chalcopyrite- and pyrrhotite-dominant veins. Late non-sulphide veins include calcite and fluorite-dominant and chlorite-dominant veins. Consistent with sph-3 situated in early ferroan-dolomite veins, sph-6 in late ferroan-dolomite veins are interpreted to have formed from the chemical remobilisation of earlier sphalerite generations upon fracturing and interstitial fluid flow. As only sph-6 that veined sph-4 breccias were analysed, we have assumed that Zn was locally derived from the sph-4 breccias.

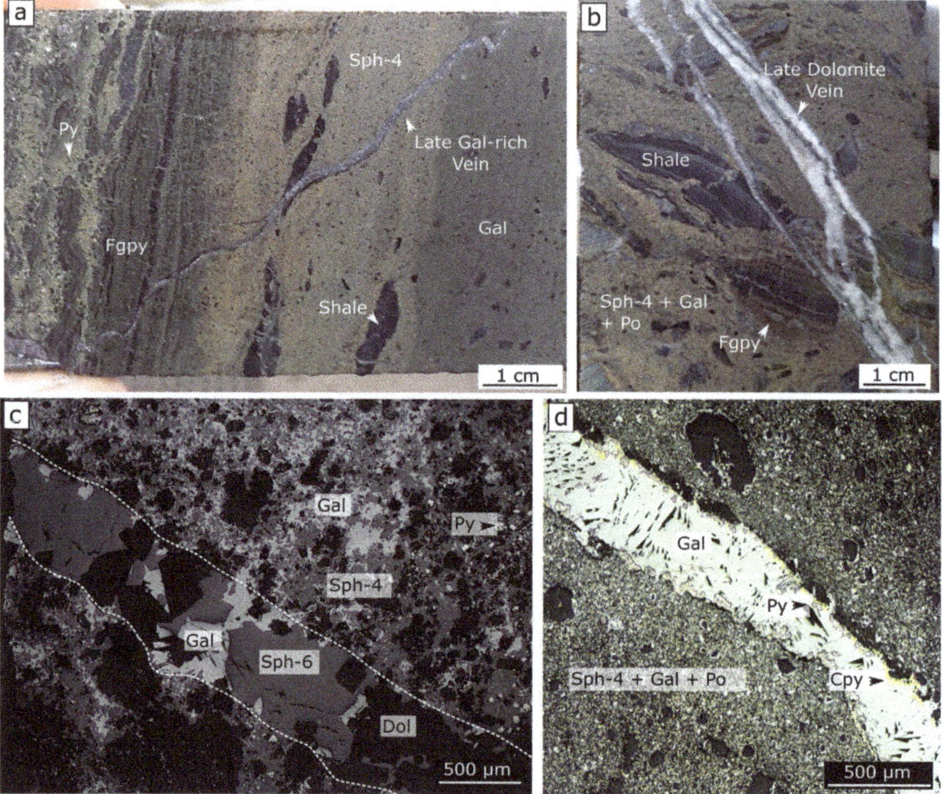

Figure 6. Sphalerite paragenesis at the Hilton Zn-Pb (Ag) deposit (continued). (**a**,**b**) Polymetallic sulphide-rich breccias cross-cut by late dolomite and sulphide-rich veins containing sph-6 and galena. (**c**,**d**) Reflected light photomicrographs of late sph-6 bearing veins cross-cutting sph-4 breccia-stage mineralisation. Abbreviations: sph = sphalerite, dol = dolomite, fgpy = fine-grained pyrite, py = pyrite, gal = galena, po = pyrrhotite, cpy = chalcopyrite.

5. Results

A total of 464 LA-ICP-MS spot analyses were conducted in this study and are reported in Supplementary Materials Table S2. This includes 109 spots on sph-1, 36 spots on sph-2, 40 spots on sph-3, 200 spots on sph-4 and 79 spots on sph-6. Due to the uncommon occurrence of sph-5, no spot analyses were acquired on this type of sphalerite. However, a trace element map was produced across a boundary between sph-4 and sph-5 to allow for a direct geochemical comparison. Sphalerite from the Hilton Zn-Pb (Ag) deposit has measurable concentrations of Mn, Co, Cu, Ga, Ge, Ag, Cd, In, Sn, Sb,

Tl, Pb and Bi (Figure 7; Supplementary Materials Table S2). The elements Mo, Se, As, Te, Ni and Au were typically below the detection limit of the ICP-MS and are therefore not used in this study.

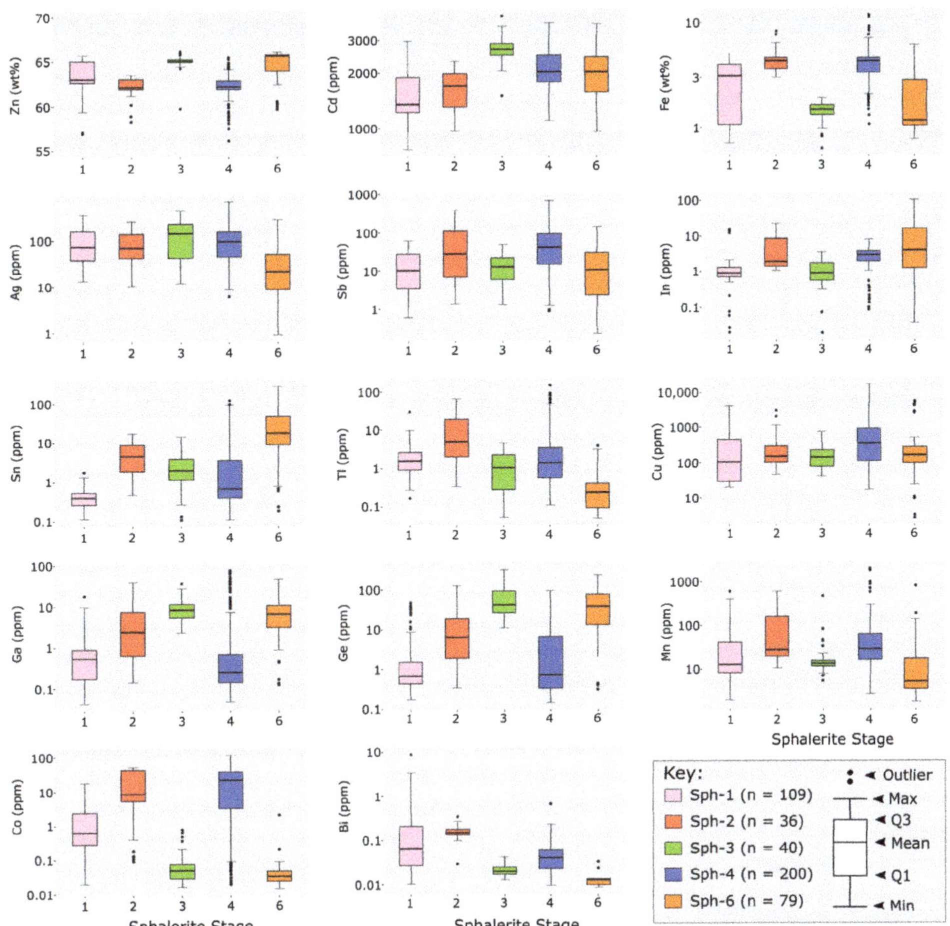

Figure 7. Box and whiskers plots of the trace elements in sphalerite analysed in this study. The box and whiskers plots are grouped into the separate sphalerite varieties.

To minimise the effect of inclusions within the sphalerite data, inclusion-free portions of the time-resolved depth profile were selected for each individual analysis. As the ablation size is significantly larger than the ablated mineral inclusions, a degree of inhomogeneity within the time-resolved depth profiles is unavoidable. The elements Zn, Fe, Cd, In and Sn consistently have smooth time-resolved depth profiles, suggesting they are homogeneously distributed over the size of the analysed spot. The elements Mn, Sb, Ag, Cu, Co, Tl, Ge and Ga show both homogeneous and heterogeneous time-resolved depth profiles, indicating their presence as both lattice-bound constituents and/or micro-scale inclusions carrying these elements. Pb and Bi always possess heterogeneous time-resolved depth profiles, indicating the presence of micron-scale inclusions of minerals carrying these elements.

5.1. LA-ICP-MS Spot Analysis

Median Fe and Zn concentrations of sph-1 are 2.73 wt% and 63.45 wt%, while sph-2 has median Fe and Zn concentrations of 4.36 wt% and 62.20 wt% (Figure 7; Table 1). Disseminated sph-2 contains higher median concentrations of Mn, Co, Ga, Ge, Cd, In, Sn, Sb and Tl than stratabound sph-1 (Figure 7; Table 1). The interquartile ranges of Co, Sn, In, Ga, Ge and Tl show minor to no overlap, indicating a significant enrichment in sph-2 relative to sph-1 (Figure 7). The concentrations of Cu and Ag do not show any considerable differences between sph-1 and sph-2.

Sph-3 in early ferroan-dolomite veins has median Fe and Zn concentrations of 1.44 wt% and 65.00 wt%, respectively (Table 1). Sph-3 has elevated concentrations of Zn, Ga, Ge, Cd and Sn, and lower concentrations of Co, Bi and Fe relative to sph-1 (Table 1; Figure 7). Ga and Ge are significantly enriched in sph-3 relative to sph-1 with median values of 9.80 ppm and 72.38 ppm, respectively (Table 1). Significant enrichments are also documented in the concentration of Cd and Sn, which do not have overlapping interquartile ranges with sph-1 (Figure 7).

Sph-6 in late cross-cutting ferroan-dolomite veins have median Fe and Zn concentrations of 2.02 wt% and 64.69 wt% while sph-4 has median concentrations of Fe and Zn of 4.15 wt% and 62.16 wt% (Table 1). Sph-6 in late cross-cutting veins has lower concentrations of Fe, Mn, Co, Ag, Sb, Tl and Bi, and higher concentrations of Zn, Ge, Ga and Sn relative to sph-4 (Table 1; Figure 7). The relative enrichments of Ga, Ge, Sn and Zn in sph-4 relative to sph-6 are consistent with the differences in the geochemistry between sph-1 and sph-3, which are interpreted to have formed from the same process. The largest depletion of trace elements in sph-6 relative to sph-4 are Tl, Bi and Co.

Table 1. The minimum, median, standard deviation and maximum values for the trace elements in sphalerite analysed in this study. All values are in ppm unless otherwise specified.

Sphalerite	Descriptives	Zn (wt%)	Cd	Fe (wt%)	Ag	Sb	In	Sn
Sph-1	Min	56.88	768	0.71	13.28	0.61	0.02	0.00
	Median	63.45	1515	2.73	104.79	16.60	2.97	0.44
	StDev	1.59	439	1.38	85.72	15.21	4.87	0.27
	Max	65.74	2968	5.09	361.09	61.96	15.53	1.75
Sph-2	Min	58.22	970	3.00	10.28	1.39	1.05	0.00
	Median	62.20	1657	4.36	91.13	73.96	5.92	5.78
	StDev	1.14	387	1.12	68.27	96.71	7.02	4.84
	Max	63.49	2306	8.29	261.32	383.10	22.50	16.78
Sph-3	Min	59.76	1501	0.63	11.77	1.32	0.00	0.11
	Median	65.00	2676	1.44	152.99	14.52	1.00	2.95
	StDev	0.93	412	0.30	116.95	10.95	0.85	2.91
	Max	66.18	4051	1.92	449.83	49.36	3.54	14.00
Sph-4	Min	54.49	1098	1.30	6.04	0.89	0.00	0.00
	Median	62.16	2132	4.15	114.50	70.67	3.00	5.52
	StDev	1.54	430	1.41	92.82	86.35	1.49	15.76
	Max	65.43	3711	11.79	692.70	673.50	8.12	119.94
Sph-6	Min	59.66	955	0.77	0.90	0.23	0.04	0.18
	Median	64.69	2025	2.02	40.67	21.08	14.21	36.55
	StDev	1.69	633	1.48	51.80	27.64	22.31	49.12
	Max	66.10	3666	6.17	283.47	140.94	103.88	272.93
Sphalerite	**Descriptives**	**Tl**	**Cu**	**Ga**	**Ge**	**Mn**	**Co**	**Bi**
Sph-1	Min	0.00	19.86	0.00	0.00	2	0.00	0.01
	Median	2.20	502.51	0.95	3.28	40	1.49	0.02
	StDev	3.13	816.98	2.07	8.53	79	2.61	0.01
	Max	29.80	4077.06	18.32	46.26	611	17.81	0.06

Table 1. *Cont.*

Sphalerite	Descriptives	Tl	Cu	Ga	Ge	Mn	Co	Bi
Sph-2	Min	0.00	46.87	0.00	0.00	11	0.00	0.02
	Median	13.25	378.55	5.04	14.40	105	16.22	0.10
	StDev	17.23	591.96	9.26	28.04	129	19.19	0.03
	Max	66.51	2984.88	39.53	124.77	613	53.31	0.16
Sph-3	Min	0.05	42.45	2.34	10.78	5	0.00	0.01
	Median	1.39	191.10	9.80	72.38	30	0.10	0.01
	StDev	1.23	167.01	6.95	74.80	74	0.17	0.00
	Max	4.55	751.87	36.71	314.16	478	0.76	0.02
Sph-4	Min	0.00	16.92	0.00	0.00	3	0.00	0.01
	Median	7.07	1017.44	4.18	5.62	80	21.14	0.03
	StDev	17.20	1824.00	11.98	16.03	218	23.28	0.02
	Max	146.81	17,137.82	75.14	94.39	1989	113.56	0.13
Sph-6	Min	0.00	2.77	0.12	0.30	2	0.00	0.01
	Median	0.35	549.30	8.59	56.66	25	0.04	0.01
	StDev	0.73	1253.63	8.79	60.14	93	0.23	0.01
	Max	3.78	5621.73	47.21	229.28	810	2.04	0.08

5.2. LA-ICP-MS Trace Element Maps

A set of trace element maps was produced between the interface of sph-4 and sph-5 (Figure 8). The trace element maps show that sph-5 has lower concentrations of Ga, Ge, Ag, In, Sn, Sb and Tl relative to sph-4. The elements Ga, Ge, In, Sn, Sb and Tl appear to be almost entirely depleted in sph-5. Contrastingly, sph-5 has comparably higher concentrations of Zn, Cd and Pb relative to sph-4. The elements Ge and Pb, and to some extent Ga are heterogeneously distributed in the analysed sph-4, indicating that high concentrations of these elements are associated with the presence of mineral inclusions (Figure 8). The highest concentration of Fe in the analysed section is consistent with the presence of pyrite and chalcopyrite.

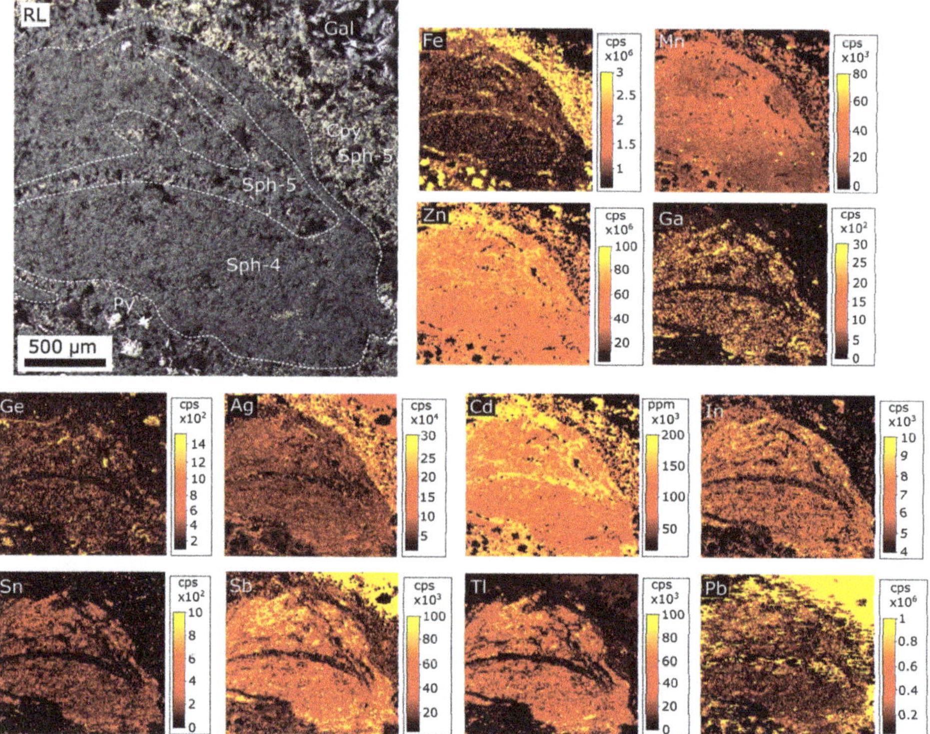

Figure 8. LA-ICP-MS trace element maps across a sample containing sph-4 and recrystallised sph-5. These maps show that sph-5 is depleted in Ge, Ga, In, Sn, Sb, Ag and Tl, and enriched in Cd and Pb relative to sph-4. Some contact zones between sph-4 and sph-5 are enriched in Tl and In. All trace element maps are in counts per second (cps).

A second set of trace element maps was produced across a section containing sph-4 and a ferroan-dolomite vein containing sph-6 (Figure 9). The trace element maps reveal that sph-6 has relatively higher concentrations of Zn, Ge, Ga, Cd, In and Sn, and a lower concentration of Co, Ag, Sb and Tl relative to sph-4. This is in agreement with LA-ICP-MS spot analyses from sph-4 and sph-6, which indicate an enrichment in Zn, Ga, Ge and Sn and a depletion in Fe, Co, Ag, Sb and Tl in the chemically remobilised sphalerite. Sph-6 is zoned in the elements Cu, Ge, Ga, Cd, In and Sn. The zonation of Cu, Ge, Ga, In and Sn appear to correlate with each other, and anticorrelate with zones of high Cd (Figure 9).

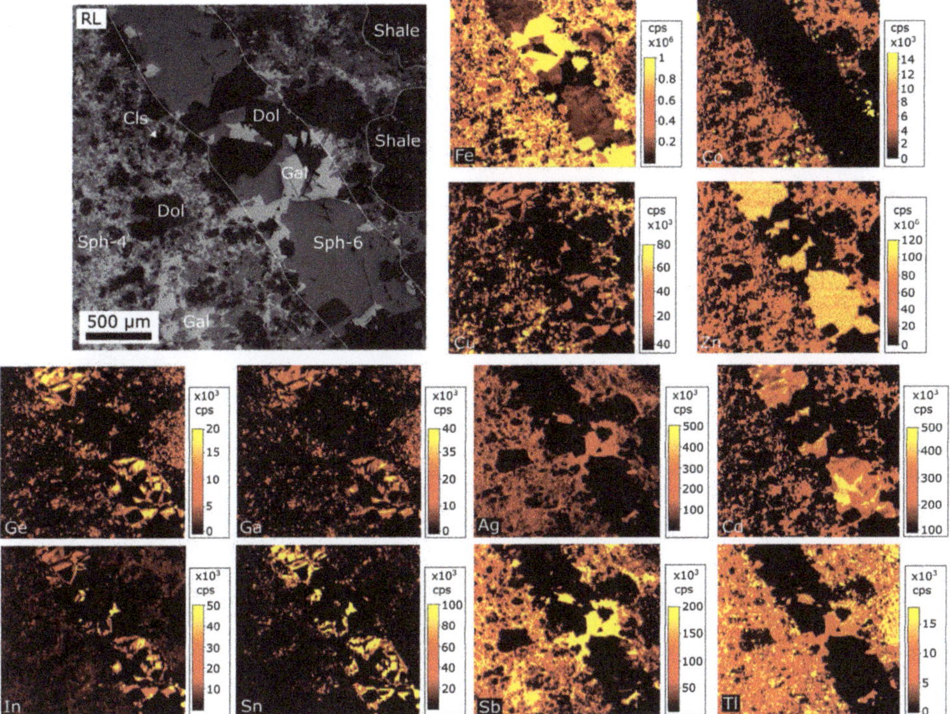

Figure 9. LA-ICP-MS trace element maps across a sample containing sph-4 and a ferroan-dolomite vein containing sph-6. The trace element maps indicate that sph-6 is enriched in Zn, Ge, Ga, Cd, In and Sn, and depleted in Fe, Co, Ag, Sb and Tl relative to sph-4. It is also seen that sph-6 is zoned in the elements Cu, Ge, Ga, Cd, In and Sn. All trace element maps are in counts per second (cps).

6. Discussion

6.1. Trace Element Substitution Mechanisms

A simple direct substitution for the bivalent elements Mn, Co and Cd with Fe is explained through the substitution mechanism (Mn^{2+}, Co^{2+}, Cd^{2+} ↔ Fe^{2+}; [7]). A positive correlation between Ag and Sb for all sphalerite varieties (Figure 10a; Supplementary Materials Table S3) is consistent with a dominant coupled substitution mechanism ($2Zn^{2+}$ ↔ Ag^{+} + Sb^{3+}) into sphalerite [13,39]. For sph-1 (R = 0.52), sph-2 (R = 0.89) and sph-4 (R = 0.77), Tl + Sn broadly correlate with the concentration of Ag + Sb (Figure 10b), indicating the presence of the substitution mechanisms $2Zn^{2+}$ ↔ (Ag^{+} or Tl^{+}) + Sb^{3+} and $3Zn^{2+}$ ↔ 2(Ag^{+} or Tl^{+}) + Sn^{4+}. The substitution of trivalent and tetravalent elements Sb^{3+}, In^{3+}, Sn^{4+}, Ge^{4+} and Ga^{3+} with varying amounts of monovalent elements Ag^{+}, Cu^{+} and Tl^{+} has been widely documented (Figure 10c; [7,40]), and is interpreted to represent the dominant substitution mechanism for the incorporation of these elements into the analysed sphalerite. A strong correlation between Ge and Cu (Figure 10d; Supplementary Materials Table S3) in sph-3 (R = 0.99) and sph-6 (R = 0.70) indicate the dominant mechanism for their substitution into sphalerite is $3Zn^{2+}$ ↔ Ge^{4+} + $2Cu^{+}$ [13], while no such correlation is shown by sph-1 (R = −0.16), sph-2 (R = 0.10) or sph-4 (R = 0.06).

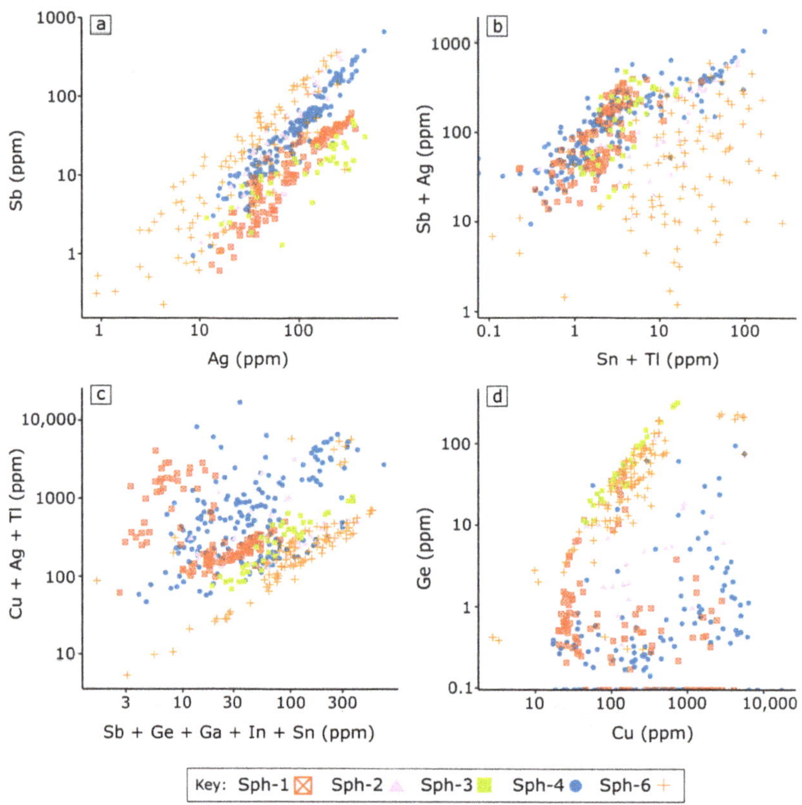

Figure 10. Binary scatter plots showing the correlation between trace elements in sphalerite and suggested substitution mechanisms. (**a**) Sb vs. Ag; (**b**) Sb + Ag vs. Sn + Tl; (**c**) Cu + Ag + Tl vs. Sb + Ge + Ga + In + Sn; (**d**) Ge vs. Cu.

6.2. Effects of Growth Conditions and Chemical Remobilisation on Sphalerite Geochemistry

6.2.1. Effect of Co-Crystallising Pyrite, Pyrrhotite and Galena

Sph-1 and sph-2 represent an early stratabound Zn mineralisation event. The timing and source of this stage of mineralisation is heavily debated and requires further investigation. Previous studies suggest that this stage of Zn-Pb mineralisation formed from an oxidised basinal brine coeval with the formation of host shale [14,41]. Studies from the nearby George Fisher Zn-Pb (Ag) deposit suggest that early Zn mineralisation post-dates the formation of the host shale, but pre-dates regional deformation [37,38]. This is challenged by work completed by Perkins and Bell [32] and Murphy [42], who favour the emplacement of early Zn-Pb mineralisation during D_4 deformation of the Isan Orogeny. Therefore, it is possible that this stage of mineralisation has undergone sub-greenschist facies metamorphism (~200 °C [30]) and subsequent deformation. Metamorphism can affect the distribution of trace elements in sphalerite, resulting in the crystallisation of discrete Ge-bearing minerals [10], and decrease the concentration of the elements Pb, Bi, Cu and Ag, which are often present in sphalerite as mineral inclusions containing these elements [43]. Detailed petrography and SEM work completed in this study did not detect any Ge-bearing minerals, and suggests that sph-1 does not exhibit any obvious textural features associated with deformation or recrystallisation, consistent with previous work by Murphy [42]. Moreover, this study assumes that if regional-scale metamorphism is a factor, it would affect sph-1 and sph-2 equally.

Sph-2 is interpreted as the alteration product of sph-1. Although these two textural varieties occur near each other and are interpreted to have precipitated from the same fluid source, sph-1 co-crystallised alongside neighbouring pyrite, pyrrhotite and galena. As sph-2 is rarely associated with neighbouring sulphide phases, it allows us to examine the effect of co-existing sulphides (pyrrhotite, pyrite and galena) on the geochemistry of sphalerite.

Sph-2 contains significantly higher concentrations of Fe, Co, Ga, Ge, In, Sn, Sb and Tl relative to sph-1. Previous studies show that pyrite and pyrrhotite are the preferred host of Co in a typical sulphide assemblage [44,45]. Pyrite can also contain considerable concentrations of Tl, Sb and In [46–48]. Galena often contains trace amounts of Sn, Ge and Ga, and is the preferred host of Tl and Sb in a typical sulphide assemblage [5,49,50]. Fe is a core component of pyrrhotite ($Fe_{(0.8-1)}S$) and pyrite (FeS_2) (Figure 8), which, if co-crystallised with sphalerite, likely decrease the available Fe. Based on the preferred sulphide host of the elements listed above, it can be inferred that the co-crystallisation of neighbouring pyrite, pyrrhotite and galena resulted in the depletion of Fe, Co, In, Sb and Tl in sph-1 relative to sph-2, and possibly contributed to the relative depletion of Ga, Ge and Sn.

While the preferred host of Sn in the sulphide system is poorly constrained and shows no systematic pattern [5], sphalerite is the primary sulphide host of Ge and Ga [5]. The elements Ge, Ga and Sn are significantly enriched in chemically remobilised sphalerite types (Figures 7 and 9), reflecting highly soluble characteristics. A relatively high solubility would allow Ge, Ga and Sn to remain in the fluid during the precipitation of sph-2 alteration selvages, resulting in an enrichment of these elements in sph-2 compared to sph-1. Alternatively, both Ge and Ga are often enriched in carbonaceous sediments with high Al content [51–53], such as the Urquhart Shale. The infiltration of fluid through the shale could have leached additional Ge and Ga, which was incorporated into sph-2 during crystallisation.

6.2.2. Effect of Recrystallisation with Abundant Neighbouring Co-Crystallising Sulphides

At the nearby George Fisher Zn-Pb (Ag) deposit, breccia-hosted sphalerite is interpreted to have formed during early-D_4 deformation of the Isan Orogeny, representing in-situ deformed sph-1 veins [37]. However, this is disputed by Murphy [42], who found only minor evidence of sphalerite recrystallisation in the Zn-dominated breccias, favouring a D_4 timing for the formation of sph-4 mineralisation. Cu-rich mineralisation is interpreted to have been emplaced from late-D_2 to post-D_4 deformation of the Isan Orogeny [17], and represents the last stage of sulphide-rich mineralisation at the Hilton Zn-Pb (Ag) deposit.

Sph-5 is interpreted to represent the recrystallisation of sph-4 during the interaction with a subsequent Cu- and Pb-rich fluid. Therefore, sph-5 recrystallised with abundant neighbouring chalcopyrite, galena, pyrrhotite and pyrite (Figure 5c,d). A direct comparison between sph-4 and sph-5 allows us to study the effect that recrystallisation alongside abundant co-crystallising sulphides has on the geochemistry of sphalerite. Figure 8 shows that sph-5 is relatively depleted in Ga, Ge, Ag, In, Sn, Sb and Tl, and is relatively enriched in Zn, Cd and Pb relative to sph-4.

The highest concentrations of Sb is associated with the presence of galena, while Ag is concentrated in both chalcopyrite and galena (Figure 8). Therefore, it is suggested that during re-crystallisation, Ag and Sb were likely removed from sph-4, and preferentially incorporated into the co-crystallising chalcopyrite and galena.

In a typical sulphide assemblage, sphalerite is the preferred host of In, followed by chalcopyrite [5,46,50]. Therefore, it can be expected that upon the recrystallisation of sph-4 to form sph-5, a majority of liberated In would be preferentially re-incorporated into sph-5, with some of the In lost to co-crystallising chalcopyrite. This could explain the relative depletion of In within sph-5 relative to sph-4. However, Figure 8 indicates that sph-5 and co-crystallising chalcopyrite possess comparably low In concentrations. This could indicate that the late Cu- and Pb-rich fluid was depleted in In, resulting in the In leached from sph-4 being redistributed over a large proportion of co-crystallising chalcopyrite and/or sph-5.

Sn does not have a predictable preferred sulphide host [5]. At Mount Isa, chalcopyrite is the preferred host of Sn, followed by galena then sphalerite [50]. Figure 8 indicates that sph-5 and co-crystallising chalcopyrite do not contain abundant Sn, and therefore it is unlikely that Sn was immediately redistributed into these neighbouring minerals. As discussed below (see Section 6.2.3.), Sn is concentrated in chemically remobilised sphalerite, indicating that it is readily expelled from the sphalerite lattice. This could suggest that Sn was transported away from the site of recrystallisation, and either incorporated into co-crystallising sulphides, such as galena and chalcopyrite elsewhere, or was lost to the late stage mineralising fluid.

Tl is also depleted in sph-5 relative to sph-4. In a typical sulphide assemblage, galena is the preferred sulphide host of Tl [5]. Figure 8 shows that sph-4 contains significantly more Tl than both sph-5 and associated co-crystallising galena. However, it is also shown that the galena contains a comparably higher concentration of Tl than sph-5 (Figure 8). Assuming the fluid associated with subsequent Cu- and Pb-rich mineralisation possessed a low concentration of Tl (as inferred by low Tl galena), and abundant neighbouring galena was present during the recrystallisation process (Figure 5c,d), it could be suggested that the Tl lost from the re-crystallisation of sph-4 was redistributed evenly among the abundant co-crystallising, low-Tl galena.

The heterogeneous distribution of Ge and Ga in sphalerite is shown in Figure 8. This suggests that high concentrations of these elements in the Hilton deposit are potentially associated with mineral inclusions. In a typical sulphide assemblage, Ga and Ge are preferentially concentrated into sphalerite, but can be also be incorporated into chalcopyrite and galena [5,50]. In the Bergslagen ore province, Sweden, it has been shown that Ga is primarily hosted by Al-silicates and oxides [54]. Additionally, both Ga and Ge can be readily incorporated into feldspars through the direct substitution with Al and Si respectively [55]. Ge and Ga are highly mobile, and can be readily expelled from sphalerite during dissolution and re-precipitation reactions (see Section 6.2.3.) and metamorphism [10]. Therefore, the recrystallisation of sph-4 to form sph-5 would have resulted in the release of Ga and Ge from sph-4 and/or the mineral inclusions holding these elements into the fluid, where they were likely redistributed into clasts of Al-rich shale, feldspars and/or partitioned into co-crystallising sulphides such as chalcopyrite or galena.

Sphalerite is the primary host of Cd in a typical sulphide assemblage [5,56]. The concentration of Cd in sphalerite can be attributed to the temperature, Cd concentration or concentration of reduced sulphur in the ore-bearing fluid [56,57]. Cd-rich sphalerite typically precipitates from low- temperature fluids [57], or from fluids with low concentrations of reduced sulphur [56]. The Cu- and Pb-rich mineralisation stage that resulted in sph-5 is associated with higher-temperature conditions compared to sph-4 [14]. Therefore, the enrichment of Cd in sph-5 relative to sph-4 may be attributed to a higher concentration of Cd, or a lower concentration of reduced sulphur in the subsequent Cu- and Pb-rich fluid, promoting Cd enrichment [56].

Figure 8 shows that Zn is enriched in sph-5 relative to sph-4. This is consistent with the enrichment of Zn in chemically remobilised sphalerite relative to its parent material (see Section 6.2.3.). These results indicate that Zn can be readily leached and redistributed from sph-4 into sph-5 upon recrystallisation. As sphalerite is the primary Zn-mineral, any additional Zn associated with the subsequent Cu- and Pb-rich fluids would also be preferentially incorporated in sph-5, contributing to the relative enrichment.

Although the Pb trace element map is somewhat chaotic, Pb is enriched in sph-5 relative to sph-4 (Figure 8). As sph-5 is coeval with the emplacement of abundant galena, this likely reflects the increased concentration of Pb within the fluid, leading to an increase in the concentration of Pb-rich mineral inclusions in the coeval recrystallised sphalerite.

6.2.3. Effect of Chemical Remobilisation

Sph-3 and sph-6 are interpreted to have resulted from the chemical remobilisation of primary sph-1 and sph-4, respectively during initial fracturing and subsequent interstitial fluid flow. This allows us to examine the effect of chemical remobilisation on the geochemistry of sphalerite. Considering the relative timing of the previous styles of mineralisation, these veins are interpreted to have formed during late D_2, or syn-D_4, and post-D_4 deformation of the Isan Orogeny, respectively. Sphalerite in both sets of carbonate veins are interpreted to have formed from the same process associated with geochemically similar fluids, and are therefore compared together in this section.

Sph-3 and sph-6 are consistently enriched in Zn, Ga, Ge, Sn and Cd, and depleted in Co, Tl, Bi and Fe relative to the primary sph-1 and sph-4. Additionally, sph-6 is also depleted in Ag, Sb and Mn relative to sph-4.

Previous studies have shown that Ge will be readily expelled from the sphalerite lattice to form discrete Ge-S minerals upon metamorphism [10,58,59]. Additionally, Ga and Ge can be readily expelled during the recrystallisation of sphalerite (see Section 6.2.2.). It has been shown that Ge and Ga are preferentially concentrated in relatively low-temperature (~120 °C), moderate- to low-salinity fluids, leading to their enrichment in low-temperature ore deposits [7,8,11]. Therefore, it is interpreted that due to their relatively high mobility, Ge and Ga will be readily leached from the primary sphalerite and/or mineral inclusions hosting these elements during interstitial fluid flow, where they are incorporated into secondary sphalerite during crystallisation. Sn is geochemically similar Ge, with differences in the behaviour of Sn attributed to its 6.63% larger ionic size and its greater tendency for divalency [51]. The similar geochemical behaviour of Sn likely contributes to its ability to be easily expelled from the sphalerite lattice, where it is re-precipitated into secondary sphalerite, a process that is consistent with recrystallisation.

The formation of Cd-rich sphalerite is favoured by low-temperature fluids and/or low concentrations of reduced sulphur within the precipitating fluid [56,57]. It is suggested that significantly less mixing with the pyrite-rich host shale relative to all other mineralisation events, and the low-temperature formation of the ferroan-dolomite veins promoted Cd enrichment in the chemically remobilised sphalerite.

Previous studies have indicated a relationship between a higher precipitation temperature and an increased concentration of Co within sphalerite [60,61]. A temperature dependent relationship could explain the inability of Co to be mobilised from sphalerite during dissolution and re-precipitation mechanisms, resulting in the chemically remobilised sphalerite being significantly depleted in Co relative to the primary sphalerite.

Galena is the primary host of Tl, Sb, Ag and Bi in a typical sulphide assemblage [5]. The depletion in these elements in chemically remobilised sphalerite types relative to the primary sphalerite can be explained by their preferential partitioning into abundant neighbouring co-crystallising galena. This is consistent with Figure 9, which shows that the highest concentration of Tl, Sb and Ag in the late veins are associated with the presence of galena. Galena is more common in late ferroan-dolomite veins than in early ferroan-dolomite veins, explaining why sph-6 has a larger depletion in Ag, Bi, Sb and Tl relative to sph-6, compared to the depletion in these elements between sph-3 and sph-1.

The chemically remobilised sphalerite sph-3 and sph-6 are depleted in Fe and enriched in Zn relative to their parent material. Differences in the Fe and Zn content of chemically remobilised sphalerite can be attributed to the rate at which Fe and Zn can be leached from surrounding sphalerite, where Zn is leached more readily than Fe [62]. This interpretation is consistent with previous research by Wagner and Cook [63], who showed that the recrystallisation of sphalerite by a low Fe/Zn fluid produced a secondary sphalerite that was enriched in Zn and depleted in Fe relative to the primary sphalerite.

Alternatively, a substantial amount of Fe was likely partitioned into the co-crystallising ferroan-dolomite that makes up the dominant mineralogical component of these veins. Evidence for this is given by the high Fe values associated with ferroan-dolomite in Figure 9, which would decrease the availability of Fe to the coeval chemically remobilised sphalerite. Moreover, competition with co-crystallising ferroan-dolomite would also explain the relative depletion of Mn in sph-6 compared to sph-4, which would also be preferentially incorporated into the ferroan-dolomite.

7. Conclusions

The identification of a complex paragenesis of six texturally distinct sphalerite types from the Hilton Zn-Pb deposit allows for evaluation of the effect that co-crystallising sulphides, recrystallisation and chemical remobilisation have on the trace element geochemistry of sphalerite.

- Sphalerite that has precipitated or recrystallised with neighbouring co-crystallising pyrite, pyrrhotite, galena and chalcopyrite is relatively depleted in the elements Fe, Co, In, Sn, Sb, Ag and Tl. This effect is due to the incorporation of these elements into the co-crystallising sulphide minerals.
- Sphalerite that is interpreted to have formed via chemical remobilisation is enriched in Zn, Ga, Ge and Sn, and depleted in Fe, Tl, Co, Bi, Ag, Sb and Mn relative to its parent material. The enrichment and depletion in trace elements from chemically remobilised sphalerite reflect their relative mobility, which allows them to be leached from the primary sphalerite into the secondary sphalerite. The relative depletion of Bi, Tl, Ag, and Sb in chemically remobilised sphalerite is associated with abundant co-crystallising galena, while the depletion in Fe and Mn may be attributed to co-crystallising ferroan-dolomite.

Overall, results from this study indicate that co-crystallising sulphides and/or chemical remobilisation greatly affect the trace element composition of sphalerite, in particular concentrating or depleting high-demand critical metals including Ge, Ga and In. We suggest that these factors are important to understanding the trace element geochemistry of sphalerite on a variety of scales and have potential implications for a range of Zn-Pb deposits worldwide.

Supplementary Materials: The following are available online at http://www.mdpi.com/2075-163X/10/9/797/s1, Table S1: EPMA vs LA-ICP-MS, Table S2: Sphalerite Geochemistry, Table S3: Element Correlation Matrix.

Author Contributions: B.C. conceived this contribution, performed analytical work, reduced the data and wrote the article. R.L. and W.H. assisted in writing, draft preparation and editing. Funding was acquired by R.L., who also assisted in study design. All authors have read and agreed to the published version of the manuscript.

Funding: This study was funded by Mount Isa Mines—a Glencore Company and forms part of the Mount Isa Research for Geology and Exploration (MIRGE) project.

Acknowledgments: I would like to thank the many geologists at the George Fisher Mine and the Mount Isa Exploration team for their warm hospitality during fieldwork and assistance during sampling. I would also like to thank Sarah Gilbert, Ben Wade and Aoife McFadden from Adelaide Microscopy for assistance with analytical procedures. Furthermore, I would also like to thank Karin Barovich, who greatly assisted in writing and drafting this manuscript. B.C. is supported by the Australian Government Research Training Program. MinEx CRC is thanked for supporting W.H. for involving this research.

Conflicts of Interest: The authors declare no conflict of interest. The funders had no role in the design of the study; in the collection, analyses, or interpretation of data; in the writing of the manuscript, or in the decision to publish the results.

References

1. Zhang, S.; Ding, Y.; Liu, B.; Chang, C. Supply and demand of some critical metals and present status of their recycling in WEEE. *Waste Manag.* **2017**, *65*, 113–127. [CrossRef] [PubMed]
2. Watari, T.; Nansai, K.; Nakajima, K. Review of critical metal dynamics to 2050 for 48 elements. *Resour. Conserv. Recycl.* **2020**, *155*, 104669. [CrossRef]
3. Cook, N.J.; Ciobanu, C.L. Mineral hosts for critical metals in hydrothermal ores. In Proceedings of the Mineral Resources in a Sustainable World, 13th SGA Biennial Meeting, Nancy, France, 24–27 August 2015; pp. 24–27.
4. Sinclair, W.D. Electronic metals (In, Ge and Ga): Present and future resources. *Acta Geol. Sin. Ed.* **2014**, *88*, 463–465. [CrossRef]

5. George, L.L.; Cook, N.J.; Ciobanu, C.L. Partitioning of trace elements in co-crystallized sphalerite–galena–chalcopyrite hydrothermal ores. *Ore Geol. Rev.* **2016**, *77*, 97–116. [CrossRef]
6. Bauer, M.E.; Seifert, T.; Burisch, M.; Krause, J.; Richter, N.; Gutzmer, J. Indium-bearing sulfides from the Hämmerlein skarn deposit, Erzgebirge, Germany: Evidence for late-stage diffusion of indium into sphalerite. *Miner. Depos.* **2019**, *54*, 175–192. [CrossRef]
7. Cook, N.J.; Ciobanu, C.L.; Pring, A.; Skinner, W.; Shimizu, M.; Danyushevsky, L.; Saini-Eidukat, B.; Melcher, F. Trace and minor elements in sphalerite: A LA-ICPMS study. *Geochim. Cosmochim. Acta* **2009**, *73*, 4761–4791. [CrossRef]
8. Frenzel, M.; Hirsch, T.; Gutzmer, J. Gallium, germanium, indium, and other trace and minor elements in sphalerite as a function of deposit type—A meta-analysis. *Ore Geol. Rev.* **2016**, *76*, 52–78. [CrossRef]
9. Ye, L.; Cook, N.J.; Ciobanu, C.L.; Yuping, L.; Qian, Z.; Tiegeng, L.; Wei, G.; Yulong, Y.; Danyushevskiy, L. Trace and minor elements in sphalerite from base metal deposits in South China: A LA-ICPMS study. *Ore Geol. Rev.* **2011**, *39*, 188–217. [CrossRef]
10. Cugerone, A.; Cenki-Tok, B.; Oliot, E.; Muñoz, M.; Barou, F.; Motto-Ros, V.; Le Goff, E. Redistribution of germanium during dynamic recrystallization of sphalerite. *Geology* **2019**, *48*, 236–241. [CrossRef]
11. Bauer, M.E.; Burisch, M.; Ostendorf, J.; Krause, J.; Frenzel, M.; Seifert, T.; Gutzmer, J. Trace element geochemistry of sphalerite in contrasting hydrothermal fluid systems of the Freiberg district, Germany: Insights from LA-ICP-MS analysis, near-infrared light microthermometry of sphalerite-hosted fluid inclusions, and sulfur isotope geochemi. *Miner. Depos.* **2019**, *54*, 237–262. [CrossRef]
12. Henjes-Kunst, E.; Raith, J.G.; Boyce, A.J. Micro-scale sulfur isotope and chemical variations in sphalerite from the Bleiberg Pb-Zn deposit, Eastern Alps, Austria. *Ore Geol. Rev.* **2017**, *90*, 52–62. [CrossRef]
13. Wei, C.; Ye, L.; Hu, Y.; Danyushevskiy, L.; Li, Z.; Huang, Z. Distribution and occurrence of Ge and related trace elements in sphalerite from the Lehong carbonate-hosted Zn-Pb deposit, northeastern Yunnan, China: Insights from SEM and LA-ICP-MS studies. *Ore Geol. Rev.* **2019**, *115*, 103175. [CrossRef]
14. Valenta, R. Deformation, fluid flow and mineralization in the Hilton area, Mt Isa, Australia. Ph.D. Thesis, Monash University, Melbourne, Australia, June 1988.
15. Wilson, A. Metal Distribution in the Hilton Mine, Mount Isa and its Genetic Implications. Honours Thesis, Monash University, Melbourne, Australia, 1992.
16. Valenta, R. Deformation of host rocks and stratiform mineralization in the Hilton Mine area, Mt Isa. *Aust. J. Earth Sci.* **1994**, *41*, 429–443. [CrossRef]
17. Valenta, R. Syntectonic discordant copper mineralization in the Hilton Mine, Mount Isa. *Econ. Geol.* **1994**, *89*, 1031–1052. [CrossRef]
18. Southgate, P.N. Carpentaria-Mt Isa Zinc Belt: Basement framework, chronostratigraphy and geodynamic evolution of Proterozoic successions. *Aust. J. Earth Sci.* **2000**, *47*, 337–340. [CrossRef]
19. Page, R.W.; Jackson, M.J.; Krassay, A.A. Constraining sequence stratigraphy in north Australian basins: SHRIMP U–Pb zircon geochronology between Mt Isa and McArthur River. *Aust. J. Earth Sci.* **2000**, *47*, 431–459. [CrossRef]
20. Southgate, P.N.; Neumann, N.L.; Gibson, G.M. Depositional systems in the Mt Isa Inlier from 1800 Ma to 1640 Ma: Implications for Zn–Pb–Ag mineralisation. *Aust. J. Earth Sci.* **2013**, *60*, 157–173. [CrossRef]
21. Domagala, J.; Southgate, P.N.; McConachie, B.A.; Pidgeon, B.A. Evolution of the Palaeoproterozoic Prize, Gun and lower Loretta Supersequences of the Surprise Creek Formation and Mt Isa Group. *Aust. J. Earth Sci.* **2000**, *47*, 485–507. [CrossRef]
22. Scott, D.L.; Rawlings, D.J.; Page, R.W.; Tarlowski, C.Z.; Idnurm, M.; Jackson, M.J.; Southgate, P.N. Basement framework and geodynamic evolution of the Palaeoproterozoic superbasins of north-central Australia: An integrated review of geochemical, geochronological and geophysical data. *Aust. J. Earth Sci.* **2000**, *47*, 341–380. [CrossRef]
23. Betts, P.G.; Giles, D.; Mark, G.; Lister, G.S.; Goleby, B.R.; Aillères, L. Synthesis of the Proterozoic evolution of the Mt Isa Inlier. *Aust. J. Earth Sci.* **2006**, *53*, 187–211. [CrossRef]
24. Neumann, N.L.; Southgate, P.N.; Gibson, G.M.; McIntyre, A. New SHRIMP geochronology for the Western Fold Belt of the Mt Isa Inlier: Developing a 1800–1650 Ma event framework. *Aust. J. Earth Sci.* **2006**, *53*, 1023–1039. [CrossRef]
25. Page, R.W.; Bell, T.H. Isotopic and structural responses of granite to successive deformation and metamorphism. *J. Geol.* **1986**, *94*, 365–379. [CrossRef]

26. Winsor, C.N. Intermittent folding and faulting in the Lake Moondarra area, Mount Isa, Queensland. *Aust. J. Earth Sci.* **1986**, *33*, 27–42. [CrossRef]
27. Connors, K.A.; Page, R.W. Relationships between magmatism, metamorphism and deformation in the western Mount Isa Inlier, Australia. *Precambrian Res.* **1995**, *71*, 131–153. [CrossRef]
28. Duncan, R.J.; Wilde, A.R.; Bassano, K.; Maas, R. Geochronological constraints on tourmaline formation in the Western Fold Belt of the Mount Isa Inlier, Australia: Evidence for large-scale metamorphism at 1.57 Ga? *Precambrian Res.* **2006**, *146*, 120–137. [CrossRef]
29. Hand, M.; Rubatto, D. The scale of the thermal problem in the Mt Isa Inlier. In *Geological Society of Australia Abstracts*; Geological Society of Australia: Hornsby, Australia, 2002; Volume 67, p. 173.
30. Chapman, L.H. Geology and Genesis of the George Fisher Zn-Pb-Ag deposit Mount Isa, Australia. Ph.D. Thesis, James Cook University, Townsville, Australia, 1999.
31. Bell, T.H.; Hickey, K.A. Multiple deformations with successive subvertical and subhorizontal axial planes in the Mount Isa region; their impact on geometric development and significance for mineralization and exploration. *Econ. Geol.* **1998**, *93*, 1369–1389. [CrossRef]
32. Perkins, W.G.; Bell, T.H. Stratiform replacement lead-zinc deposits; a comparison between Mount Isa, Hilton, and McArthur River. *Econ. Geol.* **1998**, *93*, 1190–1212. [CrossRef]
33. Danyushevsky, L.; Robinson, P.; Gilbert, S.; Norman, M.; Large, R.; McGoldrick, P.; Shelley, M. Routine quantitative multi-element analysis of sulphide minerals by laser ablation ICP-MS: Standard development and consideration of matrix effects. *Geochemistry Explor. Environ. Anal.* **2011**, *11*, 51–60. [CrossRef]
34. Norris, A.; Danyushevsky, L. *Towards Estimating the Complete Uncertainty Budget of Quantified Results Measured by LA-ICPMS*; Goldschmidt: Boston, MA, USA, 2018.
35. Paton, C.; Hellstrom, J.; Paul, B.; Woodhead, J.; Hergt, J. Iolite: Freeware for the visualisation and processing of mass spectrometric data. *J. Anal. At. Spectrom.* **2011**, *26*, 2508–2518. [CrossRef]
36. Woodhead, J.D.; Hellstrom, J.; Hergt, J.M.; Greig, A.; Maas, R. Isotopic and elemental imaging of geological materials by laser ablation inductively coupled plasma-mass spectrometry. *Geostand. Geoanalytical Res.* **2007**, *31*, 331–343. [CrossRef]
37. Chapman, L.H. Geology and mineralization styles of the George Fisher Zn-Pb-Ag deposit, Mount Isa, Australia. *Econ. Geol.* **2004**, *99*, 233–255. [CrossRef]
38. Rieger, P.; Magnall, J.M.; Gleeson, S.A.; Lilly, R.; Rocholl, A.; Kusebauch, C. Sulfur Isotope Constraints on the Conditions of Pyrite Formation in the Paleoproterozoic Urquhart Shale Formation and George Fisher Zn-Pb-Ag Deposit, Northern Australia. *Econ. Geol.* **2020**. [CrossRef]
39. Yuan, B.; Zhang, C.; Yu, H.; Yang, Y.; Zhao, Y.; Zhu, C.; Ding, Q.; Zhou, Y.; Yang, J.; Xu, Y. Element enrichment characteristics: Insights from element geochemistry of sphalerite in Daliangzi Pb–Zn deposit, Sichuan, Southwest China. *J. Geochem. Explor.* **2018**, *186*, 187–201. [CrossRef]
40. Belissont, R.; Boiron, M.-C.; Luais, B.; Cathelineau, M. LA-ICP-MS analyses of minor and trace elements and bulk Ge isotopes in zoned Ge-rich sphalerites from the Noailhac–Saint-Salvy deposit (France): Insights into incorporation mechanisms and ore deposition processes. *Geochim. Cosmochim. Acta* **2014**, *126*, 518–540. [CrossRef]
41. Cooke, D.R.; Bull, S.W.; Large, R.R.; McGoldrick, P.J. The importance of oxidized brines for the formation of Australian Proterozoic stratiform sediment-hosted Pb-Zn (Sedex) deposits. *Econ. Geol.* **2000**, *95*, 1–18. [CrossRef]
42. Murphy, T.E. Structural and Stratigraphic Controls on Mineralization at the George Fisher Zn-Pb-Ag Deposit, Northwest Queensland, Australia. Ph.D. Thesis, James Cook University, Townsville, Australia, 2004.
43. Lockington, J.A.; Cook, N.J.; Ciobanu, C.L. Trace and minor elements in sphalerite from metamorphosed sulphide deposits. *Mineral. Petrol.* **2014**, *108*, 873–890. [CrossRef]
44. Campbell, F.A.; Ethier, V.G. Nickel and cobalt in pyrrhotite and pyrite from the Faro and Sullivan orebodies. *Can. Mineral.* **1984**, *22*, 503–506.
45. Witt, W.K.; Hagemann, S.G.; Roberts, M.; Davies, A. Cobalt enrichment at the Juomasuo and Hangaslampi polymetallic deposits, Kuusamo Schist Belt, Finland: A role for an orogenic gold fluid? *Miner. Depos.* **2019**, *55*, 1–8. [CrossRef]
46. Frenzel, M.; Bachmann, K.; Carvalho, J.R.S.; Relvas, J.M.R.S.; Pacheco, N.; Gutzmer, J. The geometallurgical assessment of by-products—Geochemical proxies for the complex mineralogical deportment of indium at Neves-Corvo, Portugal. *Miner. Depos.* **2018**, *54*, 1–24. [CrossRef]

47. George, L.L.; Biagioni, C.; Lepore, G.O.; Lacalamita, M.; Agrosì, G.; Capitani, G.C.; Bonaccorsi, E.; d'Acapito, F. The speciation of thallium in (Tl, Sb, As)-rich pyrite. *Ore Geol. Rev.* **2019**, *107*, 364–380. [CrossRef]
48. Grant, H.L.J.; Hannington, M.D.; Petersen, S.; Frische, M.; Fuchs, S.H. Constraints on the behavior of trace elements in the actively-forming TAG deposit, Mid-Atlantic Ridge, based on LA-ICP-MS analyses of pyrite. *Chem. Geol.* **2018**, *498*, 45–71. [CrossRef]
49. George, L.; Cook, N.J.; Ciobanu, C.L.; Wade, B.P. Trace and minor elements in galena: A reconnaissance LA-ICP-MS study. *Am. Mineral.* **2015**, *100*, 548–569. [CrossRef]
50. Cave, B.; Lilly, R.; Barovich, K. Textural and geochemical analysis of chalcopyrite, galena and sphalerite across the Mount Isa Cu to Pb-Zn transition: Implications for a zoned Cu-Pb-Zn system. *Ore Geol. Rev.* **2020**, *124*, 103647. [CrossRef]
51. Höll, R.; Kling, M.; Schroll, E. Metallogenesis of germanium—A review. *Ore Geol. Rev.* **2007**, *30*, 145–180. [CrossRef]
52. Burton, J.D.; Culkin, F.; Riley, J.P. The abundances of gallium and germanium in terrestrial materials. *Geochim. Cosmochim. Acta* **1959**, *16*, 151–180. [CrossRef]
53. Bernstein, L.R. Germanium geochemistry and mineralogy. *Geochim. Cosmochim. Acta* **1985**, *49*, 2409–2422. [CrossRef]
54. Jonsson, E.; Högdahl, K. On the occurrence of gallium and germanium in the Bergslagen ore province, Sweden. *GFF* **2019**, *141*, 48–53. [CrossRef]
55. Goldsmith, J.R. Gallium and germanium substitutions in synthetic feldspars. *J. Geol.* **1950**, *58*, 518–536. [CrossRef]
56. Schwartz, M.O. Cadmium in zinc deposits: Economic geology of a polluting element. *Int. Geol. Rev.* **2000**, *42*, 445–469. [CrossRef]
57. Wen, H.; Zhu, C.; Zhang, Y.; Cloquet, C.; Fan, H.; Fu, S. Zn/Cd ratios and cadmium isotope evidence for the classification of lead-zinc deposits. *Sci. Rep.* **2016**, *6*, 1–8. [CrossRef]
58. Cugerone, A.; Cenki-Tok, B.; Chauvet, A.; Le Goff, E.; Bailly, L.; Alard, O.; Allard, M. Relationships between the occurrence of accessory Ge-minerals and sphalerite in Variscan Pb-Zn deposits of the Bossost anticlinorium, French Pyrenean Axial Zone: Chemistry, microstructures and ore-deposit setting. *Ore Geol. Rev.* **2018**, *95*, 1–19. [CrossRef]
59. Cugerone, A.; Cenki-Tok, B.; Muñoz, M.; Kouzmanov, K.; Oliot, E.; Motto-Ros, V.; Le Goff, E. Behavior of critical metals in metamorphosed Pb-Zn ore deposits: Example from the Pyrenean Axial Zone. *Miner. Depos.* **2020**, 1–21. [CrossRef]
60. Lee, J.H.; Yoo, B.C.; Yang, Y.-S.; Lee, T.H.; Seo, J.H. Sphalerite Geochemistry of the Zn-Pb Orebodies in the Taebaeksan Metallogenic Province, Korea. *Ore Geol. Rev.* **2019**, *107*, 1046–1067. [CrossRef]
61. Wang, Y.; Han, X.; Petersen, S.; Frische, M.; Qiu, Z.; Cai, Y.; Zhou, P. Trace Metal Distribution in Sulfide Minerals from Ultramafic-Hosted Hydrothermal Systems: Examples from the Kairei Vent Field, Central Indian Ridge. *Minerals* **2018**, *8*, 526. [CrossRef]
62. Stanton, M.R.; Gemery-Hill, P.A.; Shanks, W.C., III; Taylor, C.D. Rates of zinc and trace metal release from dissolving sphalerite at pH 2.0–4.0. *Appl. Geochem.* **2008**, *23*, 136–147. [CrossRef]
63. Wagner, T.; Cook, N.J. Sphalerite remobilization during multistage hydrothermal mineralization events—Examples from siderite-Pb-Zn-Cu-Sb veins, Rheinisches Schiefergebirge, Germany. *Mineral. Petrol.* **1998**, *63*, 223–241. [CrossRef]

Article

Metal-Selective Processing from the Los Sulfatos Porphyry-Type Deposit in Chile: Co, Au, and Re Recovery Workflows Based on Advanced Geochemical Characterization

Germán Velásquez [1,*], Humberto Estay [1], Iván Vela [2], Stefano Salvi [3] and Marcial Pablo [2]

1 Advanced Mining Technology Center (AMTC), FCFM, Universidad de Chile, Santiago 8370451, Chile; humberto.estay@amtc.cl

2 Superintendence of Geology, Los Bronces Underground Project, Anglo American Sur S.A., Santiago 7550103, Chile; ivan.vela@angloamerican.com (I.V.); marcial.pablof@angloamerican.com (M.P.)

3 Géosciences Environnement Toulouse (GET), Université de Toulouse, CNRS, GET, IRD, OMP, 14 Av. Edouard Belin, 31400 Toulouse, France; stefano.salvi@get.omp.eu

* Correspondence: gevelasqueza@gmail.com

Received: 20 May 2020; Accepted: 9 June 2020; Published: 11 June 2020

Abstract: Sulfides extracted from porphyry-type deposits can contain a number of metals critical for the global energy transition, e.g., Co and precious metals such as Au and Re. These metals are currently determined on composite mineral samples, which commonly results in their dilution. Thus, it is possible that some metals of interest are overlooked during metallurgical processing and are subsequently lost to tailings. Here, an advanced geochemical characterization is implemented directly on metal-bearing sulfides, determining the grade of each targeted trace metal and recognizing its specific host mineral. Results show that pyrite is a prime host mineral for Co (up to 24,000 ppm) and commonly contains Au (up to 5 ppm), while molybdenite contains high grades of Re (up to 514 ppm) and Au (up to 31 ppm). Both minerals represent around 0.2% of the mineralized samples. The dataset is used to evaluate the possibility of extracting trace metals as by-products during Cu-sulfide processing, by the addition of unit operations to conventional plant designs. A remarkable advantage of the proposed workflows is that costs of mining, crushing, and grinding stages are accounted for in the copper production investments. The proposed geochemical characterization can be applied to other porphyry-type operations to improve the metallic benefits from a single deposit.

Keywords: cobalt supply; rhenium; gold; by-products; pyrite processing; geo-metallurgy; porphyry-type mining; green mining

1. Introduction

Nowadays, metal resources are more crucial than ever to current global energy transition efforts [1–4], especially metals needed for the development of clean energy technologies [5,6]. Demand for these metals will increase in the coming years, with copper (Cu) being in greatest demand worldwide, estimated up to 40 times greater by 2100 [7]. This will be accompanied by other metals such as cobalt (Co), silver (Ag), tellurium (Te), rare earth elements (REEs), all of which are considered critical because of the risk they pose to supply [5,8]. Constraints on Cu supply focus on the expected decline in ore grades [9], which implies that significantly more mine material will have to be mined and processed to produce the same amount of metal. Thus, a significant challenge to the copper industry would be to move from traditional Cu (± Mo ± Au) mining to a highly efficient multi-metallic operation [10]. This would allow the exploitation of all of the metal resources contained in a single mineral deposit, while at the same time minimizing the amount of waste generated [11].

Critical and precious metals such as Co, Au, and Re, not to mention Ag, are systematically reported as by-products in porphyry-Cu-Mo deposits worldwide, with metal grades averaging at around 0.5 g/t [10,12–14]. However, an issue that must be addressed in mineral characterization protocols is that the minerals that host each specific metal of interest are not recognized, which is a crucial piece of information for metallurgical ore processing [15,16]. This issue can be explained by the following considerations: (i) mineral characterization is focused on copper ore, i.e., Cu oxides and sulfides, and does not include all metal-bearing minerals [10]; (ii) geochemical analyses are carried out by taking a sample of the whole mineralized rock, i.e., a sample containing ores and barren minerals, resulting in the dilution of some trace metals to concentrations that are below detection limits [10]; and (iii) mineral paragenesis is predicted from whole-rock geochemical data [17], instead of being determined by conventional and more-sophisticated microscopic methods.

To evaluate the economic metal content of a porphyry-type deposit, Velásquez et al. [10] proposed a high-resolution mineral and geochemical characterization, which should allow efficient determination of the grade of targeted metals and recognition of specific host minerals. The data obtained include both the metal concentration (grade) and the mode of occurrence for each metal (e.g., visible, invisible, in the structure, etc.) [10], which is valuable information for selective metal treatment.

In this paper, we present the results of a high-resolution geochemical characterization performed in situ on sulfides (i.e., bornite, chalcopyrite, molybdenite, and pyrite) from Cu-mineralized samples belonging to a porphyry-type deposit. Geochemical characterization is intended to determine and quantify the Co, Au, and Re, as well as Ag, Te, and As content of sulfides, rather than measuring trace elements in the whole rock. The resulting dataset is used to: (1) determine the preferred host sulfide for each selected trace metal; (2) highlight the difference between advanced and operational geochemical characterization; and (3) discuss how high-resolution characterization can support more appropriate Cu ore processing plant design, focused on the supply of critical and precious metals, in addition to the major compounds, i.e., Cu and Mo. The proposed metallurgical workflows can be implemented in mining operations to move from traditional Cu (± Mo) mining to high-performance multi-metallic activity in porphyry-type mines.

As a case study, we have chosen the Los Sulfatos deposit [18,19], located in the Chilean Central Andes. Managed by Anglo American Sur, S.A., Los Sulfatos is a world-class deposit (>45 Mt of contained copper) [10]. It is part of the Los Bronces–Rio Blanco–Los Sulfatos porphyry Cu-Mo system [18–20] (Figure 1), which is considered having the highest copper endowment in the Earth's crust (>200 Mt of contained copper) [18–20]. The main ore minerals found in the deposit are Cu ± Mo- sulfides, predominantly chalcopyrite, bornite, and some molybdenite [18,19]. At present, the deposit is in its preliminary development stages as the Los Bronces underground mining project. The deposit contains more than 3.9 billion tons of mine-material enriched in Cu-Mo-sulfides and grading 1.14% Cu [10], which will be extracted throughout the duration of the mine's life. The final marketing product expected to be generated during the metallurgical processing is a sulfide (bulk and selective) flotation concentrate, while the rest of the extracted mine-material will be removed to tailings [10]. The fact that Anglo American Sur S.A. is evaluating the conditions for mining operation development in the coming years for the Los Sulfatos deposit, offers a rare opportunity to propose a multi-metal metallurgical processing in a porphyry Cu-Mo mining operation by implementing the high-resolution characterization proposed by Velásquez et al. [10]. The case study is focused on porphyry-types deposits due to the increased need for copper in the coming years, which implies that the mineralized material will be extracted and processed to supply this metal and, therefore, the goal for mining geologists is to improve the metallic benefits drawn from each deposit.

In this study, we will focus on determining Co, Re, and Au occurrences in the sulfides. The specific choice of these commodities arises from the following considerations: (i) Cobalt is a key metal in the production of rechargeable lithium (Li)-ion batteries [21,22]; given that Chile is one of the world's leading suppliers of Li [23,24], the possibility of producing both Co and Li would be highly strategic for the country, as it would become a leader in the high-tech clean energy industry. (ii) Rhenium is contained in molybdenite (up to 4.7 wt.% Re), a common sulfide in porphyry-type deposits [8,13,25]. However, Re concentrations could not be assessed during operational geochemical characterization, because this procedure is performed on composite samples by whole-rock analysis. This implies that Re would not be included in the mining planning being recovered later from refineries. (iii) Gold is currently determined by bulk assays [12], but practically no mineral characterization is performed on its host mineral. As a result, the metallic Au contents would not be exploited in the metallurgical treatment and would instead be lost to tailings [10]. This is especially relevant in the case of invisible occurrences of Au [26], in which Au can occur as nanoparticles of sulfosalts or in the structure lattice.

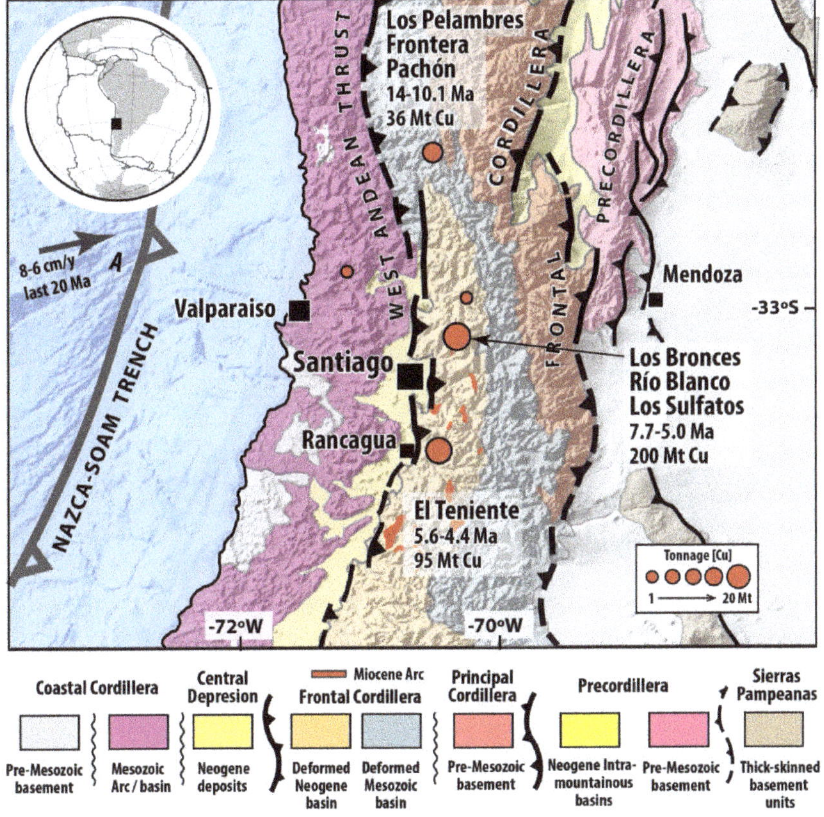

Figure 1. Physiography and first-order tectonic setting of the Central Andes showing the geological framework of the Río Blanco–Los Bronces–Los Sulfatos Cu-Mo porphyry system (modified from reference [27]).

2. Samples and Analytical Protocol

For this study, we sampled ten drill cores of mineralized material from the Los Sulfatos deposit at depths from 100 to 800 m, which are located in the expected exploitation area to be developed as the Los Bronces underground project. From these drill cores, fifteen 5 m-mineralized intervals were selected and two hand-picked samples were collected from each interval (Figure 2). It is important to note that each 5 m interval corresponds to a composite-sample analyzed by whole-rock chemical analysis, which is a routinely analytical procedure in the mining industry (Figure 2a). The goal of that correspondence is to compare the whole-rock geochemical data with the obtained results from an advanced micro-chemical characterization (Figure 2b). Thus, thirty representative samples were collected from four mineralized zones, *Min-Zones*, defined at the Los Sulfatos deposit [10], i.e., high-chalcopyrite (CPY-H), low-chalcopyrite (CPY-L), high-bornite (BRN-H), and low-bornite (BRN-L). All of these are developed in hydrothermal and magmatic breccias. Of these samples, fifteen were found to be suitable for high-resolution investigation of mineral and geochemical characterization. About one hundred sulfide crystals, including chalcopyrite, bornite, molybdenite, and pyrite (Figure 2b), were analyzed to determine their trace-metal content, focusing on critical, precious, and deleterious metals, i.e., Co, Au, Re, plus Ag, Te, and As. Mineralogical and micro-chemical studies were carried out at the Géosciences Environnement Toulouse (GET) laboratory in Toulouse, France by the application of conventional practices used for mineral protocols. These practices include petrographic and scanning electron microscope (SEM) descriptions, which were combined with more sophisticated techniques applied for micro-chemical mineral characterization [10], such as the laser ablation-inductively coupled plasma-mass spectrometry [28]. The analytical protocol is summarized below.

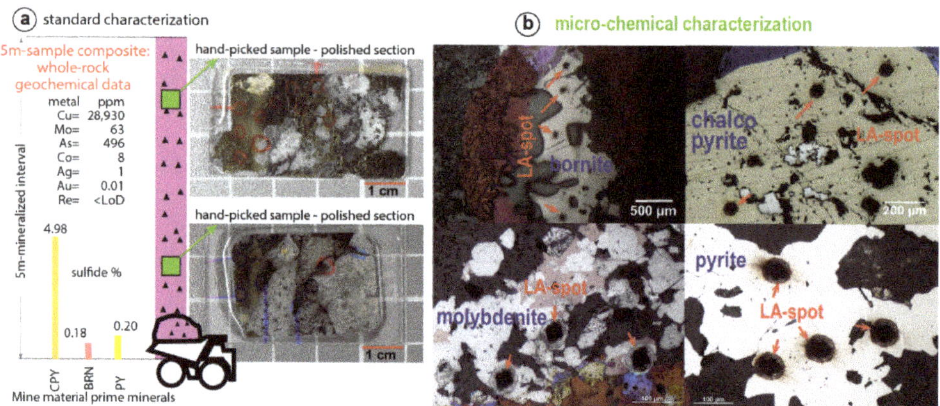

Figure 2. (**a**) Schematic diagram illustrating a selected 5 m-mineralized interval, highlighting (green filled squares) the locations of collected samples. Metal concentrations obtained from bulk analyses are indicated on the left side. Photographs of the 200 µm-thick polished sections corresponding to the hand-picked samples are shown on the right. (**b**) Photomicrographs under reflected light showing the different sulfides types analyzed by LA-ICP-MS, highlighting the locations of laser-ablation spots (LA-spot). Photomicrographs were taken after laser ablation analysis.

- Logging of a 5 m-mineralized interval of each selected drill-core, from which a core-sample was collected and prepared to design a 200 µm-thick polished section (Figure 2a);
- Identification of sulfide minerals by petrographic studies [10,29] (Figure 2a), using a polarizing microscope Nikon Eclipse LV100POL (Nikon instruments Europe, Amsterdam, The Netherlands) in reflected and transmitted light, equipped with 2×, 5×, 10×, 20× and 50× objectives;
- Characterization of metal-bearing mineral inclusions in sulfides [10,29], using a JEOL6360LV scanning electron microscopy (JEOL Ltd., Tokyo, Japan) coupled to an energy-dispersive X-ray spectrometer (SDD Bruker 129 eV) (Bruker Corporation, Billerica, MA, USA), itself equipped to acquire images in backscattered electron (BSE) mode at an acceleration voltage of 20 kV;
- Determination of trace metal concentrations in sulfides by in situ micro-chemical analyses, including electron micro-probe (EPM) and laser ablation (LA-ICP-MS) analyses [10,29], using a CAMECA SX5 microprobe (CAMECA SX5, Cameca, Gennevilliers, France), combined with a Ti: sapphire femtosecond (fs) laser couple to a quadrupole ICP-MS. To provide a sense of reproducibility of the LA-ICP-MS analysis, several spots were ablated on each single sulfide crystal, avoiding the consideration of outstanding concentration values (Figure 2b). Detection limits for trace element quantifications were calculated as three times the background standard deviation

value. At GET, the tuning routine is optimized to reduce production of molecular oxide species (typically $^{232}Th^{16}O/^{232}Th < 1\%$) and doubly charged ion species (typically $^{140}Ce{+}{+}/^{140}Ce{+} < 2\%$). This allows an analytical precision of <15% for the relative standard deviation (RSD) to be reached [30,31]. The gas blank is measured for 30 s before switching on the laser for 60 s. Raw data were processed online using the GLITTER software package (4.0, ARC National Key Centre for Geochemical Evolution and Metallogeny of Continents, Macquarie University, Sydney, Australia) (e.g., [32]), using certified reference materials, such as pyrrhotite-Po-726 [33], an in-house natural chalcopyrite, Cpy-RM [30], and NIST SRM 610 [34], as external calibrators, in bracketing mode standard-sample-standard. The following isotopes were monitored: ^{33}S, ^{34}S ^{56}Fe, ^{57}Fe, ^{59}Co, ^{63}Cu, ^{75}As, ^{95}Mo, ^{107}Ag, ^{125}Te, ^{182}W, ^{185}Re, ^{197}Au. These isotopes were selected to avoid possible argide interferences ($MeAr^{+}$), where Me corresponds to a base metal such as ^{59}Co and ^{63}Cu, in our study case. Analytical routines and conditions are described in detail in references [10,29–31].

3. High-Resolution Geochemical Characterization of Sulfides

In this study, we focus on determining the presence of critical and precious metals in sulfides, i.e., chalcopyrite, bornite, molybdenite, and pyrite from a porphyry-type deposit, taking Cu-mineralized material from drill-holes as study samples. Around one-hundred sulfide crystals were selected from the four *Min-Zones* defined at the Los Sulfatos deposit [10] and were subjected to in-depth investigations. Detailed petrographic and textural observations performed on sulfide crystals evidence two different sulfide metallogenic units in the mineralized blocks: (1) an association consisting of bornite–chalcopyrite ± molybdenite (BRN–CPY ± MOL; Figure 3a); and (2) an association formed by chalcopyrite ± pyrite (CPY ± PY; Figure 3b). The two metallogenic units are spatiality related with the *Min-Zones*, recognized in the geological 3D model for the Los Sulfatos deposit.

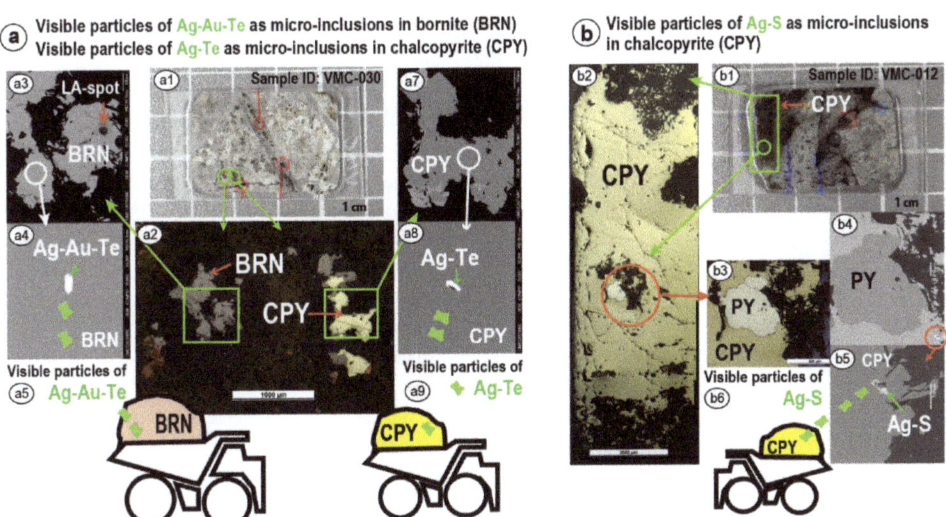

Figure 3. (**a**) Microphotograph under reflected light (a2) and BSE images (a3,a4,a7,a8) deciphering the occurrence of visible particles of Ag-Au-Te and Ag-Te, found as micron-sized inclusions in bornite and chalcopyrite, respectively. The expected final destination of these particles is also shown (a5,a9). (**b**) Microphotographs under reflected light (b2,b3) and BSE images (b4,b5) showing the occurrence of Ag-S visible particles, found as micron-sized inclusions in chalcopyrite. The estimated final destination for these metallic particles is also the flotation concentrates (b6). CPY: chalcopyrite, BRN: bornite, and PY: pyrite.

3.1. Visible Metal Content in the Sulfide Units

Visible metals correspond to the micron-sized (up to 1 μm) particles that can be determined by optical microscope or SEM. They occur as metal-bearing inclusions in the sulfides [10,29]. In the BRN–CPY ± MOL unit, recognized metallic particles include sulfosalts of Ag-Te (e.g., hessite: Ag_2Te) and Ag-Au-Te (e.g., sylvanite: (Ag, Au)Te_2). Ag-Te particles occur as inclusions in chalcopyrite, while Ag-Au-Te particles are preferentially included in bornite (Figure 3a). Studied molybdenite crystals were found to be free of visible metallic inclusions.

In the CPY ± PY unit, sulfosalts of Ag-S (e.g., acanthite: Ag_2S) were found as micron-sized inclusions in chalcopyrite (Figure 3b), whereas pyrite, which crystallizes intergrown with chalcopyrite, is almost free of visible metallic inclusions.

The association between the sulfosalt and their corresponding hosting sulfide implies that these metallic micron-sized particles (Figure 3) will accumulate in the sulfide flotation concentrates during the metallurgical processing circuits.

3.2. Invisible Trace-Metal Content in the Sulfide Units

This corresponds to the metal content which cannot be determined by SEM and EPM analyses, because the concentration is below the detection limits of these techniques. To evaluate the presence of selected trace metals, i.e., Co, Au, Re, plus Ag, Te, W, and As in the ore sulfides, 161 LA-ICP-MS analyses were performed. Of these, 54 analytical spots were ablated on sulfides from the CPY ± PY unit, and 107 spots on sulfides from the BRN–CPY ± MOL unit. The analytical routine included determination of As, because it is a deleterious metal during copper ore processing [35], and its occurrence can influence the price of sulfide flotation concentrates. Chalcopyrite, bornite, and pyrite were analyzed to determine Co, Au, Ag, Te, and As, while molybdenite was analyzed for Re, and Au, (plus W), because they represent the prime trace metals that could be concentrated in molybdenite crystals [36]. Invisible metal content can occur as mineral micro- and nano-particles or be included in the lattice structure [37].

3.2.1. Bornite–Chalcopyrite ± Molybdenite Unit

The results of LA-ICP-MS analyses (Table 1) show that Ag is preferentially contained in bornite, with an average value of 103.3 ppm (up to 316 ppm Ag) and that it is correlated with the highest concentrations of Te (up to 129 ppm), and Au (up to 1 ppm). In chalcopyrite crystals, Ag and Te contents reach up to 22 and 14 ppm, respectively, while Au concentrations are below the detection limit of the LA-ICP-MS (DL = 0.05 ppm). Chemical results are in agreement with the mineralogical results, i.e., visible micron-sized particles of Ag-Au-Te are commonly contained in bornite, and visible particles of Ag-Te are mostly found in chalcopyrite. The average concentration of selected trace metals in molybdenite ($\boldsymbol{n}$ = 5 spots) is: Re = 349 ppm (up to 514 ppm), Au = 7 ppm (up to 31 ppm), and W = 12.6 ppm (up to 31 ppm).

Time-resolved LA-ICP-MS profiles were evaluated to determine the mode of occurrence of trace Ag, Te, and Au in bornite and chalcopyrite, and to compare them with mineralogical observations. In the bornite analyses, the spectra for Ag, Au, and Te mimic each other (Figure 4a), and their signals show a spike above background. This indicates that these elements are likely to occur as micro- and nano-particles of Au-Ag-Te [10,29,37,38], in addition to the visible micron-sized particles observed as mineral inclusions. In the chalcopyrite analyses, Ag and Te show correlated signals (Figure 4b), while the Au spectra is in the background, implying the occurrence of Ag-Te micro- and nano-particles [10,29,37,38]. However, some LA-ICP-MS spectra for bornite (Figure 4c), as well as chalcopyrite, display Ag signals higher than those displayed for the other metals (e.g., Te and Au) by at least two orders of magnitude, showing that at least a part of the invisible Ag is contained also in the lattice of sulfides.

Table 1. Concentrations for Co, As, Ag, Te, and Au (in ppm) determined by in situ LA-ICP-MS analysis on selected bornite and chalcopyrite crystals.

Sample	Sulfide	Spot	Co (ppm) *DL* = 0.1	As (ppm) *DL* = 0.1	Ag (ppm) *DL* = 0.05	Te (ppm) *DL* = 0.1	Au (ppm) *DL* = 0.02
CMP-005	BRN	*n* = 19	**1.3** ± *1.4* (*<DL–6.2*)	**1.2** ± *1.0* (*1.0–4.5*)	**61.5** ± *7.6* (*60.2–80.1*)	**8.9** ± *4.8* (*12.2–16.6*)	**0.5** ± *0.3* (*<DL–1.0*)
CMP-006	BRN	*n* = 8	**0.3** ± *0.2* (*<DL–0.6*)	**1.1** ± *0.2* (*1.0–1.3*)	**50.8** ± *15.9* (*64.0–76.3*)	**5.5** ± *3.5* (*5.2–11.1*)	**0.6** ± *0.2* (*<DL–0.6*)
CMP-007	BRN	*n* = 12	**0.2** ± *0.1* (*<DL–0.4*)	**1.3** ± *0.5* (*0.8–2.5*)	**104.5** ± *14.1* (*72.5–124.4*)	**7.1** ± *6.1* (*9.0–24.6*)	**0.3** ± *0.1* (*<DL–0.3*)
CMP-007	BRN	*n* = 8	**1.1** ± *0.6* (*<DL–1.3*)	**6.1** ± *1.9* (*4.9–9.1*)	**113.8** ± *34.9* (*142.8–152.6*)	**68.9** ± *34.7* (*29.0–129.0*)	*<DL*
CMP-027	BRN	*n* = 9	**3.7** ± *2.1* (*<DL–6.2*)	**7.0** ± *4.5* (*<DL–11.7*)	**116.9** ± *60.6* (*74.1–221.1*)	**68.9** ± *39.8* (*10.0–111.0*)	*<DL*
CMP-030	BRN	*n* = 5	**0.6** ± *0.3* (*<DL–0.9*)	**1.3** ± *0.4* (*0.6–1.6*)	**307.1** ± *12.0* (*292.4–316.1*)	**34.7** ± *12.8* (*19.3–54.9*)	**0.4** ± *0.2* (*<DL–0.5*)
Bornite		***n* = 61**	**0.9**	**2.4**	**103.3**	**18.4**	**0.4**
CMP-001	CPY	*n* = 11	**2.4** ± *0.9* (*0.8–4.0*)	**1.0** ± *0.9* (*0.5–3.4*)	**7.3** ± *5.7* (*2.9–21.0*)	**3.3** ± *3.0* (*1.8–13.7*)	*<DL*
CMP-002	CPY	*n* = 4	**5.4** ± *4.6* (*0.2–11.4*)	**10.3** ± *9.3* (*1.1–23.3*)	**0.9** ± *0.7* (*0.1–1.8*)	**2.2** ± *1.5* (*0.2–3.8*)	*<DL*
CMP-003	CPY	*n* = 4	**1.0** ± *0.4* (*0.6–1.5*)	**11.3** ± *10.0* (*2.5–21.7*)	**1.7** ± *0.3* (*1.4–2.2*)	**1.2** ± *0.4* (*0.7–1.5*)	*<DL*
CMP-007	CPY	*n* = 7	**7.9** ± *5.5* (*1.8–16.9*)	**2.4** ± *1.2* (*1.4–4.4*)	**4.5** ± *4.1* (*0.2–11.7*)	**1.6** ± *1.2* (*0.3–2.5*)	*<DL*
CMP-024	CPY	*n* = 5	**2.6** ± *1.4* (*2.0–4.0*)	**4.3** ± *1.5* (*1.8–5.3*)	**1.9** ± *1.6* (*0.6–4.7*)	**2.9** ± *2.7* (*0.8–7.4*)	*<DL*
CMP-027	CPY	*n* = 5	**13.4** ± *6.7* (*8.9–15.9*)	**5.7** ± *3.3* (*4.1–6.8*)	**1.2** ± *0.9* (*0.6–2.5*)	**2.0** ± *1.3* (*1.0–3.0*)	*<DL*
CMP-030	CPY	*n* = 5	**7.5** ± *6.1* (*2.7–18.1*)	**1.1** ± *0.8* (*0.5–2.1*)	**1.7** ± *1.1* (*0.3–2.9*)	**0.8** ± *0.7* (*0.1–1.6*)	*<DL*
Chalcopyrite		***n* = 41**	**5.1**	**3.9**	**3.2**	**2.2**	*<DL*

Average values (in bold) ± standard deviation and value range (in brackets) for selected trace-element concentrations. *n*: number of spot analyses. BRN = bornite; CPY = chalcopyrite. *DL*: limit of detection. CMP: critical metal project.

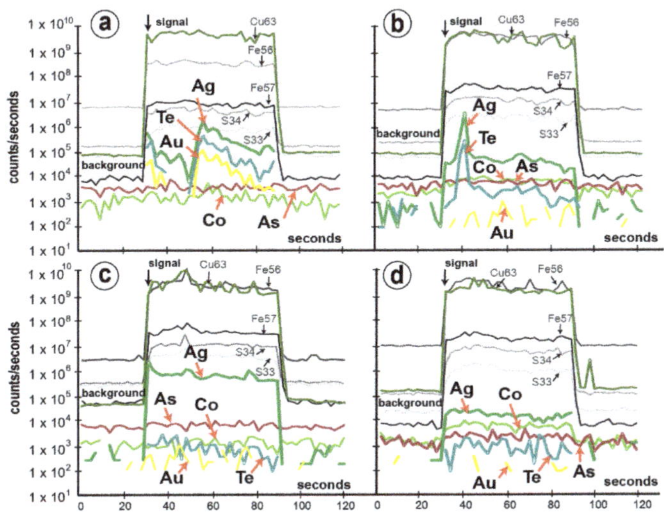

Figure 4. LA-ICP-MS patterns (in counts per second) for selected trace-element signals depicting a typical pattern for (**a**) invisible particles of Ag-Au-Te in bornite; (**b**) invisible particles of Ag-Te in chalcopyrite; (**c**,**d**) invisible silver in the mineral's lattice of bornite and chalcopyrite, respectively.

3.2.2. Chalcopyrite ± Pyrite Unit

Micro-chemical data (Table 2) show that the highest Ag values (up to 42.6 ppm) are contained in chalcopyrite. This is depleted in Te (up to 8.9 ppm) as well as Au (up to 0.07 ppm), in comparison with Cu-sulfides from the unit described previously. LA-ICP-MS profiles for the chalcopyrite analyses (Figure 4d) show a flat pattern for Ag signal well above that of Te, while the Au signal is in the background, which suggests that trace Ag is mostly contained in the mineral's lattice. Pyrite is the strategic Co-host mineral, with an average grade value of 2236 ppm and ranging of up to 24,000 ppm (Table 2). Interestingly, pyrite is also enriched in metallic Au (up to 5 ppm); however, the latter is accompanied by As with an average value of 2801 ppm (up to 6090 ppm) (Table 2). In the last case, Au and As metal content can be hosted in this sulfide's lattice [10].

Table 2. Concentrations for Co, As, Ag, Te, and Au (in ppm) determined by in situ LA-ICP-MS analysis on selected chalcopyrite and pyrite crystals.

Sample	Sulfide	Spot	Co (ppm) *DL = 0.1*	As (ppm) *DL = 0.1*	Ag (ppm) *DL = 0.05*	Te (ppm) *DL = 0.1*	Au (ppm) *DL = 0.02*
CMP-004	CPY	*n* = 11	**6.4** ± *3.7* (*<DL–10.1*)	**14.8** ± *8.3* (*3.6–24.5*)	**1.0** ± *0.4* (*0.4–1.4*)	**1.1** ± *0.9* (*0.2–2.9*)	**0.07** ± *0.02* (*<DL–0.07*)
CMP-009	CPY	*n* = 5	**12.3** ± *12.0* (*0.2–32.2*)	**3.3** ± *3.0* (*1.6–8.6*)	**12.3** ± *17.0* (*2.3–42.6*)	**3.3** ± *2.4* (*1.3–7.5*)	*<DL*
CMP-011	CPY	*n* = 7	**16.0** ± *3.9* (*10.9–22.4*)	**6.0** ± *11.0* (*1.4–31.7*)	**2.2** ± *0.3* (*1.9–2.8*)	**0.1** ± *0.2* (*<DL–0.2*)	**0.07**± *0.02* (*<DL–0.07*)
CMP-012	CPY	*n* = 10	**0.5** ± *0.3* (*<DL–0.8*)	**5.2** ± *3.1* (*<DL–8.3*)	**20.8** ± *4.2* (*15.3–27.0*)	**4.2** ± *2.2* (*1.0–8.9*)	*<DL*
CMP-013	CPY	*n* = 5	**14.1** ± *8.0* (*3.9–26.1*)	**0.3** ± *0.1* (*0.2–0.5*)	**2.5** ± *1.9* (*1.2–5.8*)	**0.5** ± *0.3* (*<DL–0.7*)	*<DL*
Chalcopyrite		***n* = 38**	**8.7**	**7.3**	**8.3**	**2.0**	**0.07**
CMP-012	PY	*n* = 6	**7700** ± *6100* (*172–24,050*)	**125.4** ± *95.6* (*1.7–14.9*)	*<DL*	**1.4** ± *0.6* (*<DL–1.8*)	*<DL*
CMP-015	PY	*n* = 12	**354** ± *240* (*142–747*)	**3702** ± *2794* (*0.7–**6090***)	**2.2** ± *1.3* (*<DL–4.0*)	**0.8** ± *0.4* (*<DL–1.6*)	**0.6** ± *0.4* (*<0.2–**5.0***)
Pyrite		***n* = 16**	**2236**	**2801**	**2.0**	**0.8**	**0.6**

Average values (in bold) ± standard deviation and value range (in brackets) for selected trace-element concentrations. ***n***: number of spot analyses. CPY = chalcopyrite; PY = pyrite. *DL*: limit of detection. CMP: critical metal project.

4. Innovation on Metal-Selective Metallurgical Processing

Based on advanced geochemical characterizations of sulfides, two geo-metallurgical units (GMU; Figures 5 and 6) are proposed. These units are focused on Cu + by-products processing: (1) a chalcopyrite–bornite ± molybdenite unit (GMU1; Figure 5), and (2) a chalcopyrite ± pyrite unit (GMU2; Figure 6); both of which correlate with the *Min-Zones* defined at the deposit. However, in the current standard characterization (Figures 5a and 6a) of copper *Min-Zones*, some trace elements are not determined as well as the specific sulfide that hosts each studied metal. From a high-resolution characterization [10] (Figures 5b and 6b), it is possible to determine the trace-metal content of a mineral, including its concentration (metal grade), the host phase, and form of occurrence (e.g., micron-sized particle, nano-particle, or in the lattice). These characteristics are depicted below for sulfides found in each GMU (Figures 5 and 6), focusing on the selected trace metals, i.e., Co, Au, and Re, plus Ag, Te, and As.

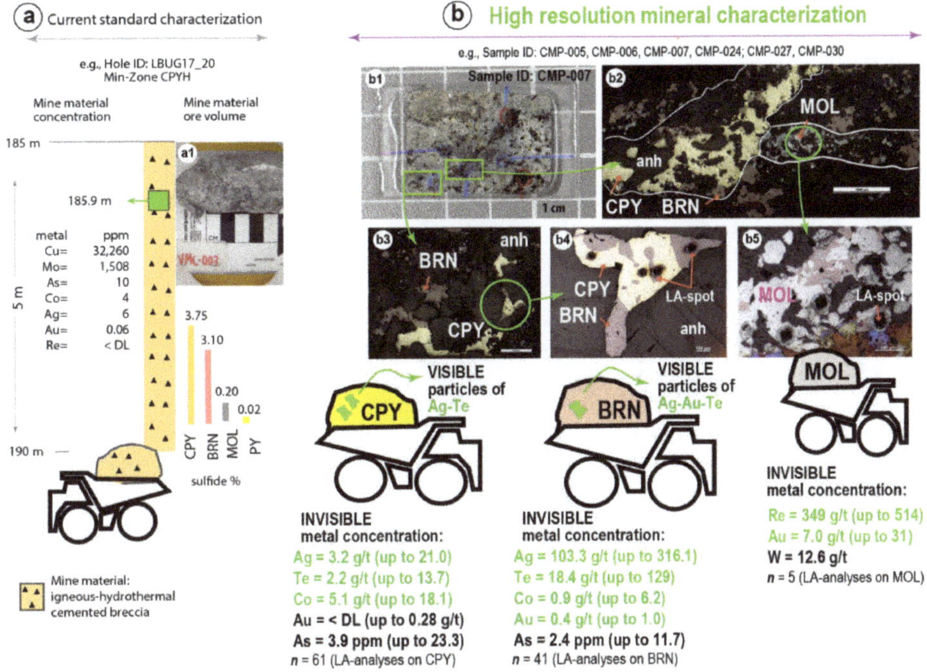

Figure 5. Schematic diagram illustrating: (**a**) current Cu-sulfide geochemical characterization, consisting of bulk geochemical analyses performed on 5 m-composite samples of mine material (on the right side a photograph is shown, a1). (**b**) The proposed high-resolution characterization for the GMU1, showing the mode of occurrence for each metal-bearing sulfide and the specific metal content, including the grade and mode of occurrence for metals.

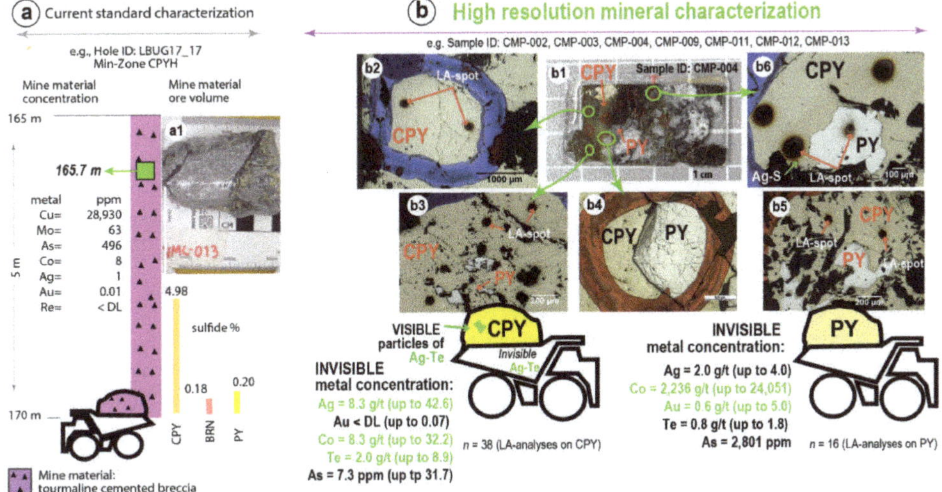

Figure 6. Schematic diagram illustrating: (**a**) the current Cu-sulfide ore characterization, which consist of bulk geochemical analyses performed on 5 m-composite samples of mine material (on the right side

photograph is shown, a1). (**b**) The proposed multi-metal high-resolution characterization for the GMU2, indicating the mode of occurrence for each metal-bearing sulfide and the specific metal content, including the grade and style of appearance of each metal.

From a standard standpoint, the GMU1 (Figure 5a) is characterized by a bulk trace-metal content of As = 10 ppm; Co = 4 ppm; Ag = 6 ppm; Au = 0.06 ppm; while Re and Te are present in concentrations below detection limits [39]. By implementing an advanced geochemical characterization (Table 1), the following strategic information can be obtained in our case study (Figure 5b): (1) bornite is the main host sulfide for trace Ag (up to 316 ppm) and Te (up to 129 ppm), and it also contains Au in concentrations of up to 1.0 ppm, results that are in agreement with the presence of micron-sized particles of Ag-Au-Te found as metallic inclusions. Bornite is almost depleted in Co and As trace elements, with concentrations of less than 6.2 ppm and 11.7 ppm, respectively; (2) chalcopyrite is enriched also with trace Ag (up to 21 ppm) and Te (up to 13.7 ppm); however, Au contents were below detection limits. In comparison with bornite, chalcopyrite is richer in trace Co (up to 18.1 ppm) and As (up to 23.3 ppm) concentrations; (3) molybdenite is a strategic host mineral for trace Au (up to 31 ppm), and also contains trace Re with an average grade of 349 ppm and ranging of up to 514 ppm. The aforementioned results represent significant metal content, which was not characterized by geochemical analysis performed on the bulk mine-material. Therefore, this metal content would not be considered in the standard ore processing design.

Standard geochemical characterization performed on the GMU2 samples (Figure 6a) shows that the bulk average for trace-elements concentrations of As = 496 ppm, Co = 8 ppm, Ag = 1 ppm; Au = 0.01 ppm and Re < DL [39]. The advanced geochemical characterization (Figure 6b; Table 2) shows that trace Co = 2236 ppm (up to 24,051 ppm), Au = 0.6 ppm (up to 5 ppm) and As = 2801 ppm (up to 6090 ppm) are preferentially contained in pyrite. In contrast, Ag is favorably concentrated in chalcopyrite (up to 42.6 ppm), being almost depleted in As = 7.3 ppm, Co = 8.3 ppm, and Au < DL, in comparison to pyrite. An important difference that arises from the advanced geochemical characterization is the mode of occurrence of As, which is not contained in chalcopyrite. However, bulk chemical results show As concentrations of around 500 ppm in the mine material, a metallic content that could erroneously penalize the Cu-grades of sulfide concentrates.

According to the standard characterization the GMU1 is richer in Ag concentration in relation to the GMU2 (GMU1 = 6 ppm Ag and GMU2 = 1 ppm Ag). However, the advanced geochemical characterization shows that chalcopyrite from GMU2 presents higher grade values for Ag (average of 8.3 ppm) in comparison with the chalcopyrite from GMU1 (average of 3.2 ppm). This silver behavior can be due to the absence of bornite in the GMU2, a mineral that preferentially concentrates Ag in its lattice [36]. Even if Ag (plus Au) is systematically dossed on prime minerals such as bornite and chalcopyrite, a remarkable data obtained from the advanced characterization is the mode of occurrence for these metals, i.e., Ag and Au, which are mainly found as Ag (±Au)-Te sulfosalts, instead of monometallic or electrum particles. This finding is fundamental to the development of metallurgical models.

4.1. Metallurgical Processing and Opportunities for Precious and Critical Metal Supply

The results presented previously reveal that the geo-metallurgical units defined for copper extraction, e.g., GMU1 and GMU2 (Figures 5 and 6), also can contain critical and precious metals in trace but profitable concentrations. For instance, in our case study: (i) the GMU1 (Figure 5) is characterized by profitable trace Au and Re contents associated with molybdenite, and (ii) the GMU2 (Figure 6) contains trace Co and Au, with the profitable metal grades associated with pyrite. These trace metals could be recovered in the same copper processing workflow during sulfide flotation processes, by modification or inclusion of some additional unit operations. Our proposed improvements are addressed to sulfide flotation workflows because sulfide concentrates will become the main marketable product produce by the copper industry in Chile over the coming years, with an estimated increase of

47.8% by 2030 [40]. If profitable by-products are not recovered during flotation processes, they will be extracted later during electro-refining and electro-winning processes, or sent to tailings [10].

The common scenario for producing sulfide flotation concentrates is a processing plant consisting of crushing, grinding, bulk-sulfide flotation, and selective Cu-sulfide flotation. However, identification of critical and precious metals associated with a specific sulfide gives rise to opportunities to propose new processing plant workflows for recovering these trace metals, in addition to copper + molybdenum, as marketable products during the flotation processes. To develop the conceptual mass balance approach, the following assumptions are considered: (i) the elemental content of each metal is obtained from the bulk geochemical characterization performed on the head ore, i.e., the current standard characterization (e.g., Figures 5a and 6a). These values are represented in the flow of the ore fed into the crushing and grinding steps; (ii) the mass balance considers the association of each metal with its specific hosting mineral (e.g., Figures 5b and 6b); (iii) the mass balance incorporates the behavior of each metal-bearing mineral in the respective unit operation. Therefore, the result of each metal of interest will be determined by the behavior of its specific mineral association.

4.1.1. Re and Au from Molybdenite

For the GMU1, the mass balance in a conventional processing plant (Figure 7) is designed to produce copper and molybdenum concentrates. In this case, a concentration factor of 10 is expected in the bulk flotation concentrate if an ideal separation between molybdenite and other sulfide minerals is reached [15].

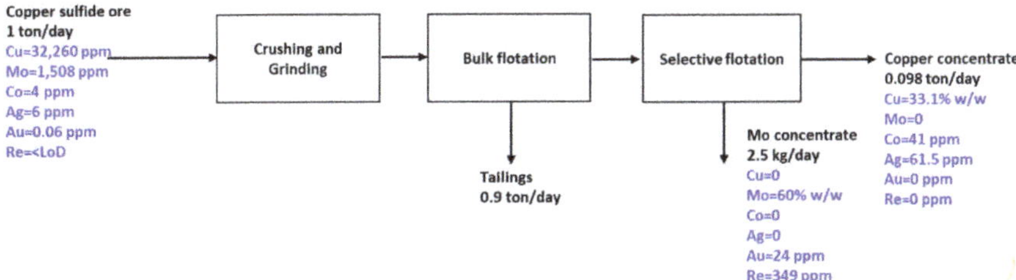

Figure 7. Schematic flow-sheet for a conventional plant design to treat porphyry ores, by processing mine material from the GMU1, to produce copper and molybdenum concentrates.

For molybdenite concentrates, only the molybdenum content is accounted for in the final price of this marketable product. To evaluate the feasibility of additional extraction of by-products, i.e., Re and Au, we present a process, typically used to recover Mo [15], and how it can be modified to also extract trace Re and particularly Au contents from molybdenite (Figure 8).

Roasting of molybdenite concentrate is implemented to obtain MoO_3 in the calcine, while Re_2O_7 is accumulated in the gas stream. The calcine, which also would contain gold, can be further treated with ammonia to dissolve the MoO_3 and transform it in the ammonium molybdate. At this stage, it is essential to consider that ammonia can dissolve gold under oxidative conditions; therefore, a rigorous control of the dissolved oxygen must be performed [15]. In spite of this, operational temperatures of this MoO_3 leaching (40–80 °C) are lower than those typically measured for ammonia leaching of gold (>80 °C) [41]. Therefore, this parameter can be operationally controlled. Finally, the leaching residue which contains gold can be treated in the proposed pathway by cyanidation and carbon adsorption to produce *dore metal* as a marketable product (Figure 8).

In the Re pathway, the ammonium molybdate solution can be crystallized to obtain ammonium dimolybdate, which is the final product typically generated from molybdenite concentrates (Figure 8). The gases containing rhenium (Re_2O_7), which are generated during roasting, can be treated in a scrubbing gas system to obtain a solution of perrhenic acid ($HReO_4$), which, in turn, are sent to an ionic exchange (IX) process to recover rhenium as ammonium perrhenate (NH_4ReO_4). The elution stage of the IX process is performed with ammonia, in which the generated solutions are crystallized to obtain the ammonium perrhenate (Figure 8), another marketable product, along with the *dore metal* produced from this proposed UGM. The processing option described in Figure 8 is not very different from that conventionally implemented by molybdenum refineries, in which rhenium and gold contents are frequently recovered from molybdenum concentrates.

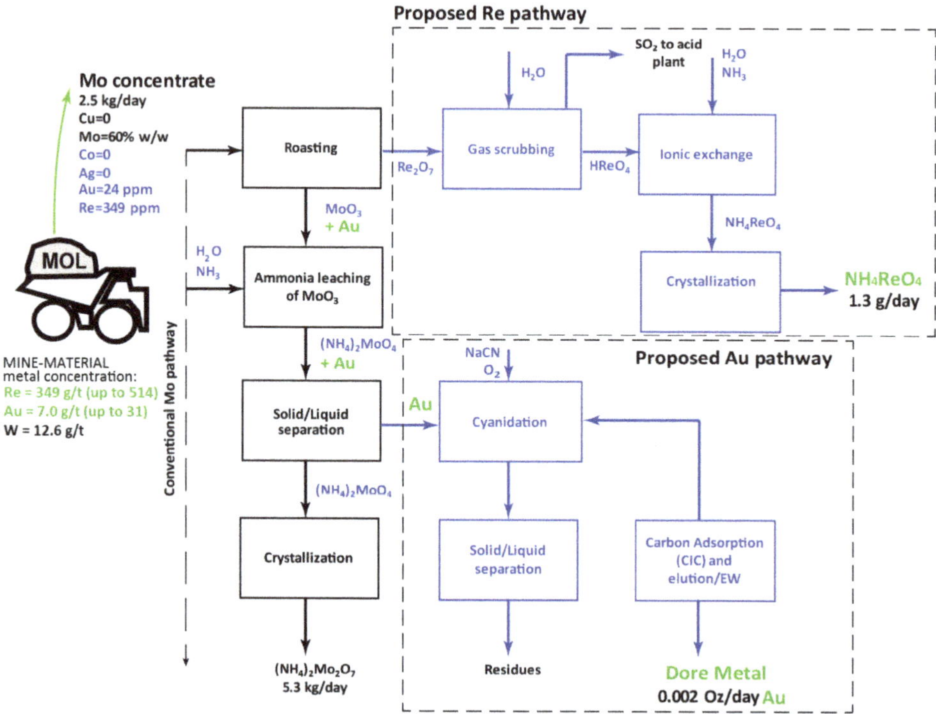

Figure 8. Schematic flow-sheet proposed to recover marketable critical and precious metals, such as Re and Au, in addition to Mo, from a molybdenite concentrate.

The presented mass balance (Figure 8) considers a unitary basis for the feed-ore, i.e., a unit estimation based on 1 ton of Cu-Mo sulfide ore processed per day, to easily scale-up to other treatment capacities. The scaling-up methodology can be expected as lineal, assuming the recoveries are not affected. For instance, in a case of a plant design with a capacity setting at 140,000 t/day of head ore treatment (a typical capacity for the big mining), the gold, rhenium, and molybdenum production as by-products could reach up to 280 Oz/day (102,200 Oz/year), 122.8 kg/day (44,822 kg/year), and 211.1 ton/day (77,000 ton/year), respectively.

4.1.2. Co and Au from Pyrite

The proposed UGM2 contains interesting grades of Co and Au, particularly concentrated in pyrite, a sulfide often not accounted for conventional process streams, which are designed to produce Cu and Mo-sulfide concentrates (Figure 9).

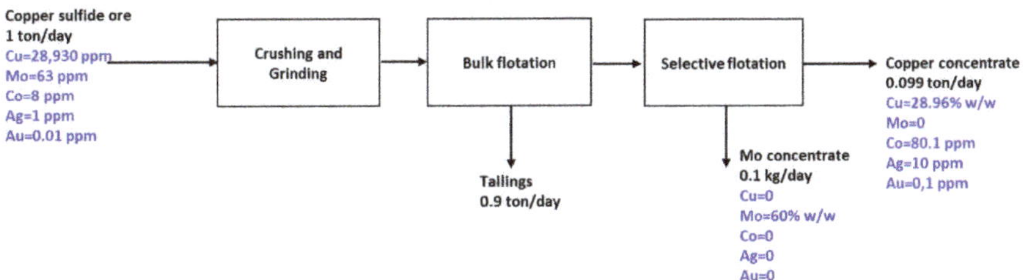

Figure 9. Schematic flow-sheet for a conventional plant design to treat porphyry ores, by processing mine material from the GMU2, to produce copper and molybdenum concentrates.

These concentrates are typically sold to a refinery furnace, which do not contain only Cu and Mo. Copper concentrates may include pyrite, which contains Co and Au (+Ag, ±Te), although generally only Au and Ag are valued in the concentrate selling price.

Considering that Co and Au trace concentrations are mainly contained in pyrite (Figure 6), there is an opportunity to modify the Cu processing design by the addition of a pyrite depression flotation stage (Figure 10). Thus, it is possible to separate pyrite from the other sulfides [42], which in our case study is chalcopyrite. Then, a pyrite concentrate can be generated and processed to recover Co and Au as marketable products (Figure 10), before sending pyrite to the tailings together with its metallic content [10]. In the proposed schematic flow-sheet (Figure 10), mass balance assumes ideal separation efficiency for each unit operation [15].

In the proposed pathway, pyrite concentrates can be treated in an oxidative environment to liberate Co and Au from the refractory sulfide matrix, considering that trace Co and Au are concentrated in nano-particles or in the mineral lattice. Here, an issue to be controlled is the percent of recoveries for Co and Au during the downstream stages. For that, it is best to incorporate an additional operational unit consisting of typical oxidative pretreatments [16], including roasting, bioleaching, pressure leaching, and oxidative acid leaching processes, which are systematically used to extract metallic Au from refractory minerals. In this way, it has recently been shown [43] that invisible Au in pyrite can be contained in an adsorbed state in the form of S-Au-S clusters, in contrast to the existing view that Au replaces Fe or S in the mineral lattice [44]. Thus, adsorbed Au is much less stable and, therefore, more easily recoverable by existing processes than the Au bound to the lattice [43]. In the case of roasting, an acid leaching of calcine must be included to dissolve the CoO formed during this process. Then, the pregnant leach solution (PLS), which will contain Co, Fe, and sulfuric acid, can be treated in a solid-liquid separation stage to remove the insoluble compounds from the PLS. Finally, insoluble compounds can be treated by cyanidation, a typical process to dissolve gold, and followed by a carbon adsorption process in pulp (CIP), including the correspondent elution and electro-winning (EW) stages [16]. Depending on the tonnage of mine material to be processed, it can also be recommended the use of an alternative carbon in column (CIC) after filtration.

In the other branch of the proposed flow-sheet, the PLS containing cobalt can be processed in a fractional stage to remove impurities, such as Fe, Cu, acid, and magnesium, and using specific pH conditions for each element [15,45,46]. Cobalt can be recovered as $Co(OH)_2$ after precipitation with MgO. Alternative processes for treating the PLS, such as solvent extraction, crystallization, or EW, can be evaluated to obtain different Co products, depending on the impurities contained and reagents price [45,46].

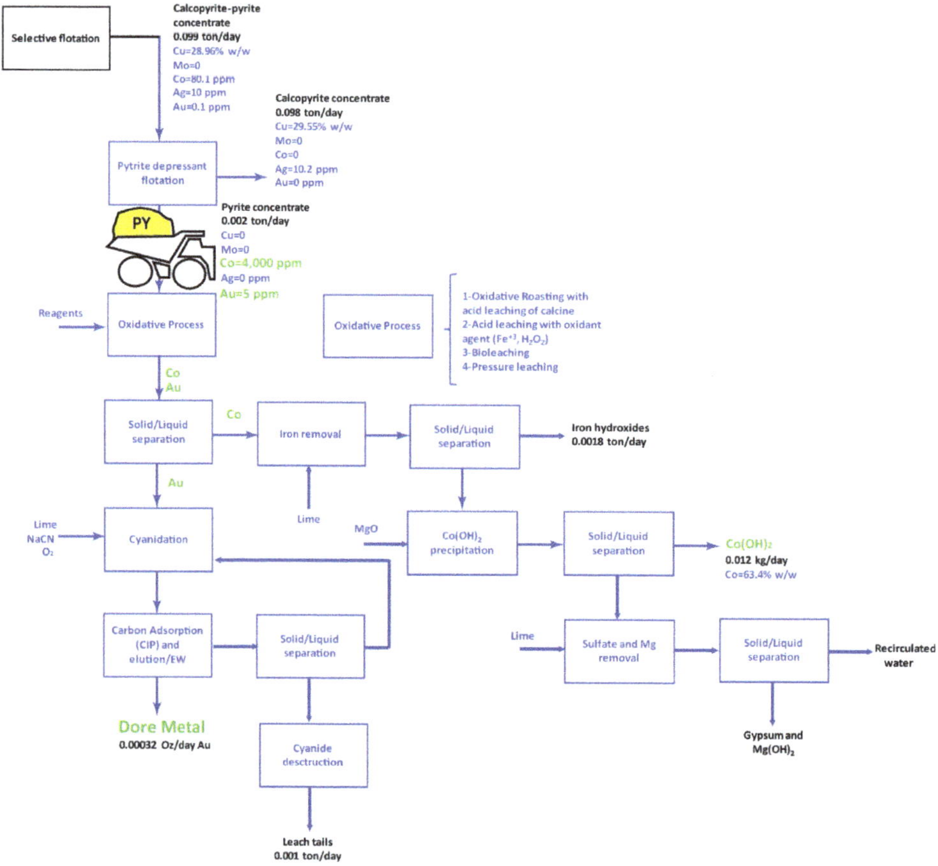

Figure 10. Schematic flow-sheet proposed to recover marketable critical metals, such as Co and Au from a pyrite concentrate, during sulfide concentrate production.

The process depicted in Figure 10 shows a significant reduction of the material volume to be treated from a pyrite concentrate (0.2% of the head ore), during the sulfide concentrate production. It has an additional positive effect, which is a remarkable increase of Co and Au grades up to 4000 and 5 g/t, respectively. Interestingly, the cost of mining, crushing, and grinding stages, developed on the bulk mine-material, are accounted into the copper production investments, which would imply that only the value for implementing additional operation units must be imputed to the by-products budget. Additionally, the reduction of the material to be processed can facilitate the installation of additional operation units in the same work areas previously designated to metallurgical plants. Considering a similar plant design with a head-ore treatment of 140,000 t/day, the by-products benefit could reach up to 44.8 Oz/day for Au (16,352 Oz/year), and 1065 kg/day for Co (389 t/year), which is a metal content not considered in the current mining plan. Therefore, future work must be focused on demonstrating, at a laboratory scale, the metal recoveries determined from a conceptual approach.

5. Concluding Remarks

In this study, we have shown that the metallic potential that could be assessed from a porphyry-type mining operation and how a high-resolution geochemical characterization of sulfides can be used to determine the concentration of trace metals as by-products. Critical and precious metals, such as Co,

Au, Re, Ag, and Te contained in ore sulfides, which are necessary for the development of society in the coming decades, are often not considered in mining planning and would be lost to tailings [10] or in the refineries wastes.

Even if the multi-metallic enrichment in porphyry-type deposits is well-known [10,12–14], very little work has been published on the metallurgical processing of these trace metals during the production of Cu- and Mo-concentrates. To our knowledge, a significant constraint to evaluate such possibility is the geochemical information obtained in operational chemical analyses, which is focused on copper contents [10]. Thus, recently, economic geologists are turning their interest in the detection of trace element concentrations directly in Cu-ore samples [47], and in evaluating possibilities to extract trace metals as by-products during Cu-sulfide processing [48] associated with sulfide concentrates production.

Here, we propose work-sheets to evaluate critical and precious metals processing during the production of sulfide concentrates, by the addition of some specific unit operations to conventional metallurgical concentrator plant designs. The advantages to be considered are as follows:

1. Millions of tons of mining material are processed each year by the porphyry mining industry, generating the possibility of beneficiation of by-product content, even if some metals are found in very low concentrations such as parts per million;
2. Investments for the extraction and processing of the mined material are reflected in the price of copper, and therefore only the additional unit operations could be taken into account in the economic assessment of the feasibility processing the by-products;
3. The identification of the specific mineral hosting each metal in trace amounts makes it possible to design a metallurgical treatment with quantities of material several times smaller than those represented by the treatment of Cu and Mo. For example, molybdenite and pyrite represent only 0.2% of the bulk material of GMU1 and GMU2, respectively. This idea implies that additional operational units can be installed in situ in the mining operations, avoiding the carrying of mined material;
4. The pyrite depression process, proposed to obtain a pyrite concentrate, can be applied both to extract the Co and Au contents of the pyrite and to reduce the amount of pyrite sent to the tailings. The latter is important from an environmental perspective because pyrite being a reactive mineral, its exposure to oxygen and water can facilitate the oxidative dissolution of the sulfide structure, resulting in acid mine drainage (AMD) [49]. The proposed workflows take into account the metallurgical processing of a recognized metal-bearing mineral, such as the processing of pyrite to obtain Co and Au, in conjunction with concentrate production. The latter idea can help reduce the amount of material to be processed, compared with a subsequent recovery of metals, such as Co, from mine tailings [49,50].

Thus, an economic evaluation can be made by considering the cost of setting up a plant similar to what we are proposing and comparing it with to the selling price of Cu and Mo concentrates, in which the trace metals contained would be valued in the final prices, i.e., Co and Au for Cu concentrates, and Re and Au for Mo concentrates.

Author Contributions: Conceptualization, G.V.; methodology, G.V. and S.S.; validation, I.V., S.S. and M.P.; formal analysis, G.V.; investigation, G.V. and H.E.; resources, G.V., I.V. and M.P.; writing-original draft preparation, G.V. and H.E.; writing-review and editing, S.S., I.V. and M.P.; visualization, G.V.; supervision, I.V.; project administration, G.V., M.P., and S.S.; funding acquisition, G.V. All authors have read and agree to the published version of the manuscript.

Funding: This research was partially funded by the National Commission for Scientific and Technological Research (CONICYT Chile) through CONICYT-PIA Project AFB180004.

Acknowledgments: We are grateful to S. Gouy and Ph. de Parseval for their help during SEM and EMP analyses at the Centre de Micro-Caractérisation Raimond Castaing (UMS 3623) in Toulouse, France, and T. Aigouy for their help during the SEM analyses at the GET Laboratory in Toulouse, France. We would also like to thank J. Chmeleff and C. Duquenoy for their technical assistance on the LA-ICP-MS analyses at the GET Laboratory. G. Velásquez and H. Estay acknowledge the National Commission for Scientific and Technological Research (CONICYT Chile), through CONICYT-PIA Project AFB180004 they were benefited to conduct part of this research. Constructive comments by two anonymous reviewers substantially improved the quality of this manuscript.

Conflicts of Interest: The authors declare no conflict of interest.

References

1. Mudd, G.M.; Weng, Z.; Jowitt, S.M. A detailed assessment of global Cu resource trends and endowments. *Econ. Geol.* **2013**, *108*, 1163–1183. [CrossRef]
2. Calvo, G.; Mudd, G.; Valero, A.; Valero, A. Decreasing ore grades in global metallic mining: A theoretical issue or a global reality? *Resources* **2016**, *5*, 36. [CrossRef]
3. Meinert, L.; Robinson, G.; Nassar, N. Mineral resources: Reserves, peak production and the future. *Resources* **2016**, *5*, 14. [CrossRef]
4. Moreau, V.; Dos Reis, P.C.; Vuille, F. Enough metals? Resource constraints to supply a fully renewable energy system. *Resources* **2019**, *8*, 29. [CrossRef]
5. Grandell, L.; Lehtilä, A.; Kivinen, M.; Koljonen, T.; Kihlman, S.; Lauri, L.S. Role of critical metals in the future markets of clean energy technologies. *Renew. Energy* **2016**, *95*, 53–62. [CrossRef]
6. Watari, T.; McLellan, B.C.; Giurco, D.; Dominish, E.; Yamasue, E.; Nansai, K. Total material requirement for the global energy transition to 2050: A focus on transport and electricity. *Resour. Conserv. Recycl.* **2019**, *148*, 91–103. [CrossRef]
7. Schipper, B.W.; Lin, H.-C.; Meloni, M.A.; Wansleebend, K.; Heijungs, R.; van der Voet, E. Estimating global copper demand until 2100 with regression and stock dynamics. *Resour. Conserv. Recycl.* **2018**, *132*, 28–36. [CrossRef]
8. McFall, K.; Roberts, S.; McDonald, I.; Boyce, A.J.; Naden, J.; Teagle, D. Rhenium enrichment in the Muratdere Cu-Mo (Au-Re) porphyry deposit, Turkey: Evidence from stable isotope analyses ($\delta^{34}S$, $\delta^{18}O$, δD) and laser ablation-inductively coupled plasma-mass spectrometry analysis of sulfides. *Econ. Geol.* **2019**, *114*, 1443–1466. [CrossRef]
9. Northey, S.; Mohr, S.; Mudd, G.M.; Weng, Z.; Giurco, D. Modelling future copper ore grade decline based on a detailed assessment of copper resources and mining. *Resour. Conserv. Recycl.* **2014**, *83*, 190–201. [CrossRef]
10. Velásquez, G.; Carrizo, D.; Salvi, S.; Vela, I.; Pablo, M.; Pérez, A. Tracking cobalt, REE and gold from porphyry copper deposits by LA-ICP-MS: A geological approach towards metal-selective mining in tailings. *Minerals* **2020**, *10*, 109. [CrossRef]
11. Nurmi, P.A. Green mining—A holistic concept for sustainable and acceptable mineral production. *Ann. Geophys.* **2017**, *60*. [CrossRef]
12. Kesler, S.E.; Chryssoulis, S.L.; Simon, G. Gold in porphyry copper deposits: Its abundance and fate. *Ore Geol. Rev.* **2002**, *21*, 103–124. [CrossRef]
13. Voudouris, P.; Melfos, V.; Spry, P.G.; Bindi, L.; Moritz, R.; Ortelli, M.; Kartal, T. Extremely Re-rich molybdenite from porphyry Cu-Mo-Au prospects in northeastern Greece: Mode of occurrence, causes of enrichment, and implications for gold exploration. *Minerals* **2013**, *3*, 165–191. [CrossRef]
14. Crespo, J.; Reich, M.; Barra, F.; Verdugo, J.J.; Martínez, C. Critical metal particles in copper sulfides from the supergiant Río Blanco porphyry Cu-Mo deposit, Chile. *Minerals* **2018**, *8*, 519. [CrossRef]
15. Habashi, F. *Handbook of Extractive Metallurgy*; Wiley-VCH: Weinheim, Germany, 1997.
16. Marsden, J.O.; House, C.I. *The Chemistry of Gold Extraction*, 2nd ed.; The Society of Mining, Metallurgy, and Exploration Inc. (SME): Colorado, CO, USA, 2006.
17. Escolme, A.; Berry, R.F.; Hunt, J.; Halley, S.; Potma, W. Predictive models of mineralogy from whole-rock assay data: Case study from the Productora Cu-Au-Mo deposit, Chile. *Econ. Geol.* **2019**, *114*, 1513–1542. [CrossRef]
18. Irarrazaval, V.; Sillitoe, R.H.; Wilson, A.J.; Toro, J.C.; Robles, W.; Lyall, G. Discovery history of a giant, high-grade, hypogene porphyry copper-molybdenum deposit at Los Sulfatos, Los Bronces-Río Blanco District, Central Chile. In *The Challenge of Finding New Mineral Resources: Global Metallogeny, Innovative Exploration, and New Discoveries*; Goldfarb, R.J., Marsh, E.E., Monecke, T., Eds.; Special Publications of the Society of Economic Geologists: Littleton, CO, USA, 2010; Volume 15, pp. 253–269.

19. Toro, J.C.; Ortúzar, J.; Zamorano, J.; Cuadra, P.; Hermosilla, J.; Spröhnle, C. Protracted magmatic-hydrothermal history of the Río Blanco-Los Bronces district, Central Chile: Development of world's greatest known concentration of copper. In *Geology and Genesis of Major Copper Deposits and Districts of the World: A Tribute to Richard H. Sillitoe*; Hedenquist, J.W., Harris, M., Camus, F., Eds.; Special Publications of the Society of Economic Geologists: Littleton, CO, USA, 2012; Volume 16, pp. 105–126.
20. Deckart, K.; Silva, W.; Spröhnle, C.; Vela, I. Timing and duration of hydrothermal activity at the Los Bronces porphyry cluster: An update. *Miner. Depos.* **2014**, *49*, 535–546. [CrossRef]
21. Bensalah, N.; Dawood, H. Review on synthesis, characterizations, and electrochemical properties of cathode materials for lithium ion batteries. *J. Mater. Sci. Eng.* **2016**, *5*, 4. [CrossRef]
22. Ghatak, K.; Basu, S.; Das, T.; Kumar, H.; Datta, D. Effect of cobalt content on the electrochemical properties and structural stability of NCA type cathode materials. *Phys. Chem. Chem. Phys.* **2018**, *20*, 22805–22817. [CrossRef]
23. Kesler, S.E.; Gruber, P.W.; Medina, P.A.; Keoleian, G.A.; Everson, M.P.; Wallington, T.J. Global lithium resources: Relative importance of pegmatites, brine and other deposits. *Ore Geol. Rev.* **2012**, *48*, 55–69. [CrossRef]
24. Munk, L.A.; Hynek, S.A.; Bradley, D.; Boutt, D.; Labay, K.; Jochens, H. Lithium brines: A global perspective. *Rev. Econ. Geol.* **2016**, *18*, 339–365.
25. Berzina, A.N.; Sotnikov, V.I.; Economou-Eliopoulos, M.; Eliopoulos, D.G. Distribution of rhenium in molybdenite from porphyry Cu-Mo and Mo-Cu deposits of Russia (Siberia) and Mongolia. *Ore Geol. Rev.* **2005**, *26*, 91–113. [CrossRef]
26. Cabri, L.J.; Chryssoulis, S.L.; de Villiers, J.P.R.; Laflamme, J.H.G.; Buseck, P.R. The nature of "invisible" gold in arsenopyrite. *Can. Mineral.* **1989**, *27*, 353–362.
27. Riesner, M.; Simoes, M.; Carrizo, D.; Lacassin, R. Early exhumation of the Frontal Cordillera (Southern Central Andes) and implications for Andean mountain-building at ~33.5° S. *Sci. Rep.* **2019**, *9*, 7972. [CrossRef] [PubMed]
28. Cook, N.; Ciobanu, C.L.; George, L.; Zhu, Z.-Y.; Wade, B.; Ehrig, K. Trace element analysis of minerals in magmatic-hydrothermal ores by laser ablation inductively-coupled plasma mass spectrometry: Approaches and opportunities. *Minerals* **2016**, *6*, 111. [CrossRef]
29. Velásquez, G.; Béziat, D.; Salvi, S.; Siebenaller, L.; Borisova, A.Y.; Pokrovski, G.S.; De Parseval, P. Formation and deformation of pyrite and implications for gold mineralization in the El Callao District, Venezuela. *Econ. Geol.* **2014**, *109*, 457–486. [CrossRef]
30. Velásquez, G.; Borisova, A.Y.; Salvi, S.; Béziat, D. In situ determination of Au and Cu in natural pyrite by near-infrared femtosecond laser ablation-inductively coupled plasma-quadrupole mass spectrometry: No Evidence for matrix effects. *Geostand. Geoanal. Res.* **2012**, *36*, 315–324. [CrossRef]
31. Borisova, A.; Thomas, R.; Salvi, S.; Candaudap, F.; Lanzanova, A.; Chmeleff, J. Tin and associated metal and metalloid geochemistry by femtosecond LA-ICP-QMS microanalysis of pegmatite-leucogranite melt and fluid inclusions: New evidence for melt-melt-fluid immiscibility. *Mineral. Mag.* **2012**, *76*, 91–113. [CrossRef]
32. Van Achterbergh, E.; Ryan, C.G.; Griffin, W.L. *Data Reduction Software for LA-ICP-MS*; Mineralogical Association of Canada Short Course Notes: Calgary, AB, Canada, 2001; Volume 29, pp. 239–243.
33. Sylvester, P.J.; Cabri, L.J.; Tubrett, M.N.; McMahon, G.; Laflamme, J.H.G.; Peregoedova, A. Synthesis and evaluation of a fused pyrrhotite standard reference material for platinum group element and gold analysis by laser ablation-ICP-MS. In Proceedings of the 10th International Platinum Symposium, Oulu, Finland, 8–11 August 2005; pp. 16–20.
34. Pearce, N.J.; Perkins, W.T.; Westgate, J.A.; Gorton, M.P.; Jackson, S.E.; Neal, C.R.; Chenery, S.P. A compilation of new and published major and trace element data for NIST SRM 610 and NIST SRM 612 glass reference materials. *Geostand. Geoanal. Res.* **1997**, *21*, 115–144. [CrossRef]
35. Long, G.; Peng, Y.; Bradshaw, D. A review of copper–arsenic mineral removal from copper concentrates. *Min. Eng.* **2012**, *36–38*, 179–186. [CrossRef]
36. Ciobanu, C.L.; Cook, N.J.; Kelson, C.R.; Guerin, R.; Kalleske, N.; Danyushevsky, L. Trace element heterogeneity in molybdenite fingerprints stages of mineralization. *Chem. Geol.* **2013**, *347*, 175–189. [CrossRef]
37. Deditius, A.P.; Utsunomiya, S.; Reich, M.; Kesler, S.E.; Ewing, R.C.; Hough, R.; Walshe, J. Trace metal nanoparticles in pyrite. *Ore Geol. Rev.* **2011**, *42*, 32–46. [CrossRef]

38. Sykora, S.; Cooke, D.R.; Meffre, S.; Stephanov, A.S.; Gardner, K.; Scott, R.; Selley, D.; Harris, A.C. Evolution of pyrite trace element compositions from porphyry-style and epithermal conditions at the Lihir gold deposit: Implications for ore genesis and mineral processing. *Econ. Geol.* **2018**, *113*, 193–208. [CrossRef]
39. Anglo American Sur, S.A. Bulk chemical assays database, until 2019. Los Sulfatos Mine-Material. The Los Bronces Underground Project. Internal Report.
40. COCHILCO-Comision Chilena del Cobre. Proyección de la Producción de Cobre en Chile 2019–2030. DEPP 15/2019. Registro Propiedad Intelectual N° 310805: Santiago, Chile, 2019; p. 43. Available online: https://www.cochilco.cl/Listado%20Temtico/Proyecci%C3%B3n%20de%20la%20producci%C3%B3n%20esperada%20de%20cobre%202019%20-%202030%20Vfinal.pdf (accessed on 20 March 2020).
41. Aylmore, M.G. Alternative Lixiviants to Cyanide for Leaching Gold Ores. Chapter 27. In *Gold Ore Processing: Project Developments and Operations*, 2nd ed.; Adams, M.D., Ed.; Elsevier: Amsterdam, The Netherlands, 2016.
42. Mu, Y.; Peng, Y.; Lauten, R.A. The depression of pyrite in selective flotation by different reagent systems —A Literature review. *Miner. Eng.* **2016**, *96–97*, 143–156. [CrossRef]
43. Pokrovski, G.S.; Kokh, M.A.; Proux, O.; Hazemann, J.-L.; Bazarkina, E.F.; Testemale, D.; Escoda, C.; Boiron, M.-C.; Blanchard, M.; Aigou, T.; et al. The nature and partitioning of invisible gold in the pyrite-fluid system. *Ore Geol. Rev.* **2019**, *109*, 545–563. [CrossRef]
44. Simon, G.; Huang, H.; Penner-Hahn, J.E.; Kesler, S.E.; Kao, L. Oxidation state of gold and arsenic in gold-bearing arsenian pyrite. *Am. Mineral.* **1999**, *84*, 1071–1079. [CrossRef]
45. Sole, K.C.; Parker, J.; Cole, P.M.; Mooiman, M.B. Flowsheet options for cobalt recovery in African copper-cobalt hydrometallurgy circuits. *Min. Proc. Extr. Metall. Rev.* **2019**, *40*, 194–206. [CrossRef]
46. Shengo, M.L.; Kime, M.-B.; Mambwe, M.P.; Nyembo, T.K. A review of the beneficiation of copper-cobalt-bearing minerals in the Democratic Republic of Congo. *J. Sustain. Min.* **2019**, *18*, 226–246. [CrossRef]
47. Rollog, M.; Cook, N.J.; Guagliardo, P.; Ehrig, K.; Ciobanu, C.L.; Kilburn, M. Detection of Trace Elements/Isotopes in Olympic Dam Copper Concentrates by nanoSIMS. *Minerals* **2019**, *9*, 336. [CrossRef]
48. Schmandt, D.S.; Cook, N.J.; Ehrig, K.; Gilbert, S.; Wade, B.P.; Rollog, M.; Ciobanu, C.L.; Kamenetsky, V.S. Uptake of trace elements by baryte during copper ore processing: A case study from Olympic Dam, South Australia. *Miner. Eng.* **2019**, *135*, 83–94. [CrossRef]
49. Falagán, C.; Grail, B.M.; Barrie-Johnson, D. New approaches for extracting and recovering metals from mine tailings. *Miner. Eng.* **2017**, *106*, 71–78. [CrossRef]
50. Ahmadi, A.; Khezri, M.; Akbar-Abdollahzadeh, A.; Askari, M. Bioleaching of copper, nickel and cobalt from the low grade sulfidic tailing of Golgohar Iron Mine, Iran. *Hydrometallurgy* **2015**, *154*, 1–8. [CrossRef]

Article

Study of Copper Leaching from Mining Waste in Acidic Media, at Ambient Temperature and Atmospheric Pressure

Juan María Terrones-Saeta *, Jorge Suárez-Macías, Francisco Javier Linares del Río and Francisco Antonio Corpas-Iglesias

Department of Chemical, Environmental, and Materials Engineering, Higher Polytechnic School of Linares, University of Jaen, Scientific and Technological Campus of Linares, 23700 Linares, Spain; jsuarez@ujaen.es (J.S.-M.); fjlr0015@red.ujaen.es (F.J.L.d.R.); facorpas@ujaen.es (F.A.C.-I.)

* Correspondence: terrones@ujaen.es; Tel.: +34-675-20-939

Received: 24 August 2020; Accepted: 29 September 2020; Published: 1 October 2020

Abstract: Mining activity produces a series of wastes that must be treated to avoid environmental pollution. In addition, some of these mining wastes still contain metallic elements that are interesting for their extraction with new less expensive techniques and that can work with low mineral grades, such as hydrometallurgy. This study evaluates the suitability of Copper recovery in mining wastes, coming from waste dump, with a high percentage of metal oxides and granite. This recovery is carried out through leaching in 0.05, 0.10, 0.15 and 0.20 molar Sulphuric Acid solutions, at ambient temperature and atmospheric pressure. The exposure of the waste to the solution was made for 96 h, taking measurements of the leaching and evaluating the increase in Copper concentration every 24 h. The results reflected a good Copper recovery rate with concentrations up to 1.9 g/L. The best results were obtained for the 0.20 molar Sulphuric Acid solutions, producing a stability in the Copper concentration after 72 h. Other elements in smaller proportion as the Zinc were also recovered. Therefore, a process of recovery of Copper was obtained with a robust, versatile and economic technique in mining residues that currently represent an environmental pollution.

Keywords: hydrometallurgy; leachate; ICP-MS; polymetallic sulphides; granite; copper; mining waste; waste rock; recovery rate; sustainability

1. Introduction

Copper is one of the most demanded materials nowadays. This fact is mainly due to its use in several areas, such as in information, energy, electronics, construction, the military, shipping, railway, etc. [1]. The use in different areas is due to its properties and versatility, presenting very good electrical conductivity, thermal conductivity, malleability, resistance to chemical agents and, in addition, being easily combined with other elements to obtain alloys with very particular characteristics, widely used in construction.

Therefore, Copper is a strategic material for different countries, as there is a significant imbalance between supply and demand. An example is China, which for years is the largest consumer of Copper in the world, [2], obtaining this material mainly from Copper Sulphide minerals. These Copper Sulphide minerals are treated in 98% with pyrometallurgical techniques [3]. However, the protection of natural resources, low mineral grades and an excessive increase in demand means that these techniques are no longer as profitable as in the past. On the other hand, Copper is one of the 100% recyclable materials, so countries such as China are opting for the treatment of industrial waste [4], scrap metal and electronic waste to obtain Copper [5]. However, these recycling processes involve a significant consumption of energy and natural resources, and therefore cause a considerable environmental impact, as well as that caused by the usual pyrometallurgical techniques for Copper processing.

Worldwide, 80% of copper is obtained by pyrometallurgical processes and only 20% by hydrometallurgical techniques, derived from processing mainly Copper Sulphide mineral [6]. However, due to the existence of decreased mineral laws, more restrictive environmental regulations and significant pollution due to the processes, makes pyrometallurgical techniques unprofitable. It is therefore necessary to develop hydrometallurgical processes that obtain Copper with lower energy consumption and in materials with much lower grades, such as industrial by-products, waste, mining waste, etc. In other words, to obtain Copper mainly from different raw materials impossible to concentrate by pyrometallurgical techniques [7], either because Copper is found in such low proportions that it is not economically viable or because it contains chemical elements that restrict its concentration, such as Arsenic, or because they are polymetallic minerals impossible to treat by traditional methods [8].

Therefore, it is the hydrometallurgical techniques that must be developed and implemented to provide the solutions to the problem of Copper demand. These techniques are economically viable and, as associated with their slow and low recovery, there are much lower processing costs than in pyrometallurgical techniques. However, they have not yet been commercially applied for landfills for ore or industrial waste [9].

There are new hydrometallurgical techniques for recovering Copper from Copper Sulphide minerals [10,11]—mainly chalcopyrite, with acidic solutions and at atmospheric pressure [12]. These new techniques have some essential advantages, on the one hand, their economy, as they are much slower processes and require less technology, so the cost of the equipment is reduced considerably; on the other hand, their versatility makes them applicable to almost any type of mineral independent of the impurities that they contain already, such as Zinc, Lead, Arsenic, Antimony, Mercury, etc. [13]. Another advantage is its simplicity, a series of parameters must be controlled periodically for its success but its implementation is simple and easy to maintain [14]—in short, its robustness as a technique, and being able to process diverse minerals with different chemical elements and coming from diverse formations. It is therefore a technique that is currently being evaluated in depth in the laboratory but that must be implemented commercially [15].

At the same time, it must be recognized that it has a series of disadvantages with respect to pyrometallurgical processes that need to be mentioned and evaluated [16]. In terms of production, and based on different case studies, it is difficult to fully recover Copper and the precious mineral; in terms of product quality, the leachate may not always contain a high proportion of Copper depending mainly on the solution, time and mineral used; variation in recovery rates is common, depending obviously on the minerals processed and their activity. Regarding the production of waste, it is therefore necessary to parallel research projects to this technique that are able to process the waste in order to obtain a process as sustainable as possible. Finally, there is the scarcity of equipment adapted to this technique compared to pyrometallurgical techniques, mainly due to its poor implementation. However, the advantages mentioned above represent a very wide field of research that may lead to an important development of hydrometallurgical techniques in the future, with its consequent implementation. In addition, as mentioned above, the low quality of the leached product, in some cases, is solved by the low cost of the initial material, the low cost of the process and the existing demand for Copper.

The recovery rate for hydrometallurgical techniques depends on several factors. These factors must be controlled in the process to optimize production and evaluate viability, even more so in the recovery of copper from landfills, which is the fundamental idea on which this project is based. The factors on which the Copper recovery rate depends are the following:

- The Copper minerals present. Obtaining Copper from mining waste by hydrometallurgical techniques is obviously influenced by the percentage of Copper minerals present. Usually they will be in combination with other types of minerals without value in greater or lesser proportions. At the same time, not all Copper minerals produce the same recovery rates, so it is interesting for its classification and quantification. Impurities will also influence the recovery of Copper, although never to the same extent as in pyrometallurgical processes.

- The mineral associations in the existing rocks in the landfills and the easy release of the Copper minerals. There must be an initial study of the landfill to evaluate this factor directly related to that mentioned above.
- The particle size to be processed [17]. The absence of a concentrate in the mining waste and the combination with other non-valuable minerals means that leaching does not occur at its maximum recovery rate if these Copper minerals are not in contact with the acidic solution. Therefore, a crushing of the landfill materials will influence a higher probability of contact of the Copper ores with the acidic solution and, consequently, will result in a higher recovery rate. The crushing process is therefore essential if the aim is to obtain the highest recovery rate in the shortest time [18]. At the same time, it must be taken that milling very fine crushing would damage the industrial process, due to problems in the recirculation of the leachate and the creation of contaminating sludges. These sludges produced should be evaluated and reused in other industrial processes to avoid their deposition in landfills and the associated environmental pollution. In other words, it is important to evaluate the appropriate particle size to obtain an adequate recovery rate, a lower amount of contaminating waste and a more energetically optimized leaching process [19,20].

On the basis of the above, it is necessary to develop new hydrometallurgical techniques, at atmospheric pressure and in an acidic media [21], with different types of raw materials for various reasons. These reasons are the satisfaction of the demand, the implementation of new strategic industries, the search for particular solutions, the reduction in the environmental impact associated with the traditional processes and the optimization of the natural resources.

In addition, the use of waste as a material for the extraction of elements, such as those as important today as Copper, makes the development of a sustainable industry possible [22]. This industry reduce pollution from the deposition of these wastes, reduces the extraction of new virgin materials and provides economically viable solutions [23]. In this study, therefore, a process is developed within the circular economy.

With the detailed purpose, this work studies the recovery of Copper by hydrometallurgical techniques through the leaching in Sulphuric Acid solutions of materials coming from Lead mining landfills. To this end, the mine that is producing these dumps is initially characterized, the mineralogy of the existing landfills is evaluated, and representative samples are taken. These samples were chemically analyzed for the presence of Copper, as well as other chemical elements that could damage the recovery rate. Later, the samples were crushed to obtain an adequate particle size and it was subject to different solutions of low molar Sulphuric Acid for 4 days. Leachate samples were taken from all the solutions every 24 h and the recovery rate of Copper was evaluated as a function of time and the molar of the solution. In turn, were analyzed the presence in the leachate of the other secondary chemical elements or harmful chemical elements.

The results showed a concentration of Copper in the 0.20 molar solution of Sulphuric Acid maximum which was stable at 72 h. The recovery rate corresponding to this concentration value is approximately 80%. At the same time, interesting concentrations of other elements, such as Zinc, were obtained, with almost total recovery rates. The elements harmful to the process were in low proportion, among them Arsenic.

2. Materials and Methods

This section describes the material used for the extraction of Copper from Lead and Silver mine waste dumps, as well as the methodology followed to evaluate the suitability of Copper recovery by hydrometallurgy, based on leaching in acidic media of the minerals, at atmospheric pressure and ambient temperature.

2.1. Materials

The material evaluated for the extraction of Copper by leaching in acidic solutions comes directly from waste dumps in the mining district of Linares, Jaén. The mining activity of Linares was mainly focused on the extraction of Lead and in a smaller proportion the Silver from the galena argentiferous mineral present in the area. This mining activity was maintained for hundreds of years when Lead was an essential material for the population and there was a high demand for it. In the last few decades of the 20th century, the brutal fall in Lead prices, as well as the need for research into new exploitable seams, led to the total closure of mining activity. Therefore, there are different mining wastes from this intense mining activity that can be found throughout the mentioned territory, waste from washing plants, flotation or the waste landfill under study.

These mining wastes cause significant environmental pollution, as previous mining legislation did not provide for adequate environmental measures to be taken to avoid this. Therefore, it is usual that there is contamination of underground water and surface water by the Lead present in the waste, as well as other chemical elements, such as Arsenic, Sulfur, Zinc, etc. In turn, waste with smaller particle size causes a direct effect on vegetation. The impact of these enormous waste deposits notably affects the image of the territory, even though this has already been accepted by the majority of the population.

In these wastes, as mentioned, there is a high proportion of Lead. There are also contaminating elements that should be retained and, at the same time, minerals rich in chemical elements that can be profitable and which are currently in great demand—in this particular case, Copper.

In particular, the residue evaluated for the extraction of Copper belongs to the waste dump of the Lead extraction from the seam called "EL COBRE". This reef of about 4500 m in length crosses the towns of Guarroman, Linares and Bailen. The seam is located in the direction of the most generalized fractures of the rock that forms it, this being mainly granite and, in the lesser half, very metamorphosed shale. This seam, from which the Lead mineral was extracted in the past, has an arrangement close to vertical with runs of up to 3000 m. The thickness of the same Lead usually oscillates between centimeters and several meters, it being usual that this was between one or two meters. The main minerals existing in the seam are the argentiferous galena, blende, Iron Sulphides (mainly pyrite) and Copper Sulphides, mainly chalcopyrite. This last mineral is the one of interest in the study. Figure 1 shows the image of the waste dump, as well as the particles that form it; their diversity and different sizes.

Figure 1. Image of the "EL COBRE" seam waste dump. (**a**) General image of the waste dump and (**b**) detailed image of the different particles that make it up.

The seam of "EL COBRE", from which the study of its waste dumps is based and which has the name of the element to be extracted, has never been used for the extraction of Copper, only Lead and Silver were extracted in a much smaller proportion. Therefore, in its waste dumps it is easy to find the Copper Sulphides that have not been processed. These Copper Sulphides have been transformed in most cases into Copper Oxides, due to the treatments carried out and the exposure over decades to the atmospheric conditions. The grades obtained by the mining company during the extraction and processing of this vein are variable, given its depth and thickness, being between 4.5% and 6% for Lead. In turn, the Silver contents were modest, being these between 240 and 360 grams of metal per ton of Lead recovered.

The exploitation of the "EL COBRE" seam at present, as has been commented on, is totally paralyzed, with almost all the reserves having been extracted. Underground mining was used and the mentioned minerals were processed to extract Lead later by gravimetric methods. The waste rock materials from lead extraction were deposited in large waste dumps, which are distributed along the course of the seams. In these waste dumps are found the Copper Oxides from the seam combined with the other minerals.

The waste dumps have a fairly random particle size, with particles of 10 to 15 cm predominating. The existing mineralogy is easily observable in these particles, this being mainly granite (rock that forms the seam), Copper Oxides, Iron and Lead Oxides. Different representative samples were taken from these waste dumps so that they could later reflect, in the laboratory, the composition and the real process that could be carried out in industry.

The samples taken from the "EL COBRE" seam waste dumps were first dried in an oven to remove any moisture they might contain and thus not influence the subsequent process with uncontrolled variables. However, the existence of humidity in the industry would not be detrimental to the hydrometallurgical process, but should be taken into account so that the acidic solutions were suitable and marked by research.

Later, the samples will be prepared differently for the leaching process or for the initial characterization, detailing such processes in the methodology.

2.2. *Methodology*

Once the origin of the waste to be analyzed for the extraction of Copper by hydrometallurgical methods was determined, the tests carried out to assess the suitability of this study were detailed. The methodology was a series of logical, structured and ordered tests capable of objectively evaluating the final conclusion of the study, as well as determining those key points on which greater attention must be paid in order to obtain acceptable results.

Once the mineralogy of the waste dump under study had been evaluated in situ, and the main minerals detailed in the previous section had been identified, the procedure was that representative samples were taken. The samples from the waste dump belonging to the mining waste mentioned above were dried for 24 h at a temperature of 105 ± 2 °C to eliminate the humidity.

Subsequently, the chemical characterization of the waste was necessary to study its chemical composition, as well as the Copper content present in the samples. The first test carried out was the elemental analysis for the detection of the chemical elements Carbon, Nitrogen, Hydrogen and Sulfur. Secondly, the loss on ignition test was performed, and finally, the X-ray fluorescence test, for the determination of the chemical elements with the highest atomic weight, including Copper.

After evaluating the chemical composition and determining the percentage of Copper present, leaching was carried out in acid solution. For this purpose, recirculation equipment was prepared in which the sample was introduced, and the watering of the same sample was executed continuously. The solution of Sulphuric Acid was achieved with four low molars, in order to avoid high molarities that could lead to greater environmental pollution. In the different recirculation equipment, with different Sulphuric Acid solutions, measurements were taken every 24 h up to a total of 96 h. All leachate samples of the waste, in the four solutions and for the different times, were analyzed in the equipment, called the ICP-MS Instrument. In this way, the percentage of Copper present in all the samples was studied quantitatively, as well as that of other chemical elements that could be interesting.

With the results obtained, the ideal time of leaching and the adequate solution of Sulphuric Acid was determined, determining the viability of the study.

All the above-mentioned tests are described in two large blocks, analysis of the chemical composition of the waste and the leaching process and measurement of the concentration. These blocks are described in more detail below.

2.2.1. Analysis of the Chemical Composition of the Waste

The analysis of the chemical composition of the waste to be processed by leaching techniques in acidic media is essential. The aim is to determine the percentage of Copper present in the sample tested, as well as the existence of other chemical elements that may be harmful to this leaching process. The chemical composition of the dump waste will confirm that the chemical elements present coincide with the minerals identified.

Based on the above, the first of the tests is the elemental analysis test, performed with TruSpec Micro equipment (LECO, St. Joseph, MI, USA) from the LECO brand. This test consists of the combustion of the sample at a temperature of 1000 ± 10 °C and the analysis of the gases produced, analyzing the chemical elements of Carbon, Nitrogen, Hydrogen and Sulfur. The residue sample, as it has a mainly inorganic composition, will present relatively low percentages of these elements. However, the percentage of Carbon will reflect the percentage of organic matter and carbonates. On the other hand, the Sulfur present will show the percentage of the sulphurous minerals. However, not all the percentage of Sulphur determined by elemental analysis will be the existing one, since combustion takes place up to 1000 ± 10 °C, existing at a higher percentage determined by X-ray fluorescence (Thermo Fisher Scientific, Waltham, MA, USA).

In turn, the loss on ignition test reflects the change in weight after the sample has been subjected to the temperature of 1000 ± 10 °C. This loss of weight, at the above mentioned temperature, corresponds mainly to the loss of weight by the percentage of organic matter, the loss of weight by the percentage of carbonates, the transformation of some chemical compounds as well as the oxidation of other chemical elements. It is therefore difficult to assess which aspect they correspond to; however, this test provides sufficient information in correlation with the other tests.

Finally, the X-ray fluorescence test on the waste sample from the "EL COBRE" seam waste dump will provide the composition of the remaining chemical elements not analyzed in the elemental analysis. This test is essential to determine the percentage of Copper in the sample, as well as to corroborate the chemical composition predicted by the mineralogical study determined in the previous section. That is, the chemical composition must correspond to the chemical elements of the minerals present in the waste sample. In turn, it will determine the percentage of contaminating or conflicting elements for the leaching process that could damage the complete and rapid leaching of the Copper. Among these chemical elements are Arsenic, Zinc, Lead, Iron, etc.

2.2.2. Leaching Process and Concentration Measurement

After analyzing the chemical composition of the waste dump from the "EL COBRE" seam, we proceeded to study the leaching of Copper with different solutions of Sulphuric Acid, at different times of leaching and with atmospheric pressure.

First the samples were crushed until a particle size between 6 mm and 10 mm was obtained. The crushing of the sample and obtaining this size is based on several factors. Firstly, this process is carried out in such a way that the Copper ore is in contact with the solution on the greatest possible surface, since, as has been mentioned, this ore appears associated with other minerals and even with the conforming rock that is granite. If the mineral is encapsulated, the leaching process in an acidic media is slower and not as effective since the solution must penetrate to leach the existing Copper. Therefore, initial particle sizes around 10 cm and 15 cm are discarded. On the other hand, a very fine particle size would favor the leaching process but would create sludges, this fact being a detriment to the continuous recirculation of the solution and causing a higher energy expenditure. At the same time, these sludges produced should be separated and finally treated, since it is a polluting waste that cannot be directly deposited in a landfill. In short, the particle size selected is that which will provide an adequate recovery rate and, in turn, very low percentages of waste sludges.

The mining waste samples with particle size between 6 mm and 10 mm were divided into subsamples of similar composition to perform the most objective leaching analysis possible. The UNE-EN 932-2 standard was used for the subdivision. This process is very important, since obtaining similar samples will significantly influence the quality of the results.

Once the sample had been crushed and the desired particle size had been obtained, the leaching test was carried out in an acidic media and with different solutions. The samples selected for leaching had a mass of 10 ± 0.1 g and were previously dried to remove humidity.

On the other hand, the Sulphuric Acid solutions were 0.05, 0.10, 0.15 and 0.20 molar. The use of these low molar solutions is motivated by different factors; firstly, the use of low molar Sulphuric Acid solutions will create a more easily treatable residue, respecting the sustainability of the process in a global way and avoiding unnecessary contaminating residues. On the other hand, higher molars could lead to excess sludge creation without providing a higher extraction rate. Finally, solutions with higher concentrations of Sulphuric Acid would create a greater leaching of other chemical elements that could damage and disturb the subsequent processes for the extraction of Copper, increasing the percentage of Iron, Magnesium, Arsenic, etc. in the leachate.

The leaching process was carried out in equipment specially prepared for this study. This very simple equipment consisted of an airtight container wherein the sample was poured. Below the sample there were two filters—one of 1 mm, to avoid release particles larger than that size during the leaching process, and another of 0.1 mm, to try to retain the sludge derived from the leaching process. At the same time, there was a lower intake from which the solution was taken and pumped to the top for the continuous irrigation of the waste sample. This process is kept constant during the 96 h of the test, without interruption and controlling the temperature at 25 ± 1 °C. The container was large enough to hold the samples and 200 ± 1 mL of each of the solutions. The image of the equipment can be seen in Figure 2.

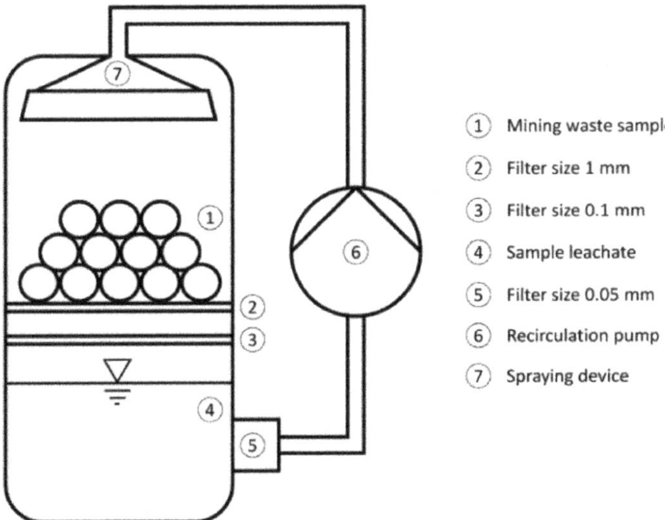

Figure 2. Equipment for leaching mining waste in an acidic media, ambient temperature and atmospheric pressure. Own elaboration.

Once the four solutions were made, the four processes were started with the four similar samples of waste from the "EL COBRE" seam. Every 24 h, we proceeded to take samples of leaching of the four solutions of Sulphuric Acid, as well as the continuous corroboration of the temperature. In this test, a total of 4 repetitions was carried out for each solution.

The samples obtained were analyzed in the ICP-MS of the commercial brand Agilent model 7900 (Agilent Technologies, Santa Clara, CA, USA). For the quantification of the percentage of Copper and the other chemical elements with higher precision, patterns of Copper and the other elements under study were used.

3. Results and Discussions

This section describes the results of all the tests mentioned in the methodology. From each test, partial conclusions were drawn that to help reach a final conclusion. The final conclusion is the evaluation of the suitability of leaching of mining waste from waste dumps by acidic media, at ambient temperature and atmospheric pressure, for the recovery of Copper. The above-mentioned tests are presented in the two large blocks mentioned.

3.1. Analysis of the Chemical Composition of the Waste

The chemical composition of the waste of the dump waste under study must be characterized for various reasons. On the one hand, for the knowledge of the percentage of Copper existing in the sample; on the other hand, for the evaluation of the chemical elements that can damage the leaching process; finally, the study of other interesting elements to leach in a smaller proportion.

For this purpose, the study was performed with the sample prepared according to the methodology—the elemental analysis test. The elemental analysis test is reflected in Table 1.

Table 1. Elemental analysis Carbon, Hydrogen, Nitrogen and Sulfur for mining waste.

Samples	Nitrogen, %	Carbon, %	Hydrogen, %	Sulfur, %
Mining waste	0.001 ± 0.002	3.735 ± 0.051	0.393 ± 0.014	0.647 ± 0.024

The elemental analysis of the mining waste sample shows a very low percentage of Carbon. This low percentage of Carbon confirms the low proportion of organic matter and carbonates. By this analysis, it is difficult to identify specifically what it is; however, the value is adequate and does not present problems. The low percentage of Hydrogen may be due to the transformation of hydrated compounds or to the unavoidable residual humidity from the test process. However, as with Carbon, this value is quite small. Finally, it can be seen that there is a percentage of Sulfur evaluated by means of this test in the gases analyses of the combustion of the sample at 1000 ± 10 °C. The existence of this low percentage of Sulphur reflects the transformation of primary minerals, Polymetallic Sulphides, into oxides. However, it should be noted that the percentage of Sulfur determined by elemental analysis is not the total of that existing in the sample. Subsequently, we evaluated the percentage of Sulfur remaining by the X-ray fluorescence method.

The loss of ignition test reflects the change in weight of the sample when it is subjected to a temperature of 1000 ± 10 °C. This variation in weight is due to different reasons. On the one hand, the loss of Carbon from organic matter and carbonates; on the other hand, the loss of the most volatile chemical elements, such as Hydrogen, Nitrogen and Sulfur. Finally, it is due to the transformation of some chemical compounds and the oxidation of the chemical elements. The loss of ignition of the mining waste sample under study is 19.28 ± 0.36%. This loss of ignition, after observing the reduced percentages shown by the elemental analysis of the elements analyzed, is mainly due to the transformation of minerals and their oxidation. This process that was carried out was the one commonly used in hydrometallurgy for leaching the treatment of chalcopyrite in acidic media, also called roasting. The value of loss on ignition is usual and similar to those obtained in tests of this type for residues.

X-ray fluorescence determined the percentage of the chemical elements the highest atomic weight. For this purpose, the test was carried out on the calcined sample at a temperature of 1000 ± 10 °C. This procedure was carried out because the equipment was not capable of detecting chemical elements, such as Carbon, so if it was not carried out with the calcined sample, the X-ray fluorescence equipment would distribute the percentage of chemical elements detected without taking this it into account, and therefore with an error. The results of the X-ray fluorescence test performed on the mining waste sample after the loss on ignition test are shown in Oxygen compounds, because they have been transformed. The results of the X-ray fluorescence test are shown in Table 2.

Table 2. X-ray fluorescence of the mining waste sample.

Compound	wt, %	Est. Error
CaO	31.40	0.23
SiO_2	19.49	0.20
Fe_2O_3	9.01	0.14
Al_2O_3	6.10	0.12
CuO	5.85	0.12
MgO	2.73	0.08
K_2O	1.52	0.06
MnO	1.37	0.06
PbO	1.06	0.05
ZnO	0.989	0.049
S	0.440	0.022
TiO_2	0.317	0.016
BaO	0.209	0.010
P_2O_5	0.0704	0.0035
NiO	0.0354	0.0018
As_2O_3	0.0353	0.0026
Co_3O_4	0.0179	0.0013
SnO_2	0.0159	0.0014
SrO	0.0149	0.0007
Cr_2O_3	0.0140	0.0014
ZrO_2	0.0117	0.0009
Y_2O_3	0.0114	0.0008
Rb_2O	0.0111	0.0006
CdO	0.0088	0.0010
V_2O_5	0.0042	0.0015
Ag_2O	0.0024	0.0009

The results in Table 2 reflect the chemical composition of the mining waste in oxide compounds; therefore, we proceeded to eliminate the Oxygen and recalculate the percentage of the chemical elements present in the mining waste sample, obviously taking into account the percentage of Oxygen. The elemental composition of the mining waste obtained by X-ray fluorescence is shown in Table 3.

Table 3 shows the percentage that the different chemical elements analyzed present in the sample, as well as the percentage of Oxygen due to the oxides present in the mining waste after oxidation in the calcination process. The sum of all the elements, plus the Oxygen and the loss of ignition reflects 100% of the sample; therefore, this method is the most reliable for the determination of the real percentage of each of the chemical elements that exist in the sample, independent of their combination in different chemical compounds.

Table 3. Elementary chemical composition obtained by X-ray fluorescence from the mining waste sample.

Element	wt, %	Est. Error
Ca	22.45	0.17
Si	9.11	0.09
Fe	6.30	0.10
Al	3.23	0.06
Cu	4.67	0.09
Mg	1.65	0.05
K	1.26	0.05
Mn	1.06	0.05
Pb	0.98	0.05
Zn	0.794	0.040
S	0.440	0.022
Ti	0.1900	0.0095
Ba	0.1870	0.0093
Px	0.0307	0.0015
Ni	0.0278	0.0014
As	0.0267	0.0020
Co	0.0132	0.0009
Sn	0.0125	0.0011
Sr	0.0126	0.0006
Cr	0.0096	0.0010
Zr	0.0087	0.0006
Y	0.0090	0.0006
Rb	0.0101	0.0005
Cd	0.0077	0.0009
V	0.0023	0.0008
Ag	0.0022	0.0009
Total Weight% Oxygen	28.23	0.33

In view of the results, it can be seen that the percentage of Copper, the main element of this study, is 4.67 ± 0.09%. This Copper is the one that is intended to be extracted by leaching in acidic media at atmospheric pressure and ambient temperature. The percentage of Copper is contemplated as a viable percentage for its recovery; that is to say, it is not a reduced percentage, but it would be if presenting problems to be extracted by pyrometallurgical means. This fact is due to the diversity of existing minerals, the combination of them and the existence of some harmful elements for the pyrometallurgical process that will be detailed later. In short, the present percentage of Copper for the development of this new hydrometallurgical technique is correct and be expected from the material to which it corresponds.

Calcium represents a high percentage in the composition, corresponding to feldspars and granite mica, as does Aluminium. It should be remembered that granite is the main rock in which metallic sulphide deposits have been formed, so its presence in the waste dump is important. However, Calcium and Aluminum are not chemical elements that harm the hydrometallurgical leaching process, so they are not a problem. Similarly, silicon is present in a significant percentage as granite is a siliceous rock.

The Magnesium, Manganese and Potassium, as in the previous cases, come directly from the granite, and it is not unusual to find them in the percentages detailed in this rock.

Lead comes from the galena, a mineral mainly extracted in the mine workings of the seam. The percentage of Lead is reduced, as the waste belongs to the waste dumps of the waste rock materials of the mine, so the mining process drastically reduced the percentage of Lead in the waste. In a much smaller proportion, Silver appears, associated to Lead in the argentiferous galena mineral.

Zinc, belonging to the blende mineral and recognized in the process of mineralogical study, appears in low proportion in the composition. Therefore, it is not the main element of extraction by leaching in acidic media, but it can be an extracted in lower recovery rate after obtaining Copper, also being an element quoted by the industry.

The other chemical elements are in very small proportion, making wet process tests necessary to quantify them under higher precision. It is common to find, in these types of mining formations, Arsenic Sulphides that would seriously damage the pyrometallurgical process; however, in hydrometallurgy, its influence is much smaller and the percentage it represents in the mining waste is very low. It should be noted that the low percentage of Sulphur reflects the transformation and oxidation of the Polymetallic Sulphides mentioned by the action of the atmospheric conditions.

In short, the percentage of Copper existing in the sample, and the inexistence of chemical elements that could damage the process in high percentages, make possible and viable the subsequent leaching section of the sample in acidic media at atmospheric pressure and ambient temperature, based on hydrometallurgical techniques.

3.2. Leaching Process and Concentration Measurement

The chemically characterized mining waste sample was then divided into several subsamples. These subsamples were subjected to the leaching process with the equipment detailed in Figure 2, with different solutions of Sulphuric Acid, 0.05, 0.10, 0.15 and 0.20 molar. The process was carried out at ambient temperature (25 ± 1 °C) and atmospheric pressure for 96 h, taking samples of the leachate every 24 h. It should be noted that, at the end of the leaching process, no sludge was found at the bottom of the vessel, confirming the suitability of the sample grading for this purpose.

Firstly, and as a main element of study, the concentrations of Copper in the leachate for the four Sulphuric Acid solutions and at different leaching times are detailed in Figure 3.

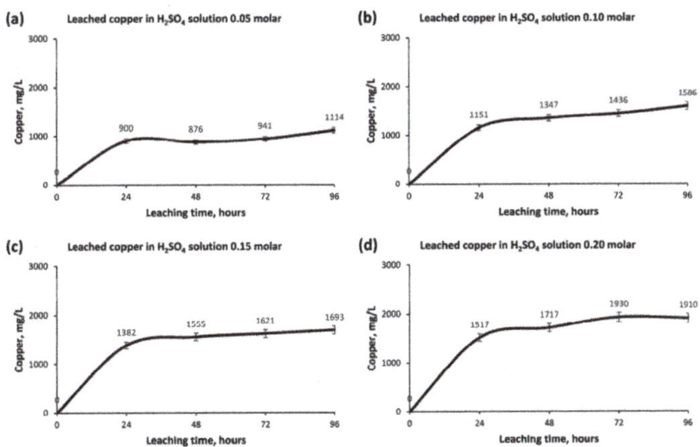

Figure 3. Analysis of Copper concentration in the leaching process at atmospheric pressure, ambient temperature and in different Sulphuric Acid solutions: (**a**) 0.05 molar H_2SO_4 solution, (**b**) 0.10 molar H_2SO_4 solution, (**c**) 0.15 molar H_2SO_4 solution, (**d**) 0.20 molar H_2SO_4 solution.

As can be seen in Figure 3, the concentration of Copper in the leachate is higher for the 0.20 molar Sulphuric Acid solution, the concentration being progressively higher with the increasing molar of the solution. Sulphuric Acid solutions of 0.05 molar, 0.10 molar and 0.15 molar, reflect the fact that there is no stabilization in the leach after 96 h, with a reduced upward slope in all three solutions. Therefore, more time leaching is necessary to achieve the results of the 0.20 molar of Sulphuric Acid solution. On the other hand, in the solution of 0.20 molar of Sulphuric Acid, if there is a stabilization of the concentration after 72 h, it shows very similar values of Copper concentration after a longer times.

In turn, most of the Copper leaching occurs in the first 24 h of the process and for the four Sulphuric Acid solutions. The percentage increase in Copper concentration after the first 24 h is about 20%.

The results reflect a better behavior of the 0.20 molar Sulphuric Acid solution because it needs a shorter leaching time, as well as a stabilization of the concentrations at 72 h. Taking into account the mass of the sample tested, the volume of the solution and the percentage of Copper in the sample, it can be stated that, under the prescribed conditions and with the detailed equipment, there is an approximate recovery of 82% of the Copper in the sample with the 0.20 molar solution of Sulphuric Acid. It is therefore a good result of copper extraction given the material it comes from (mining waste), the leaching conditions (atmospheric pressure and ambient temperature), the low molar of Sulphuric Acid of the solution and the more economical extraction process.

Higher values of Copper recovery rates would require higher concentrations of Sulphuric Acid and the smaller particle sizes of the sample. This would negatively affect the principles of this study, since we would create a leachate with higher Sulphuric Acid and it would be more difficult to treat, as well as a higher percentage of sludge that would damage the leaching process and subsequently be very difficult to reuse due to its environmental pollution. In addition, it is important to note that a concentration of almost 2 g/L of Copper was obtained, this being an acceptable value for classical hydrometallurgical techniques with Copper Oxide minerals and viable for subsequent treatment of the extraction.

The results of the Copper concentrations of the four Sulphuric Acid solutions, at different times and in the mentioned leaching process, are reflected for comparison in Figure 4.

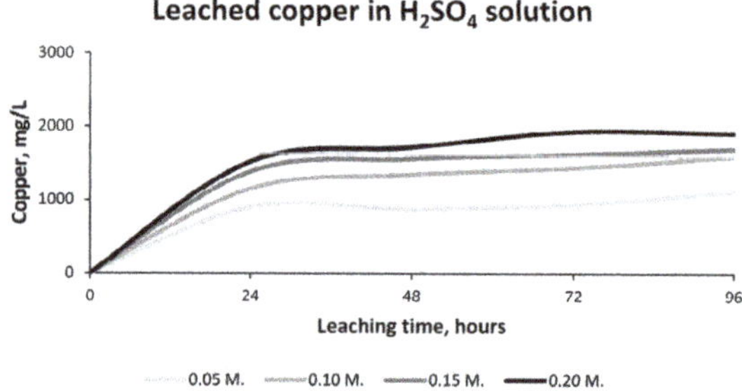

Figure 4. Comparison of the Copper concentrations in the four Sulphuric Acid solutions with different leaching times and leaching process at ambient temperature and atmospheric pressure.

Figure 4 shows the comparison of the Copper concentration in the three leachates, reflecting the existence of a great similarity between the 0.10, 0.15 and 0.20 molar Sulphuric Acid concentrations, taking into account the detail commented on before the non-stabilization of the Copper concentration in the time evaluated for the 0.10 and 0.15 molar Sulphuric Acid solutions. They are, therefore, the three acceptable solutions of Sulphuric Acid for its dissolution for obtaining similar concentrations of Copper. The leaching of the other chemical elements will determine the advantages and disadvantages of each solution. For this purpose, we proceed to describe the concentration in the leachate of the metallic chemical elements that exist in the sample in greater proportions.

The Iron concentration for the four acidic solutions, at different times of the leaching process at ambient temperature and atmospheric pressure, are detailed in Figure 5.

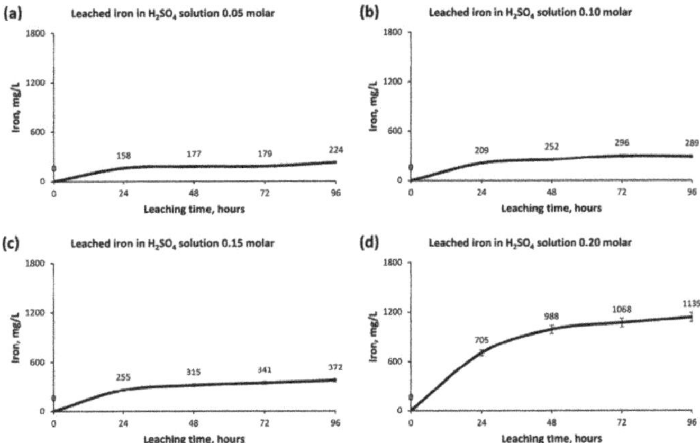

Figure 5. Analysis of the Iron concentration in the leaching process at atmospheric pressure, ambient temperature and in different Sulphuric Acid solutions: (**a**) 0.05 molar H_2SO_4 solution, (**b**) 0.10 molar H_2SO_4 solution, (**c**) 0.15 molar H_2SO_4 solution and (**d**) 0.20 molar H_2SO_4 solution.

The concentrations of Iron in the leachates of the mining waste sample for the four Sulphuric Acid solutions clearly reflect a higher suitability of the 0.20 molar Sulphuric Acid solution, while its concentration in the other solutions is very low. Nevertheless, and based on what has been commented on, the solutions do not present the stabilization of the concentrations of Iron. This fact clearly reflects the need for a longer leaching time, as well as solutions with a greater molarity of Sulphuric Acid for leaching. The maximum percentage of recovery is approximately 36% of the Iron present in the sample.

However, the Iron is not the element that intended to be extracted in this process, focusing the extraction on Copper and secondary elements of greater interest at present. Therefore, it is not recommended to use a higher molarity in the solution of Sulphuric Acid or a longer leaching time, since it would induce a higher extraction of Iron that would complicate the main process of Copper extraction. Therefore, the Iron extraction values for these Sulphuric Acid solutions can be considered acceptable, not being high and not harming the extraction of Copper as the main element.

In turn, the concentration values of Magnesium in the leachates of the different solutions in the process at ambient temperature and atmospheric pressure are detailed in Figure 6.

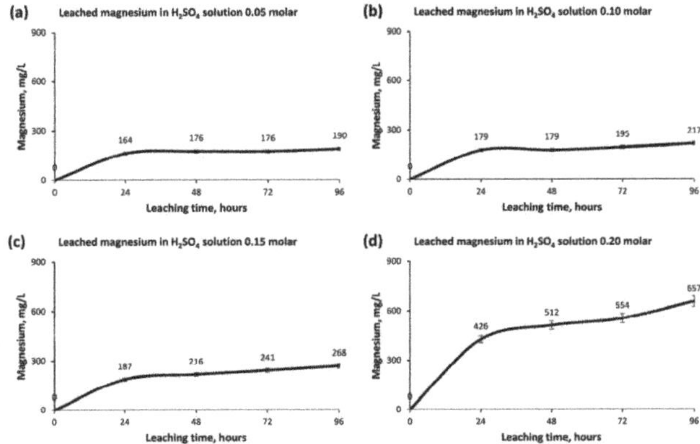

Figure 6. Analysis of the concentration of Magnesium in the leaching process at atmospheric pressure, ambient temperature and in different solutions of Sulphuric Acid: (**a**) 0.05 molar H_2SO_4 solution, (**b**) 0.10 molar H_2SO_4 solution, (**c**) 0.15 molar H_2SO_4 solution and (**d**) 0.20 molar H_2SO_4 solution.

In the concentration of Magnesium in the different Sulphuric Acid solutions, it can be seen, as in the case of Iron, that the leaching process cannot extract the right percentage of Magnesium. The ascending trend of the 0.20 molar Sulphuric Acid solution clearly reflects this fact. A higher concentration of Sulphuric Acid or a longer leaching time is therefore necessary to exceed the 80% recovery rate of Magnesium. However, the recovery reflects a high value, so Magnesium is not the main element of extraction and to obtain an adequate recovery value, it is not recommended to vary the molarity of the solution nor increase the leaching time, as happens in the case of the Iron. At the same time, the low percentage of recovered Magnesium does not negatively influence the Copper recovery process, this being a hydrometallurgical process which is highly robust and with capacity for the differentiated extraction of the different elements.

The concentration in the Zinc leachates for the different Sulphuric Acid solutions in the process at ambient temperature, atmospheric pressure and different leaching times, are detailed in Figure 7.

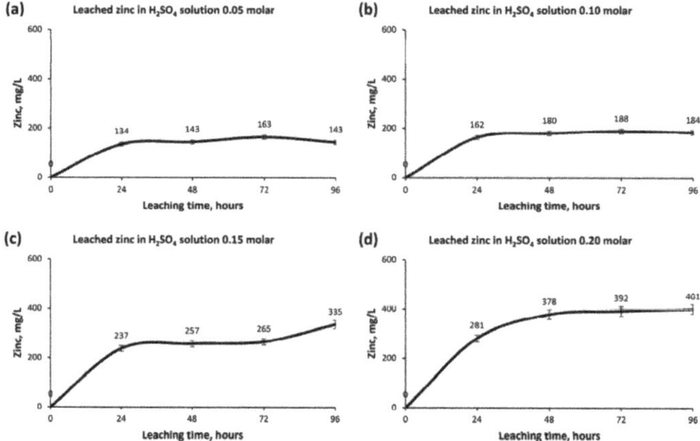

Figure 7. Analysis of Zinc concentration in the leaching process at atmospheric pressure, ambient temperature and in different Sulphuric Acid solutions: (**a**) 0.05 molar H_2SO_4 solution, (**b**) 0.10 molar H_2SO_4 solution, (**c**) 0.15 molar H_2SO_4 solution and (**d**) 0.20 molar H_2SO_4 solution.

In the case of Zinc, behaviors similar to the process of Copper, but even better behaved, occur. The recovery rate of Zinc for the solution of Sulphuric Acid with 0.20 molarity is practically total, obtaining an excellent behavior of the solution. The solutions with smaller molarity of Sulphuric Acid produce a smaller concentration of Zinc, showing better quality results in the 0.15 molar solution than the two solutions of the smaller molarity.

In the solution of Sulphuric Acid with 0.20 molarity, the results of the concentration of Zinc are stabilized, being practically maximum in the first 48 h of the leaching process.

Zinc, in a low proportion in the sample, is a strategic element for obtaining it by hydrometallurgical methods. Therefore, obtaining this secondary element in the leachate makes its later extraction possible, since the cost of the process is assumed by the recovery of Copper and Zinc, which is one more incentive for its economic viability.

Based on the above, the solutions and exposure times are adequate, a higher molarity of Sulphuric Acid or a longer leaching time for the concentration of Zinc not being necessary.

On the other hand, the results of the Manganese concentration in the different Sulphuric Acid solutions and in the different exposure times for the leaching process at ambient temperature and atmospheric pressure are detailed in Figure 8.

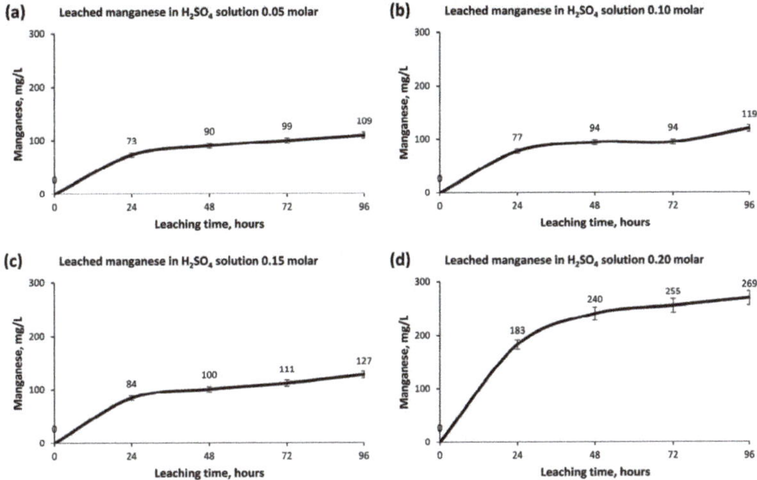

Figure 8. Analysis of the concentration of Manganese in the leaching process at atmospheric pressure, ambient temperature and in different solutions of Sulphuric Acid: (**a**) 0.05 molar H_2SO_4 solution, (**b**) 0.10 molar H_2SO_4 solution, (**c**) 0.15 molar H_2SO_4 solution and (**d**) 0.20 molar H_2SO_4 solution.

In the case of Manganese, something similar to Magnesium and Iron occurs. There is an inefficiency for the complete extraction of Manganese in the solutions and in the proposed leaching times, the recovery rate being approximately 50%. However, this element is not in the scope of this study and therefore its extraction is unnecessary.

Figure 9 shows the results of the leaching process for Nickel.

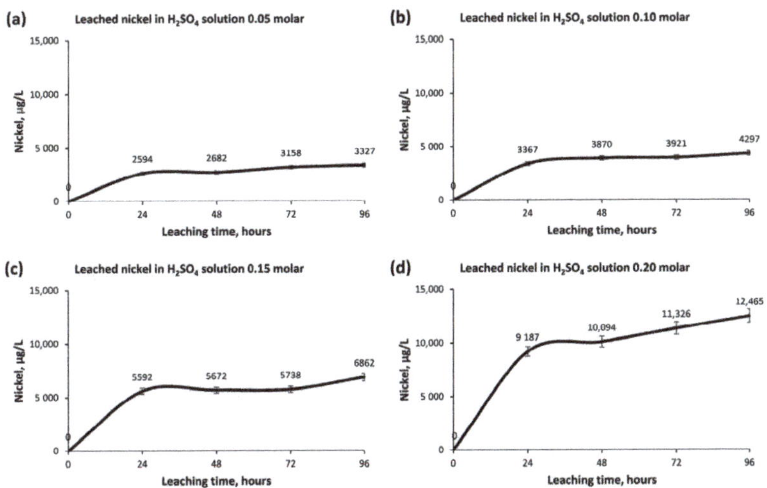

Figure 9. Analysis of the Nickel concentration in the leaching process at atmospheric pressure, ambient temperature and in different Sulphuric Acid solutions: (**a**) 0.05 molar H_2SO_4 solution, (**b**) 0.10 molar H_2SO_4 solution, (**c**) 0.15 molar H_2SO_4 solution and (**d**) 0.20 molar H_2SO_4 solution.

The results of the Nickel concentration in the different solutions and for the different times of the above-mentioned leaching process reflect a similar behavior to Zinc and Copper. The solution of Sulphuric Acid with 0.20 molarity represents the best results, obtaining recovery rates of up to 90%. The other solutions of lower molarity of Sulphuric Acid show incomplete leaching, without the stabilization of concentrations in the stipulated times. The element Nickel is found in low proportion in the sample; however, it is interesting to note a practically full recovery. It is worth noting that it does not influence the Copper extraction process.

The concentrations of Lead in the leachates of the four solutions at different times and in the leaching process are reflected in Figure 10.

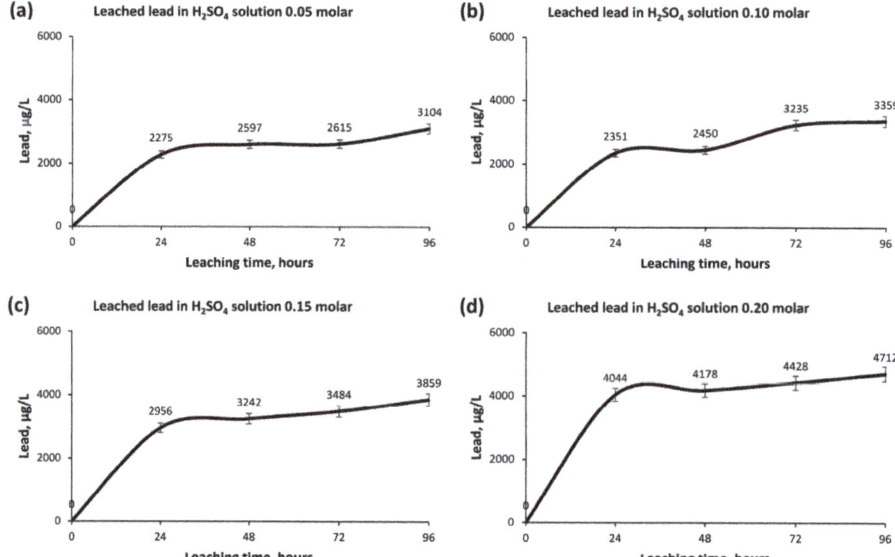

Figure 10. Analysis of Lead concentration in the leaching process at atmospheric pressure, ambient temperature and in different Sulphuric Acid solutions: (**a**) 0.05 molar H_2SO_4 solution, (**b**) 0.10 molar H_2SO_4 solution, (**c**) 0.15 molar H_2SO_4 solution and (**d**) 0.20 molar H_2SO_4 solution.

The concentration of Lead in the leachate of the different solutions presents a minimum recovery rate, so this hydrometallurgical method is not suitable for the recovery of Lead. The low concentration of Lead in the leachates, the problems that it can cause in the extraction of Copper, the reduced percentage in the original sample, as well as the inactivity of Lead before the leaching process, make it possible to affirm that the results are acceptable. This assertion is based on the fact that no unnecessary element is leached in the process that later remains in the leachate and is difficult to treat. This is a good result considering the low interest of Lead in the present study.

On the other hand, the results of the Titanium concentration in the leaching solutions and in the different times, are shown in Figure 11.

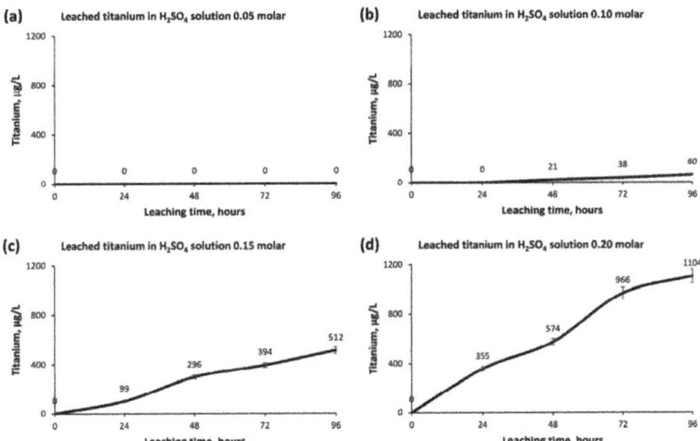

Figure 11. Analysis of the Titanium concentration in the leaching process at atmospheric pressure, ambient temperature and in different Sulphuric Acid solutions: (**a**) 0.05 molar H_2SO_4 solution, (**b**) 0.10 molar H_2SO_4 solution, (**c**) 0.15 molar H_2SO_4 solution and (**d**) 0.20 molar H_2SO_4 solution.

Titanium, as reflected by the x-ray fluorescence test, represents a very low percentage in the mining waste sample and its extraction is not interesting. However, it shows a very interesting behavior to determine by comparison the quality of the Copper leaching process. Unlike the results obtained from the Copper concentrations, the concentrations in the Titanium solutions perfectly show the inactivity of the 0.05 and 0.10 molar Sulphuric Acid solutions. The solutions with molarities 0.15 and 0.20 reflect, on the other hand, the beginning of activity, showing an upward trend without reaching the stabilization in the time and an important difference between the concentrations of the last two solutions. This fact reflects the inadequacy of this leaching process for Titanium, which is not in the scope of the study, and the notable difference with the Copper recovery process is, therefore, determined objectively by the quality of the leaching process for copper recovery from mining waste.

Finally, Figure 12 shows the concentrations of Arsenic in the different solutions and in the different leaching times.

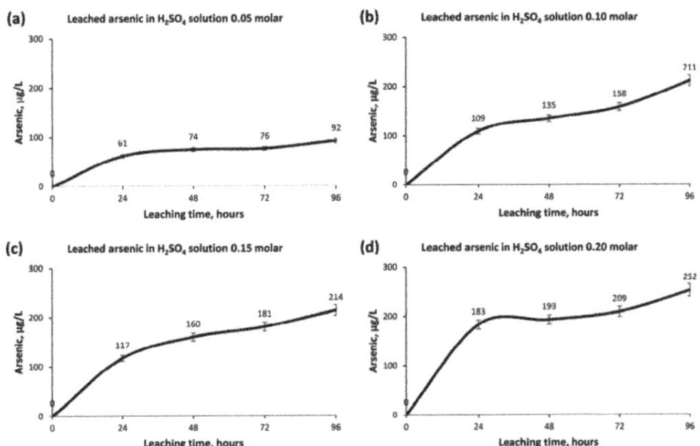

Figure 12. Analysis of Arsenic concentration in the leaching process at atmospheric pressure, ambient temperature and in different Sulphuric Acid solutions: (**a**) 0.05 molar H_2SO_4 solution, (**b**) 0.10 molar H_2SO_4 solution, (**c**) 0.15 molar H_2SO_4 solution and (**d**) 0.20 molar H_2SO_4 solution.

Arsenic is not an element of interest in recovery by this hydrometallurgical process for different reasons; among them, its low percentage in the sample and the environmental problem that it represents. However, it is an element that seriously damages the pyrometallurgical process and, to a lesser extent, the hydrometallurgical process. Therefore, the low existence of this element in the sample and the low recovery that has been obtained in the leaching, approximately 2%, confirms the quality of the leachate obtained for the extraction of Copper. Furthermore, it confirms the adequacy of the molarity of the Sulphuric Acid in the solutions and the leaching times, since a higher molarity for the extraction of a percentage of Copper would cause an unnecessary leaching of the elements commented as well as of the Arsenic, damaging the process and not obtaining substantial differences in the extraction of Copper due to the increase of the molarity of the Sulphuric Acid solution.

Figure 13 shows the maximum leaching values obtained for the elements detailed above with the 0.20 molar solution of Sulphuric Acid, at ambient temperature and atmospheric pressure.

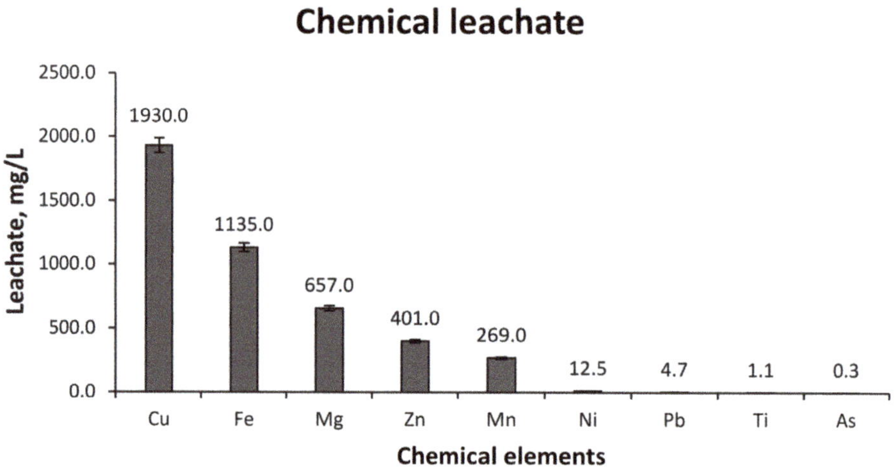

Figure 13. Comparison of the maximum concentration values of the different elements analyzed in the 0.20 molar Sulphuric Acid solution. Leaching process at ambient temperature and atmospheric pressure.

Figure 13 shows how Copper is the element with the highest concentration in the leachate, reaching values of 1.930 ± 0.007 g/L. This value represents an adequate concentration for the viability of the process, causing a rate of recovery of the Copper of the mining waste of approximately 82%. Excellent values if the origin of the raw material, the simplicity of the process and the energy consumption associated with it are taken into account. On the other hand, the element subsequently leached and unavoidable is Iron, in a much lower concentration and with a low recovery rate. This fact is very interesting, since it is not an element of interest for its extraction and its extraction in large proportions can be a problem. The same happens with Magnesium, in much lower concentration than Copper.

The compatibility of the hydrometallurgical process at ambient temperature, atmospheric pressure and low molar acidic media in the extraction of Zinc should be highlighted. The concentration of Zinc in the leachate is even higher than that of Manganese, the latter being an element in greater proportion than Zinc in the sample. The rate of extraction of Zinc from mining waste is almost complete, obtaining low values of concentration in the leachate, but it is interesting to value its extraction as a secondary element.

Contrary to what has been commented on, Lead is not leached in great proportion by the hydrometallurgical process, a fact that benefits the subsequent process of Copper extraction. It is not an element of interest and remains in the waste sample after the leaching process.

Nickel and Titanium appear in very low concentrations in the leachate and, therefore, their percentages are practically negligible. It is worth mentioning the good compatibility of Nickel with the leaching process used, achieving a recovery rate of about 90% and, in turn, achieving a higher concentration in the leachate than Lead and Titanium.

Finally, the low percentage of Arsenic in the leachate should be highlighted. This is an interesting fact since there is a very low recovery rate and it will not condition the hydrometallurgical process, unlike pyrometallurgical techniques in which Arsenic is totally limiting.

4. Conclusions

The partial conclusions obtained from the results of all the tests mentioned in the methodology will be the fundamental basis for the corroboration of the final hypothesis, the leaching of Copper in acidic media, at atmospheric pressure and ambient temperature from mining waste samples for extraction. The importance of using a mining waste, deposited in the vicinity of the seam, without current use and with problems of contamination of underground water and surface water due to its chemical composition, should be highlighted in this study.

The partial conclusions are detailed below.

- The mine waste dump from which the sample was taken for analysis contained mainly granite, Iron Sulphides, Copper Sulphides, Lead Sulphides and, to a lesser proportion, Zinc Sulphides. The Polymetallic Sulphides had been transformed into Oxides by their continuous exposure to the atmospheric conditions and their treatment.
- The chemical composition of the sample under study reflected a low percentage of carbonates and organic matter, with the presence of Copper in a proportion of 4.67%, being a useful sample for the extraction of Copper by hydrometallurgical techniques. There were also percentages of Iron, Lead and Zinc, as well as a low proportion of Arsenic.
- The leaching process was done at ambient temperature so as not to increase the production costs of the process.
- The low molarities of the Sulphuric Acid solutions were selected to avoid subsequent environmental problems with the leachate after the extraction of Copper and elements of interest. In addition, higher molarities of Sulphuric Acid in the solution would have directly caused higher leaching of other elements that are not of interest to extract and would impair the process of extracting Copper from the leachate.
- The particle size between 6 mm and 10 mm of the mine waste sample did not produce sludge at the end of the leaching process for any of the Sulphuric Acid solutions evaluated.
- The leaching times were limited to 96 h, not being necessary longer times as in other hydrometallurgical processes, existing at present in which the leaching time is of weeks and even months.
- The 0.20 molar Sulphuric Acid solution yielded results at 72 h of Copper concentrations of 1.930 ± 0.007 g/L and with recovery rates of approximately 80% of the Copper in the sample. This value is a good result for the subsequent hydrometallurgical stages
- The 0.15 and 0.10 molar Sulphuric Acid solutions give similar results to those of the 0.20 molar Sulphuric Acid solution; nonetheless, which is higher.
- The greatest leaching of Copper occurs in the first 24 h, producing a small increase in concentration afterwards until its stabilization after 72 h.
- Zinc with the 0.20 molar Sulphuric Acid solution has an almost complete recovery rate, obtaining maximum concentrations of 0.401 ± 0.002 g/L. It is therefore a secondary element that can be economically evaluated the interest of its extraction. A similar process occurs with Nickel, but its concentration is very low, so its extraction is of no interest.
- Lead reflected a total incompatibility with the leaching process; therefore, it achieved the objective of not obtaining its leaching and that it remained in the mining sample after the leaching process.

- The other elements, Magnesium, Manganese and Titanium, did not obtain important rates of recovery, a fact that favors the hydrometallurgical process in its later extraction and allows us to recognize the aptitude of the Copper leaching.
- Arsenic, a very harmful element in Copper pyrometallurgical processes and, to a lesser extent, in hydrometallurgical processes, obtained a very low concentration in the leachate and a very low recovery rate. This fact benefits the hydrometallurgical technique.

Based on this, it can be stated that the extraction of Copper from the mining waste in the waste dump belonging to the detailed seam and called "EL COBRE" is feasible, as well as the leaching process in acidic media with low molarity, atmospheric pressure and ambient temperature.

It is worth noting that adequate and comparable Copper leaching values have been obtained to those of other traditional hydrometallurgical processes, through the use of a mining waste (which is not used and which implies environmental pollution) and with a simple, robust and versatile leaching process. Due to these characteristics, this process has a lower economic cost than traditional processes and consumes less energy than other processes, as it is carried out at atmospheric pressure, ambient temperature and in shorter leaching times. At the same time, the low molarity and the absence of sludge after the leaching process make the process more sustainable. These are the competitive advantages of this study compared to pyrometallurgical techniques and other hydrometallurgical processes.

On this basis, it is possible to verify the usefulness of the procedure and of the starting hypotheses for the solution of the problems existing at present in the extraction of Copper, mainly due to the lower grades of the minerals. This is because the lower economic cost of the initial material and low process costs make copper mineral grades viable.

Author Contributions: Conceptualization, F.A.C.-I., F.J.L.d.R., J.M.T.-S. and J.S.-M.; methodology, F.A.C.-I., F.J.L.d.R., J.M.T.-S. and J.S.-M.; software, J.M.T.-S. and J.S.-M.; validation, F.A.C.-I. and F.J.L.d.R.; formal analysis, F.A.C.-I. and F.J.L.d.R.; investigation, J.M.T.-S. and J.S.-M.; resources, F.A.C.-I.; data curation, F.J.L.d.R.; writing—original draft preparation, J.S.-M.; writing—review and editing, J.M.T.-S.; visualization, J.M.T.-S.; supervision, F.A.C.-I.; project administration, J.S.-M.; funding acquisition, F.A.C.-I. All authors have read and agreed to the published version of the manuscript.

Funding: This research received no external funding.

Acknowledgments: Technical and human support provided by CICT of Universidad de Jaén (UJA, MINECO, Junta de Andalucía, FEDER) is gratefully acknowledged.

Conflicts of Interest: The authors declare no conflict of interest.

References

1. Li, L.; Pan, D.; Li, B.; Wu, Y.; Wang, H.; Gu, Y.; Zuo, T. Patterns and challenges in the copper industry in China. *Resour. Conserv. Recycl.* **2017**, *127*, 1–7. [CrossRef]
2. Wang, M.; Chen, W.; Zhou, Y.; Li, X. Assessment of potential copper scrap in China and policy recommendation. *Resour. Policy* **2017**, *52*, 235–244. [CrossRef]
3. Wang, H.T.; Liu, Y.; Gong, X.Z.; Wang, Z.H.; Gao, F.; Nie, Z.R. Life Cycle Assessment of Metallic Copper Produced by the Pyrometallurgical Technology of China. *Mater. Sci. Forum* **2015**, *814*, 559–563. [CrossRef]
4. Bonnin, M.; Azzaro-Pantel, C.; Domenech, S.; Villeneuve, J. Multicriteria optimization of copper scrap management strategy. *Resour. Conserv. Recycl.* **2015**, *99*, 48–62. [CrossRef]
5. Shuva, M.A.H.; Rhamdhani, M.A.; Brooks, G.A.; Masood, S.; Reuter, M.A. Thermodynamics data of valuable elements relevant to e-waste processing through primary and secondary copper production: A review. *J. Clean. Prod.* **2016**, *131*, 795–809. [CrossRef]
6. Copper: World Mine Production, By Country. Available online: https://www.indexmundi.com/en/commodities/minerals/copper/copper_t20.html (accessed on 19 May 2020).
7. Kowalczuk, P.; Bouzahzah, H.; Kleiv, R.; Aasly, K. Simultaneous Leaching of Seafloor Massive Sulfides and Polymetallic Nodules. *Minerals* **2019**, *9*, 482. [CrossRef]
8. Watling, H.R. Chalcopyrite hydrometallurgy at atmospheric pressure: 1. Review of acidic sulfate, sulfate-chloride and sulfate-nitrate process options. *Hydrometallurgy* **2013**, *140*, 163–180. [CrossRef]

9. Meshram, P.; Prakash, U.; Bhagat, L.; Abhilash; Zhao, H.; van Hullebusch, E.D. Processing of Waste Copper Converter Slag Using Organic Acids for Extraction of Copper, Nickel, and Cobalt. *Minerals* **2020**, *10*, 290. [CrossRef]
10. Hernández, P.C.; Dupont, J.; Herreros, O.O.; Jimenez, Y.P.; Torres, C.M. Accelerating copper leaching from sulfide ores in acid-nitrate-chloride media using agglomeration and curing as pretreatment. *Minerals* **2019**, *9*, 250. [CrossRef]
11. Hernández, P.; Taboada, M.; Herreros, O.; Graber, T.; Ghorbani, Y. Leaching of Chalcopyrite in Acidified Nitrate Using Seawater-Based Media. *Minerals* **2018**, *8*, 238. [CrossRef]
12. Chen, J.; Wang, Z.; Wu, Y.; Li, L.; Li, B.; Pan, D.; Zuo, T. Environmental benefits of secondary copper from primary copper based on life cycle assessment in China. *Resour. Conserv. Recycl.* **2019**, *146*, 35–44. [CrossRef]
13. Jones, D.J. CESL Copper Process. In Proceedings of the Alta Copper Hydrometallurgy Forum, Brisbane, Australia, 14–15 October 1996.
14. Watling, H.R.; Elliot, A.D.; Maley, M.; van Bronswijk, W.; Hunter, C. Leaching of a low-grade, copper-nickel sulfide ore. 1. Key parameters impacting on Cu recovery during column bioleaching. *Hydrometallurgy* **2009**, *97*, 204–212. [CrossRef]
15. Wang, S. Copper leaching from chalcopyrite concentrates. *JOM* **2005**, *57*, 48–51. [CrossRef]
16. Dreisinger, D. Copper leaching from primary sulfides: Options for biological and chemical extraction of copper. *Hydrometallurgy* **2006**, *83*, 10–20. [CrossRef]
17. van Staden, P.J. The Mintek/Bactech copper bioleach process. In Proceedings of the ALTA Copper Hydrometallurgy Forum, Brisbane, Australia, 19–21 October 1998.
18. Gericke, M.; Govender, Y.; Pinches, A. Tank bioleaching of low-grade chalcopyrite concentrates using redox control. *Hydrometallurgy* **2010**, *104*, 414–419. [CrossRef]
19. Hourn, M.; Halbe, D. The Nena Tech Process: Results on Frieda River copper gold concentrates. In Proceedings of the International Conference of Randol Copper Hydromet Roundtable 1999, Phoenix, AZ, USA, 10–13 October 1999; pp. 97–102.
20. Hourn, M.M.; Turner, D.W.; Holzberger, I.R. Atmospheric Mineral Leaching Process. U.S. Patent 5,993,635, 30 November 1999.
21. Cerda, C.; Taboada, M.; Jamett, N.; Ghorbani, Y.; Hernández, P. Effect of Pretreatment on Leaching Primary Copper Sulfide in Acid-Chloride Media. *Minerals* **2017**, *8*, 1. [CrossRef]
22. Arroyo, F.; Fernández-Pereira, C.; Bermejo, P. Demonstration Plant Equipment Design and Scale-Up from Pilot Plant of a Leaching and Solvent Extraction Process. *Minerals* **2015**, *5*, 298–313. [CrossRef]
23. Peacey, J.; Guo, X.J.; Robles, E. Copper hydrometallurgy—Current status, preliminary economics, future direction and positioning versus smelting. *Trans. Nonferrous Met. Soc. China (Engl. Ed.)* **2004**, *14*, 560–568.

Article

Petrogenetic Constraints of Early Cenozoic Mafic Rocks in the Southwest Songliao Basin, NE China: Implications for the Genesis of Sandstone-Hosted Qianjiadian Uranium Deposits

Dong-Guang Yang [1], Jian-Hua Wu [1,2,*], Feng-Jun Nie [1,2,*], Christophe Bonnetti [1], Fei Xia [1,2], Zhao-Bin Yan [1], Jian-Fang Cai [3], Chang-Dong Wang [3] and Hai-Tao Wang [3]

1 State Key Laboratory of Nuclear Resources and Environment, East China University of Technology, Nanchang 330013, China; yangdg@ecut.edu.cn (D.-G.Y.); christoph.bonnetti@gmail.com (C.B.); fxia@ecit.cn (F.X.); zbyan@ecit.cn (Z.-B.Y.)

2 College of Earth Science, East China University of Technology, Nanchang 330013, China

3 No. 243 Geological Party, The China National Nuclear Corporation, Chifeng 024006, China; awfi0013@126.com (J.-F.C.); 18747688243@163.com (C.-D.W.); Simply721@126.com (H.-T.W.)

* Correspondence: jhwu@ecit.cn (J.-H.W.); niefj@ecit.cn (F.-J.N.); Tel.: +86-0791-83897801 (J.-H.W.)

Received: 14 October 2020; Accepted: 12 November 2020; Published: 14 November 2020

Abstract: The tectonic inversion of the Songliao Basin during the Cenozoic may have played an important role in controlling the development of sandstone-type uranium deposits. The widely distributed mafic intrusions in the host sandstones of the Qianjiadian U ore deposits provided new insights to constrain the regional tectonic evolution and the genesis of the U mineralization. In this study, zircon U-Pb dating, whole-rock geochemistry, Sr-Nd-Pb isotope analysis, and mineral chemical compositions were presented for the mafic rocks from the Qianjiadian area. The mafic rocks display low SiO_2 (44.91–52.05 wt.%), high ${}^{T}Fe_2O_3$ contents (9.97–16.46 wt.%), variable MgO (4.59–15.87 wt.%), and moderate K_2O + Na_2O (3.19–6.52 wt.%), and can be subdivided into AB group (including basanites and alkali olivine basaltic rocks) and TB group (mainly tholeiitic basaltic rocks). They are characterized by homogenous isotopic compositions (ε_{Nd} (t) = 3.47–5.89 and $^{87}Sr/^{86}Sr$ = 0.7032–0.7042) and relatively high radiogenic $^{206}Pb/^{204}Pb$ (18.13–18.34) and Nb/U ratios (23.0–45.6), similar to the nearby Shuangliao basalts, suggesting a common asthenospheric origin enriched with slab-derived components prior to melting. Zircon U-Pb and previous Ar-Ar dating show that the AB group formed earlier (51–47 Ma) than the TB group (42–40 Ma). Compared to the TB group, the AB group has higher TiO_2, Na_2O, K_2O, P_2O_5, Ce, and HREE contents and Ta/Yb and Sr/Yb ratios, which may have resulted from variable depth of partial melting in association with lithospheric thinning. Combined with previous research, the Songliao Basin experienced: (1) Eocene (~50–40 Ma) lithospheric thinning and crustal extension during which mafic rocks intruded into the host sandstones of the Qianjiadian deposit, (2) a tectonic inversion from extension to tectonic uplift attributed to the subduction of the Pacific Plate occurring at ~40 Ma, and (3) Oligo–Miocene (~40–10 Ma) tectonic uplift, which is temporally associated with U mineralization. Finally, the close spatial relation between mafic intrusions and the U mineralization, dike-related secondary reduction, and secondary oxidation of the mafic rocks in the Qianjiadian area suggest that Eocene mafic rocks and their alteration halo in the Songliao Basin may have played a role as a reducing barrier for the U mineralization.

Keywords: Qianjiadian uranium deposit; Songliao Basin; mafic rocks; tectonic inversion; reducing barrier; U mineralization

1. Introduction

The tectonic evolution of sedimentary basins from extension to compression has been regarded as the key factor to the formation of sandstone-type uranium deposits [1–6]. For example, tectonic movements controlled the deposition of sediments, fluid flow, and climatic variations within the basin [7], and the timing of uranium mineralization is generally related to tectonic uplifts [5]. Moreover, the relationship between basic rocks emplaced during extension and uranium deposition has always attracted much attention. Previous studies mainly focused on the effects of mafic dikes on uranium mineralization in volcanic- and granite-related uranium deposits [8–12]. Except for U mineralization associated with mafic intrusions in Proterozoic sandstones [13], the roles of mafic dikes in the genesis of sandstone-hosted uranium deposits have been poorly constrained.

The Qianjiadian uranium deposit has been widely regarded as a typical interlaminar oxidation zone-type uranium deposit in the southern part of the Songliao Basin (Figure 1) [14–16]. Many publications have documented the sedimentary facies, depositional environment, sandstone petrography and geochemistry, mineral paragenesis and textures, mineralization geochronology, and ore genesis model for the deposit [17–21]. In the Qianjiadian area, uranium is mainly adsorbed or reduced on carbonaceous debris or migrated oil, and U minerals such as pitchblende and coffinite are intimately associated with iron disulphides [22–27]. Several studies were conducted on the micro-morphological observations, in situ sulfur isotope analysis of pyrites, and carbon isotope composition of calcite cement, emphasizing the role of bacterial sulphate reduction (BSR) in the genesis of the U mineralization in the Qianjiadian area [28–33]. Nevertheless, the tabular U ore bodies in the Qianjiadian deposit are spatially associated with secondary reduced gray sandstones or bleached white sandstones [34] (Figure 2) together with the widely distributed mafic rocks [17], which significantly differ from the two predominant models of redox front sandstone-hosted U deposits presented in [33]. In addition, detailed petrographic studies performed on carbonaceous debris from this deposit showed significant increase of the vitrinite reflectance, suggesting that the U mineralization is spatially associated with the alteration halo of the diabase dikes [20]. However, the emplacement age and geochemical signatures of these mafic rocks have not been carefully constrained, and their possible role in the genesis of the Qianjiadian deposit has to be clearly characterized.

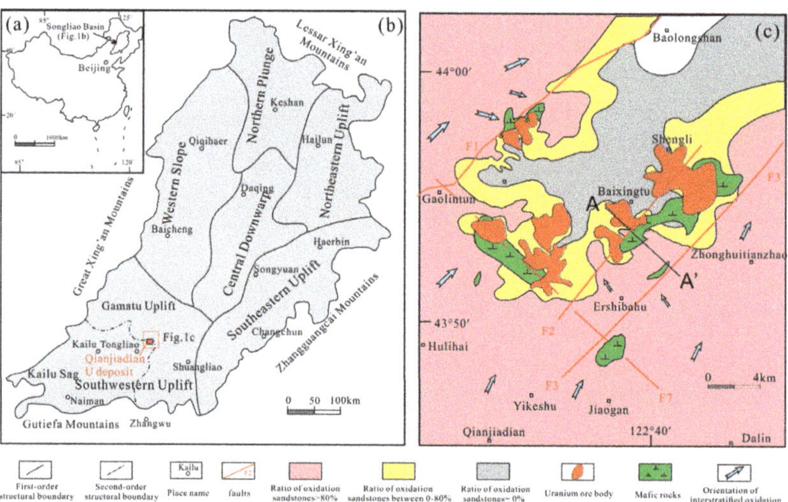

Figure 1. (**a**) A sketch map showing the location of the Songliao Basin. (**b**) Tectonic units of the Songliao Basin (modified from [35]). (**c**) Spatial distribution of the interstratified oxidation and mafic rocks in the Qianjiadian uranium deposit, modified after [19,34].

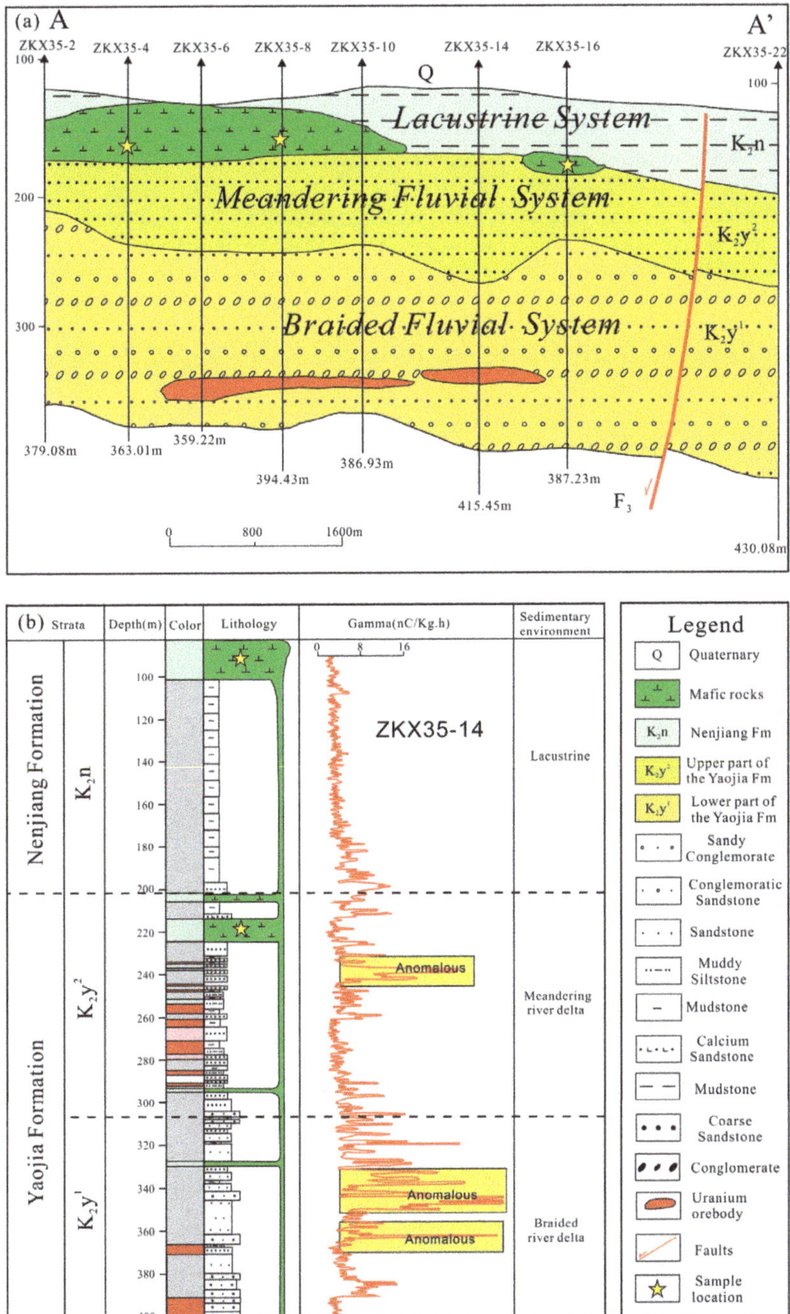

Figure 2. Geological cross-section (**a**) and lithological column of drill hole ZKX35-14 (**b**) in the Qianjiadian uranium deposit showing the relationship between diabase, faults, and uranium mineralization (the location is shown in Figure 1b).

Moreover, the apatite and zircon fission-track data of the Baixingtu ore deposits recorded two distinct stages of rapid cooling after the Late Cretaceous, with stage I at ~80–50 Ma and stage II at ~40–10 Ma, respectively [5]. Based on whole rock U-Pb isotopic dating and geochemical features of the Qianjiadian deposit, Luo et al. [18] proposed a multi-stage genetic model, with an early stage of synsedimentary U mineralization (96 ± 14 Ma), later overprinted by oil and gas migration (67 ± 5Ma) and hydrothermal fluids percolation (40 ± 3 Ma). Recently, U-Th-Pb chemical ages obtained by electron probe micro-analyzer (EPMA) revealed two younger and more dominant stages of mineralization at 43–28 Ma and 19–3 Ma, respectively [36]. Most of the uranium mineralization occurs around the tectonic uplifts with ages younger than 40 Ma [36], and these tectonic uplifts and sandstone-type uranium deposits show close spatial and temporal relationships [5].

Moreover, Cheng et al. [6] proposed that Miocene tectonic uplift (~40–10 Ma) of the basin caused large-scale uranium mineralization. Although the tectonic uplift events have been extensively studied, when and how the tectonic inversion from extension to uplift occurred has not been carefully constrained. Mafic rocks that generally occur in extension environment are widespread in the sandstone-hosted uranium ore deposits in the Qianjiadian area, Songliao Basin (Figure 1c), and have intruded into the sandstones of the Yaojia and Nenjiang formations (Figure 2). Based on whole-rock Ar–Ar dating, Xia et al. [15] proposed that the diabase dikes were emplaced at 49.5 ± 5 Ma. However, LA-ICP-MS analysis of zircons from mafic dike in Baixingtu ore deposit yielded an age of 70.0 ± 3.0 Ma [37]. Therefore, detailed geochronology and geochemistry of the mafic rocks could provide important information for regional tectonic evolution of the basin.

Here, zircon U-Pb ages, whole-rock geochemistry, Sr-Nd-Pb isotopic compositions, and mineral chemical compositions were presented for the mafic rocks in the Qianjiadian U deposit in the Songliao basin, Northeastern China. These new data combined with previous low-temperature thermochronologic data are then used to discuss regional uplift and extension events, and their geodynamic mechanism and implications for uranium mineralization.

2. Geological Setting

The Songliao Basin is a large oil, gas, and uranium producer and covers an area of 260,000 km^2 in NE China [35,38]. It is bounded by the Zhangguangcai Range to the east, the Great Xing'an Range in the northwest, the Lesser Xing'an Range to the northeast, and the North China Craton in the south [39] (Figure 1b). This Mesozoic–Cenozoic sedimentary basin trends north-northeast to south-southwest and was formed on a folded basement. This basement is mainly composed of Paleozoic to Mesozoic metamorphic and igneous rocks [40]. This basin began with syn-rift volcanogenic successions during the period from 150 to 105 Ma, followed by post volcanic thermal sagging between 105 and 79.1 Ma, and ended with regional uplift and basin inversion from 79.1 to 40 Ma [41]. Sediments from the Upper Jurassic and Lower Cretaceous Huoshiling (J_3h), Shahezi (K_1s), and Yingcheng (K_1y) formations were deposited in a syn-rifting tectonic setting. The Upper Cretaceous Quantou (K_2q), Qingshankou (K_2qn), Yaojia (K_2y), and Nenjiang (K_2n) formations developed during the post-rift thermal subsidence of the basin [35,38]. The Yaojia Formation is mainly constituted of fluvial and deltaic coarse-grained deposits, whereas the Nenjiang Formation is primarily composed of lacustrine fine-grained deposits (Figure 2).

The Qianjiadian uranium deposit is hosted within the Yaojia Formation in the southwestern part of the Songliao Basin and is located at a depth between 200 and 400 m below the surface. Uranium ore bodies with an average grade of 0.0104% to 0.0287% are mainly tabular or lenticular in shape [42]. The ore-hosting sediments are mainly composed of fine-grained sandstone, siltstone, and mudstone with pitchblende and coffinite as the predominant uranium minerals [15–20]. Intrusive mafic rocks are also widespread in sandstones of the Yaojia Formation in the Qianjiadian area (Figure 1c). They are crosscut by relatively abundant carbonate veins and their alteration halo displays a secondary reduced green-white zone characterized by a pervasive green alteration in reduced grey sandstones as described by Bonnetti et al. [27] and the bleaching of oxidized sandstones (Figure 3a). The mineral assemblage of the green alteration includes newly precipitated chlorite, epidote, and carbonate [27,28] that mainly occur as cement filling sandstone porosity. Based on grain sizes, these mafic rocks show fine-grained and medium- to coarse-grained textures. The fine-grained mafic rocks mainly consist of diabase (Figure 3b) and medium-coarse grained mafic rocks mainly composed of gabbro (Figure 3d). The diabase exhibits an ophitic texture and massive structure and is mainly composed of plagioclase (60–65 vol.%), clinopyroxene (30–35 vol.%), and small amounts of olivine (5–10 vol.%) and most grains have a maximum size less than 1 mm. The gabbro is massive and has a fine- to medium-grained granular texture, and consists of plagioclase (45–50 vol.%), clinopyroxene (40–45 vol.%), and small amounts of olivine (5–8 vol.%) and magnetite (3–5 vol.%).

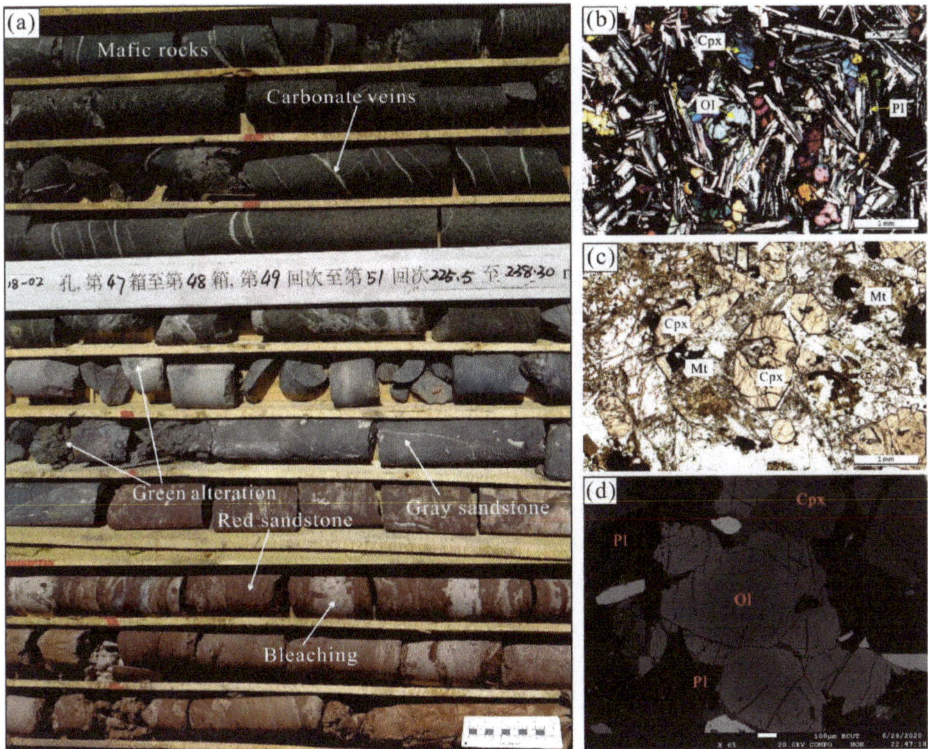

Figure 3. Field photographs and photomicrographs of selected samples from the Cenozoic mafic rocks in the Qianjiadian region. (**a**) Field photographs show the contact relationship of mafic rocks, grey sandstone with green alteration, and secondary reduced (bleaching) white/pink siltstone. (**b**) Microscope photograph of diabases with Ol, Cpx, and Pl. (**c**) Diabases with Mt and Cpx. (**d**) Back scattered electron (BSE) image of diabase with Ol, Cpx, and Pl. Cpx= clinopyroxene; Ol = olivine; Pl = plagioclase; Mt = magnetite.

3. Analytical Methods

3.1. U-Pb Zircon Dating

Zircon separation was undertaken using heavy liquid and magnetic techniques and was further purified by hand picking under a binocular microscope. Separated grains were cast into epoxy resin discs and polished approximately to expose the grain centers. Prior to isotopic analyses, the microstructures of zircon grains were performed using cathodoluminescence (CL) imaging. Zircon U–Pb isotopic analysis was conducted using an Agilent 7500a inductively coupled plasma-mass spectrometer (ICP-MS) instrument equipped with a ComPex102 193 nm ArF excimer laser ablation system at the Key Laboratory of Mineral Resources Evaluation in Northeast Asia, Ministry of Natural Resources of China, Jilin University, China. The detailed description of the analysis technique was described by Liu et al. [43]. Harvard zircon 91500 was used as an external standard to normalize isotopic fractionation, and a standard silicate glass (NIST SRM 610) was used as an external standard to calculate the U, Th, and Pb concentrations of unknowns.

3.2. *Major and Trace Element Determinations*

Bulk rock abundances of major and trace elements were conducted at the Beijing Createch Testing Technology Co. Ltd, Beijing, China. Major element compositions were analyzed by X-ray fluorescence spectrometer (XRF, Rigaku RIX 2100), with analytical uncertainties of 1%–5%. Trace elements were conducted using a Perkin-Elmer Elan 6000 ICP-MS after acid digestion of the samples in high-pressure Teflon bombs, with analytical uncertainties between 1% and 3%. The National standards GSR1, GSR3, BCR-2, and GSP-2 were chosen to calibrate element abundances of the analyzed samples.

3.3. *Whole-Rock Sr–Nd–Pb Isotope Analysis*

Whole-rock Sr–Nd–Pb isotopic analyses were undertaken using a high-resolution multiple-collector (MC)-ICP-MS (Thermo-Fisher Neptune Plus) at the Beijing Createch Testing Technology Co. Ltd., Beijing, China. Prior to the separation of Sr, Nd, and Pb by ion-exchange techniques, approximately 50~100 mg powder for each sample was dissolved in a PFA beaker with HF + HNO_3 at 100 °C for 6 days. Repeated standard analyses yielded $^{87}Sr/^{86}Sr$ of 0.710247 ± 13 (NBS-987) and $^{143}Nd/^{144}Nd$ of 0.512192 ± 15 (GSB Nd), respectively. For Pb isotope analyses, the CAGS Pb was used as the standard yielded average for $^{206}Pb/^{204}Pb$ = 17.9698 ± 0.0090 (2 SD, n = 11), $^{207}Pb/^{204}Pb$ = 15.5616 ± 0.0008 (2S D, n = 11), and $^{208}Pb/^{204}Pb$ = 38.4036 ± 0.0019 (2 SD, n = 11), respectively.

3.4. *Olivine Composition Analyses*

The major and minor elements of olivine were determined by EPMA (JEOL-JXA 8230) at the State Key Laboratory of Nuclear Resources and Environment, East China University of Technology, Nanchang, China. The operating conditions were conducted using 20 kV accelerating voltage, 300 nA beam current, 2 μm beam diameter, and 20 s peak counting time for major elements (Mg, Fe, Si), 40s peak counting time for Ca and Mn, and 90 s peak counting time for Ni. In details, Si, Mg, Al, and Ca were normalized to olivine, Fe and Mn to diopside, and Ni to ZBA Oxide NiO. The original data reduction was performed using the ZAF correction.

4. Results

4.1. *U-Pb Zircon Geochronology*

The analyzed zircons were clear euhedral–subhedral prisms and showed fine-scale and striped absorption oscillatory zoning (Figure 4). They had high Th/U ratios of 0.51–1.82 (Table S1), indicating their magmatic origin [44]. Four samples yielded ages of 46.6–40.3 Ma that were significantly younger than the stratigraphic ages of the Yaojia Formation (~91–88.5 Ma) [45]. Generally, silica-rich mafic rocks with coarse-grained texture have more potential to crystallize their own zircons. In this study, three samples of medium- to coarse-grained mafic rocks (sample BLS12-1, BLS13-5, and BLS13-7) yielded ages of 42.3–40.3 Ma; nevertheless, only two zircon grains from fine-grained mafic rocks (sample BLS14-2) yielded apparent ages of 46.5 and 46.6 Ma (Table S1).

A total of 20 concordant zircon U–Pb analyses were obtained from sample BLS12-1, yielding apparent ages ranging from 40 ± 1 to 3549 ± 17 Ma. Twelve zircon grains yielded apparent ages between 40 ± 1 and 44 ± 1 Ma, with a weighted mean $^{206}Pb/^{238}U$ age of 42 ± 1 Ma (N = 12, mean square weighted deviation (MSWD) = 4.6; Figure 5a). The remaining zircons yielded apparent ages between 254 and 3549 Ma (Table S1). The younger age of 42 ± 1 Ma represents the crystallization age of the diabase, with the older ages representing the crystallization ages of captured zircon grains entrained by the diabase.

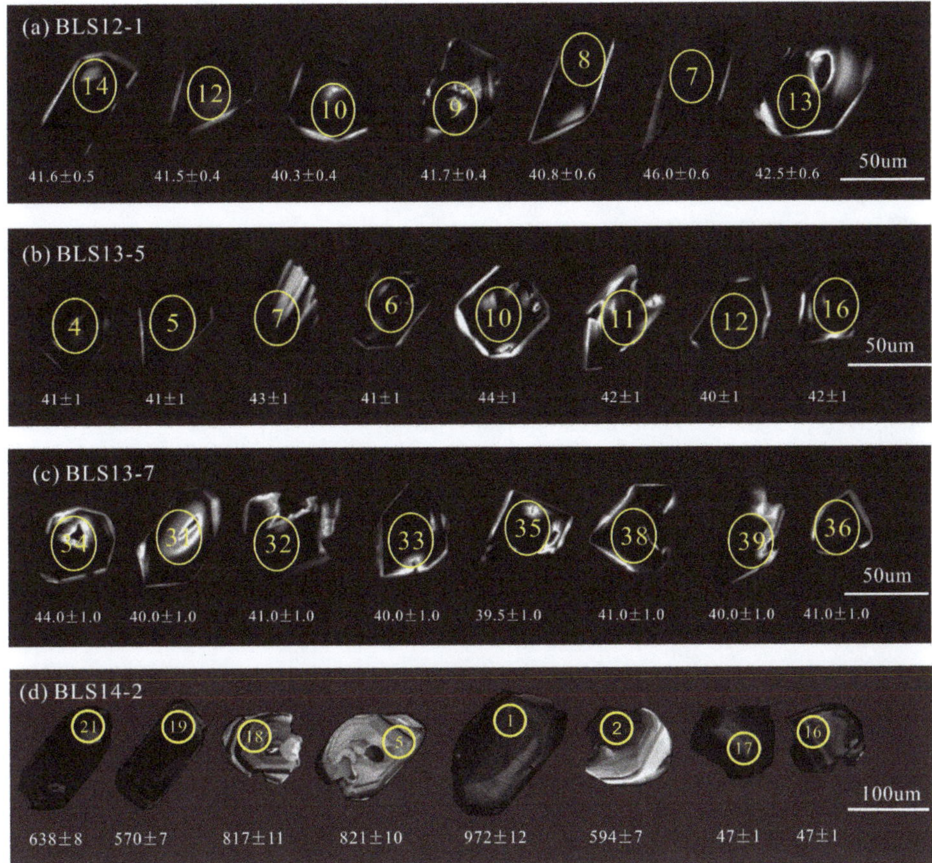

Figure 4. Representative cathodoluminescence (CL) images of selected zircons from the Cenozoic mafic rocks in the Qianjiadian region.

A total of 15 concordant zircon U–Pb analyses were obtained from sample BLS13-5, yielding $^{206}Pb/^{238}U$ ages ranging from 38 to 42 Ma, with a weighted mean $^{206}Pb/^{238}U$ age of 40 ± 1 Ma (N = 16, MSWD = 2.0; Figure 5b), representing the crystallization age of the diabase porphyrite.

A total of 20 concordant zircon U–Pb analyses were obtained from sample BLS13-7, yielding $^{206}Pb/^{238}U$ ages ranging from 40 to 44 Ma with a weighted mean $^{206}Pb/^{238}U$ age of 42 ± 1 Ma (N = 20, MSWD = 1.1; Figure 5c), representing the crystallization age of the diabase.

A total of 25 U–Pb analyses were obtained from sample BLS14-2 (Table S1), and most grains showed apparent ages ranging from 46.5 ± 0.9 to 972 ± 12 Ma (Figure 5d). Two zircon grains yielded $^{206}Pb/^{238}U$ ages of 46.5 and 46.6 Ma, with a weighted mean $^{206}Pb/^{238}U$ age of 47 ± 1 Ma (N = 2, MSWD = 0.01; Figure 5d). The remaining zircons yielded apparent ages between 570 and 898 Ma. These older ages represent the times of crystallization of captured zircons entrained by the diabase, whereas the younger age (47 Ma) was interpreted as the crystallization age of this diabase, which is very similar to the Ar-Ar age of the nearby Shuangliao alkali basalts (48.5~51 Ma) [46].

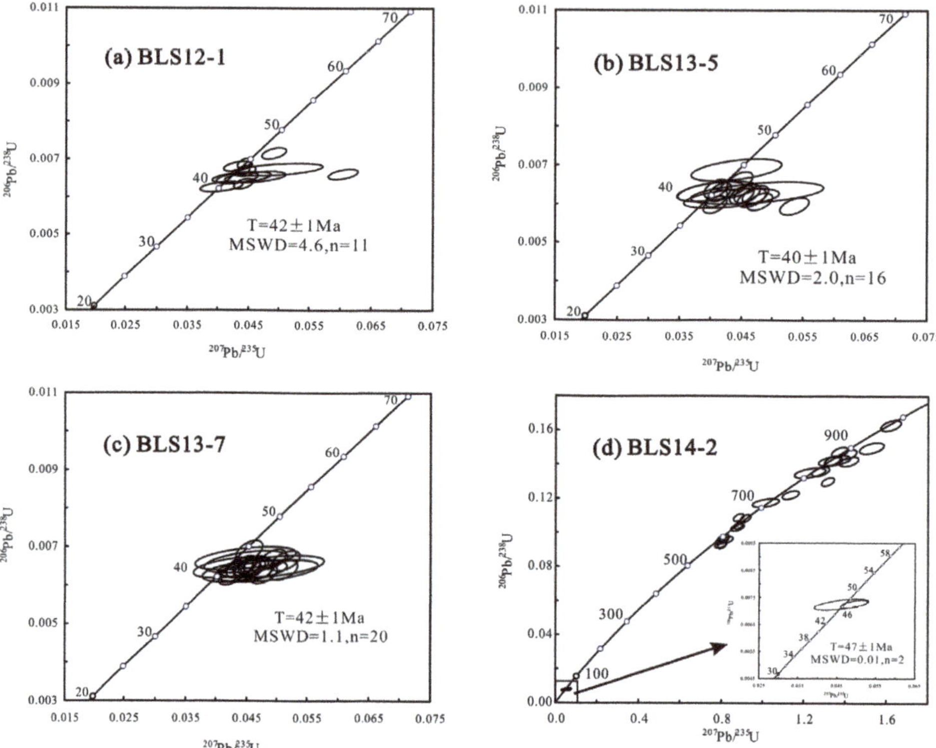

Figure 5. (**a**–**d**) U–Pb concordia diagrams for zircons from samples of Cenozoic mafic rocks in the Qianjiadian area.

4.2. Major and Trace Elements

Major element data for a total of 27 rock samples are given in Table S2. The studied samples fall into two categories in the SiO_2 vs. Na_2O+K_2O (Figure 6a) and Zr/TiO_2 vs. Nb/Y diagrams (Figure 6b): The AB group is predominantly alkali basaltic rocks and the TB group is tholeiitic basaltic rocks (Figure 6d). The CIPW normative calculation further indicates that the AB group is mostly alkali olivine basaltic rocks and basanite, whereas the TB group is predominantly olivine tholeiitic basaltic rocks with minor quartz tholeiitic basalts (Table S2). Both AB and TB groups have high $^{T}Fe_2O_3$ (9.97–16.46 wt.%) and TiO_2 (1.44–2.20 wt.%) contents and high $FeO^t/(FeO^t + MgO)$ ratios (0.65–0.82), indicating a ferroan nature. These rocks also have higher Na_2O contents compared to the Wudalainchi rocks and most of the Keluo-Nuomin samples (Figure 6c). In Harker diagrams (Figure 7), the AB group has higher TiO_2, Na_2O, K_2O, and P_2O_5 contents than the TB group. The TB group has higher SiO_2, Ni, Co, and Sc contents and Al_2O_3/TiO_2 ratios relative to the AB group samples. In general, the increase of Ni and Co contents is positively correlated with the increase of MgO in AB and TB Groups (Figure 7).

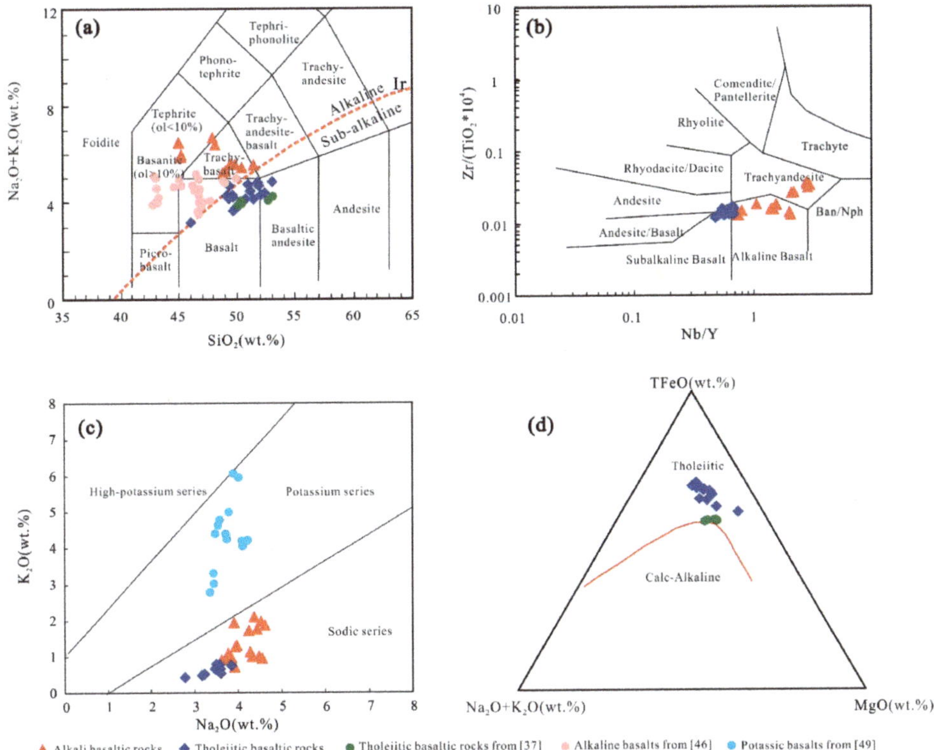

Figure 6. Plots of (**a**) Na_2O+K_2O vs. SiO_2 (after Irvine and Baragar [47]), (**b**) $Zr/(TiO_2 \times 10^{-4})$ vs. Nb/Y (after Winchester and Floyd [48]), (**c**) K_2O (wt.%) versus Na_2O (wt.%) [49], and (**d**) TFeO-Na_2O+K_2O-MgO triangular diagram, the boundary line between alkaline series and the tholeiitic series after Irvine and Baragar [47], A = Na_2O + K_2O (wt.%); F = ^{T}FeO (wt.%); M = MgO (wt.%). The chemical data for the alkaline basalts of Shuangliao [46], Tholeiitic basaltic rock of Qianjaidian areas [37], and potassium basalts of Greater Khingan Range [49] are shown for comparison.

Trace element data are presented in Table S2. For comparison, data for Eocene alkali basalts from Shuangliao and tholeiitic diabases from Qianjiadian are also plotted in Figure 8. All samples are enriched in large-ion lithophile elements (LILE) and light rare earth elements (LREEs), with obvious positive Ba and Sr anomalies and negative Th, U, and P anomalies (Figure 8a). In general, the AB group has higher LREE contents (La = 18.42–53.71 ppm) and LREE/HREE ratios ($(La/Yb)_N$ = 10.53–35.91) than the TB group (La = 7.52–15.29 ppm, $(La/Yb)_N$ = 3.80–7.41). In addition, the samples of AB group show positive Nb–Ta anomalies, which are similar to oceanic island basalt (OIB) (Figure 8b), whereas the TB samples display lower abundances of most trace elements similar to those of enriched mid-oceanic ridge basalt (E-MORB).

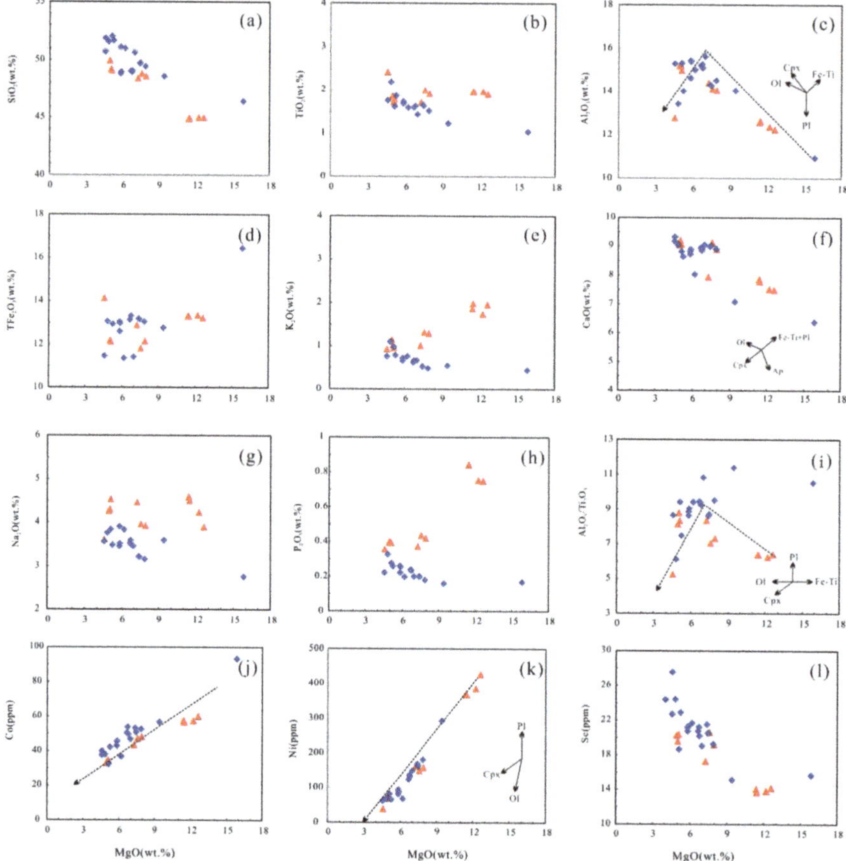

Figure 7. MgO versus major (**a–i**) and trace elements (**j–l**) of the Cenozoic mafic rocks in the Qianjiadian area.

4.3. Sr–Nd–Pb Isotopes

The Sr–Nd–Pb isotope compositions are presented in Table S3. The isotope data of basalts from Hannuoba, Wudalianchi, southeast China, are also plotted for comparison (Figure 9). The Hannuoba and Wudalianchi basalts represent typical indicators of two end members mixing between the depleted mantle (DM) and the enriched mantle 1 (EM1) [50,51] and the basalts from southeast (SE) China record source mixing between DM and EM 2 components [52]. The Qianjiadian mafic rocks exhibit a limited range in Sr-Nd isotopic composition ($^{87}Sr/^{86}Sr$ = 0.7033–0.7043, ε_{Nd} (t) = 3.47–5.89) and plot to the right of the mantle array in the $^{143}Nd/^{144}Nd$ vs. $^{87}Sr/^{86}Sr$ diagram (Figure 9a). This horizontal displacement may reflect the hydrothermal interaction of seawater with oceanic crust [53]. The TB group displays lower ε_{Nd} (t) (3.47–4.32) and higher $(^{87}Sr/^{86}Sr)_i$ (0.7040–0.7042) than AB group (ε_{Nd} (t) = 5.51–5.89; $(^{87}Sr/^{86}Sr)_i$ = 0.7032–0.7033) (Figure 9a,b), indicating more enriched components in their mantle source. The studied samples also exhibit a relatively homogeneous Pb isotopic composition ($^{206}Pb/^{204}Pb$ = 18.13–18.34, $^{207}Pb/^{204}Pb$ = 15.46–15.50, and $^{208}Pb/^{204}Pb$ = 38.18–38.58, Table S3). In the $^{208}Pb/^{204}Pb$ vs. $^{206}Pb/^{204}Pb$ isotopic correlation diagrams, the samples exhibit higher Pb isotopic ratios than those of Hannuoba and Wudalianchi, but lower than those of Cenozoic basalts in southeast China, and plot close to the region the Cenozoic basalts in Indian MORB (Figure 9d).

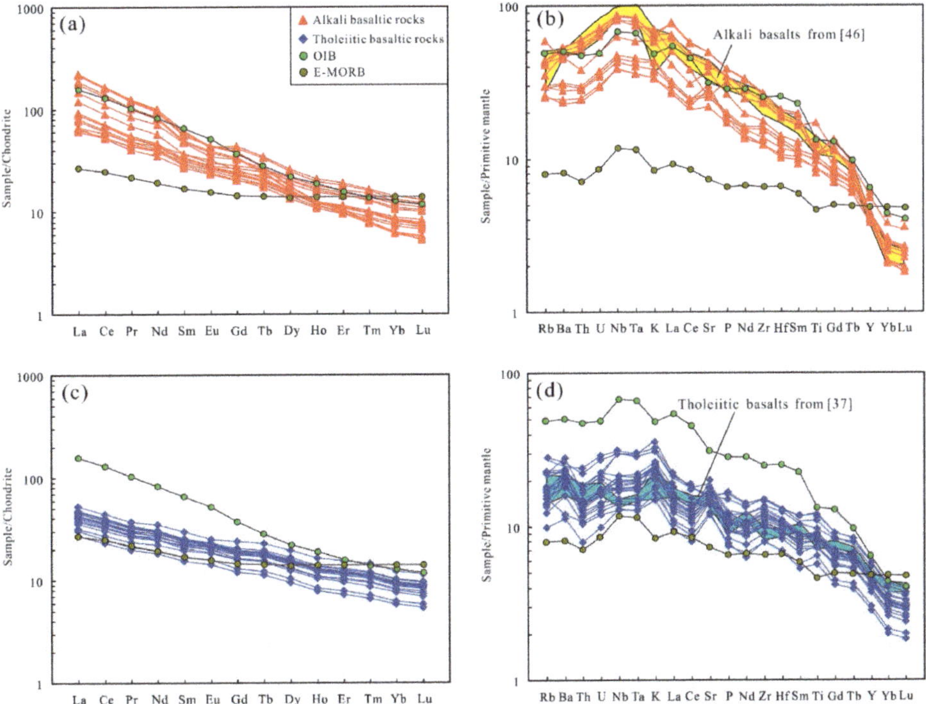

Figure 8. Chondrite-normalized REE patterns (**a**,**c**) and primitive mantle normalized trace element diagrams (**b**,**d**) for the samples from the Qianjiadian region. The values of chondrite and primitive mantle are from Boynton [54] and Sun and McDonough [55], respectively.

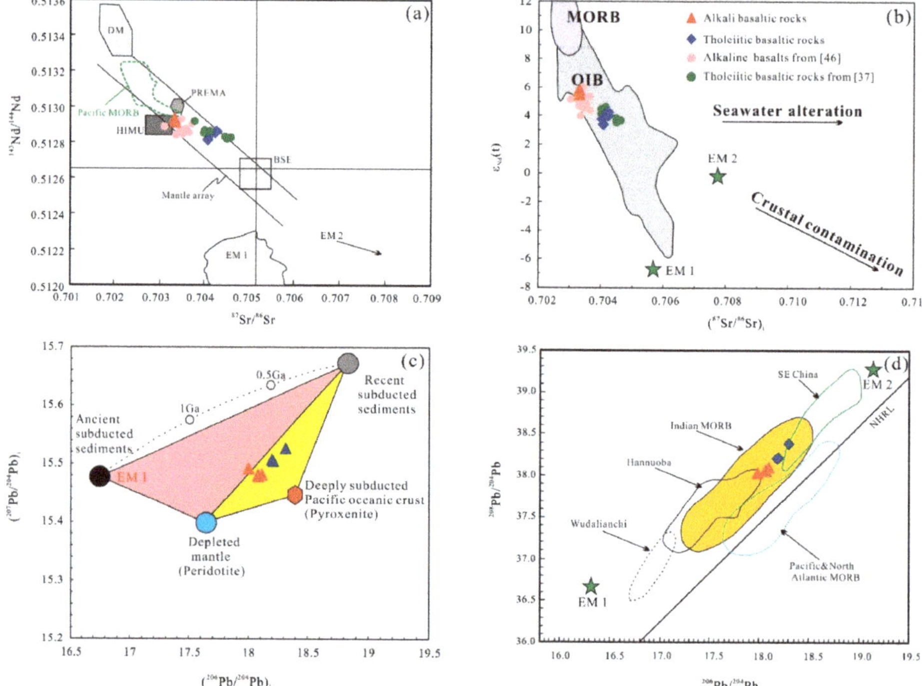

Figure 9. Whole-rock Sr–Nd–Pb isotopic compositions for the Cenozoic Qianjiadian mafic rocks. (**a**) $^{87}Sr/^{86}Sr$ vs. $^{143}Nd/^{144}Nd$ diagram; (**b**) $(^{87}Sr/^{86}Sr)_i$ vs. ε_{Nd} (t) diagram; (**c**) $^{206}Pb/^{204}Pb$ vs. $^{207}Pb/^{204}Pb$; (**d**) $^{206}Pb/^{204}Pb$ vs. $^{208}Pb/^{204}Pb$. DM, depleted mantle; EM-1 and EM-2, enriched mantle-1 and enriched mantle-2; HIMU, mantle with high U/Pb ratio; PREMA, frequently observed prevalent mantle; BSE, bulk silicate Earth; MORB, mid-ocean ridge basalt. The fields of the above mantle reservoirs are from Zindler and Hart (1986) [56]. Pb isotope mixing model is from Kuritani et al. [57]. Fields for the Southeastern (SE) China, Indian Ocean MORB, and Pacific and North Atlantic MORBs are from Chung et al., Barry and Kent (1998), and Zou et al. [52,58,59]. The Northern Hemisphere Reference Line (NHRL) is after Hart [60].

4.4. Mineral Compositions

The chemical compositions of olivine from alkali basaltic rocks (sample BLS01-3) and tholeiitic basaltic rocks (sample BLS07-2) were presented in Table S4 and plotted in Figure 10. The olivine exhibits granular in texture (Figure 10) and is characterized by higher CaO contents (> 0.26 wt.%), but lower NiO (< 0.23 wt.%) and Fo (<75.82) (Figure 11a,b) than the olivine xenocrysts of the peridotite xenoliths entrained from the Shuangliao basalts [61]. These characteristics are similar to magmatic phenocrysts, differing from fragmented mantle-derived xenocrysts.

The zoned olivine phenocrysts from alkali basalts and tholeiitic mafic rocks show consistent variations from core to rim (Figure 10a,b). In general, the rims have relatively lower NiO and Fo and higher CaO, FeO, and MnO contents than the cores (Figure 10c,f), which correspond to light margin and dark interior in BSE images (Figure 10a,b), respectively, indicating the effect of chemical diffusive re-equilibrium between olivine and melt in the process of crystallization [62,63]. We, therefore, favor that only the composition of the cores of olivine phenocrysts can reflect the natural origin of their mantle source.

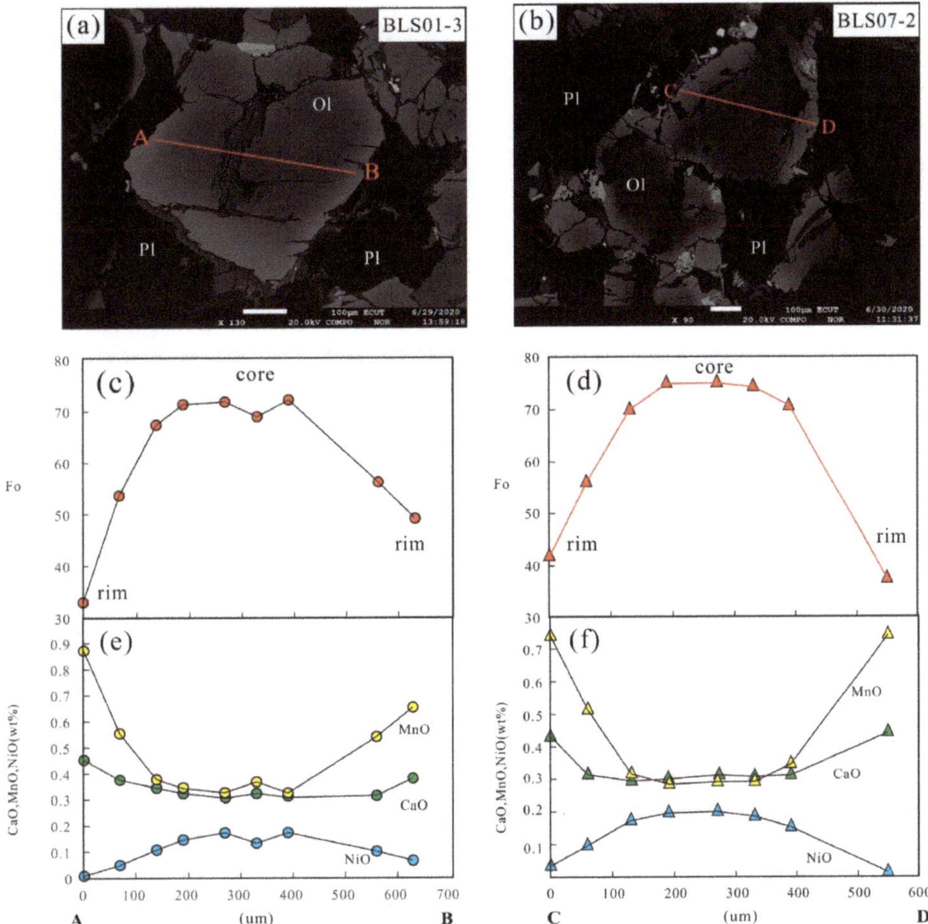

Figure 10. Core-rim compositions of two representative zoned olivine from samples BLS01-3 and BLS07-2 of the Qianjiadian basaltic rocks. Back-scattered electron images show the compositional traverses (**c–f**) marked by red lines (**a**,**b**). The cores have higher Fo and NiO and lower MnO and CaO content than the rims. Fo = 100 × Mg/(Mg + Fe + Mn). For mineral abbreviations, see Figure 3.

4.5. Characteristics of the Host Sandstones and Secondary Alteration Related to Mafic Intrusions

The sediments of the Yaojia Formation mainly consist of sandy conglomerate, very coarse- to fine-grained sands, siltstone, and mudstone. Volumetrically, sandy lithofacies predominate the stratigraphic column (Figure 2b). Sandy facies show a large grain-size distribution from very coarse to very fine sands and display constant stratigraphic occurrence of phytoclasts within primary reduced grey lithofacies [27]. The sandstones hosting the U mineralization in the Qianjiadian area are globally of arkosic sandstone-to-sandstone composition and are mainly grain-supported, except for fine-grained facies or altered lithologies, which are more matrix-supported or show secondary cementation. The primary reduced host sandstones are mainly composed of lithic fragments and individual detrital minerals such as quartz, K-feldspar, plagioclase, biotite, muscovite, and chlorite. In the Qianjiadian area, the U mineralization occurs at the redox interface between the primary reduced and the secondary oxidized sandstones (Figure 2b) or close to a secondary reduced green-white zone [18,27,31] related to mafic rock intrusions that are spatially distributed along the redox boundary

(Figures 1c and 2b). Besides, the content of U from Yaojia Formation sandstones is inversely correlated with REEs' content [17]. Compared with the sandstones far from the mafic intrusions, the sandstones near the mafic intrusions are obviously depleted in REEs [17]. Regionally, sediments of the Yaojia Formation were deposited under a hot and dry climate; hence, resulting in the primary oxidation of part of these sediments (Figure 3a) [64]. Then, the sediments of the Yaojia Formation experienced a secondary reduction related to mafic intrusions, which are characterized by a pervasive green alteration in permeable sandstone (e.g., as shown in [27]) and the bleaching of primary oxidized lithologies (e.g., white sandstone in Figure 3a). The green alteration occurring within the alteration halo of these mafic rocks is mainly characterized by pore filling of the host sandstone with newly formed chlorite, epidote, and carbonate [27]. However, the spatial extension of this secondary reduction remains relatively limited and mostly restricted to the vicinity of the mafic rocks.

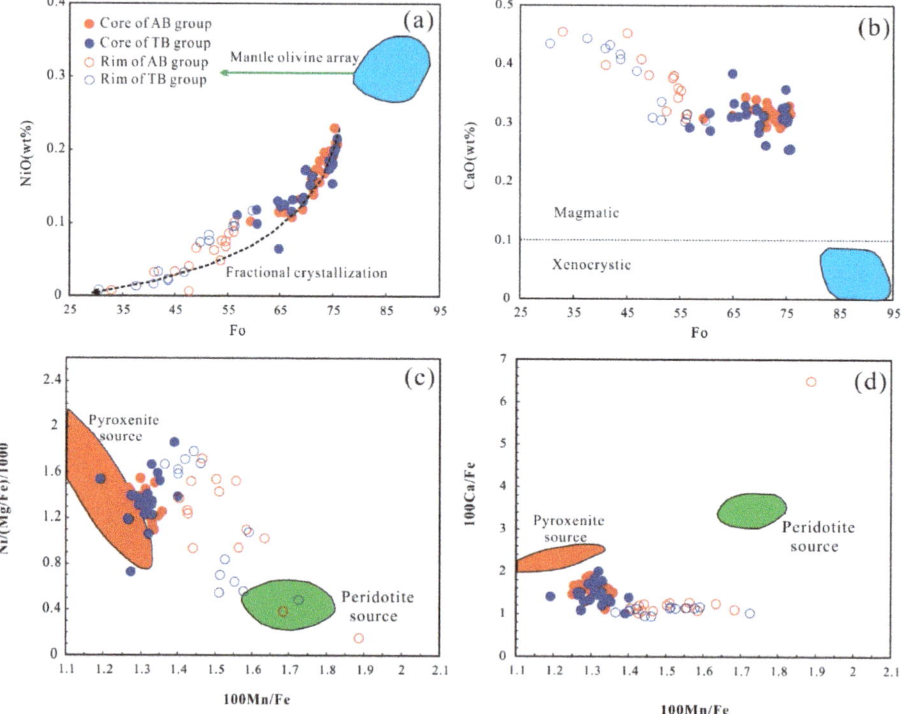

Figure 11. The composition variation of olivines from the Qianjaidian mafic rocks. (**a**) The mantle olivine array and fractional crystallization trend are from [65]. (**b**) The dashed line separates the composition of magmatic and xenocrystic olivine on the basis of CaO [66]. The fields and lines in plots (**c**,**d**) are from [67]. All plots concur in a dominant pyroxenite source for the Qianjaidian mafic rocks.

5. Discussion

5.1. Crystal Fractionation and Crustal Contamination

Both fractional crystallization and crustal contamination could have significantly modified the composition of basalt and, thus, hamper our interpretation of the mantle source. The existence of inherited zircon grains (Figure 5d) suggests that a crustal assimilation occurred during magma ascent or emplacement. However, most mafic rocks have relatively low SiO_2 and high MgO contents (Table S2). In the $^{87}Sr/^{86}Sr$ and ε_{Nd} (t) vs. MgO diagrams (Figure 12), no clear linear relationship is observed, indicating insignificant crustal contamination was involved in the mantle source for the AB and TB group.

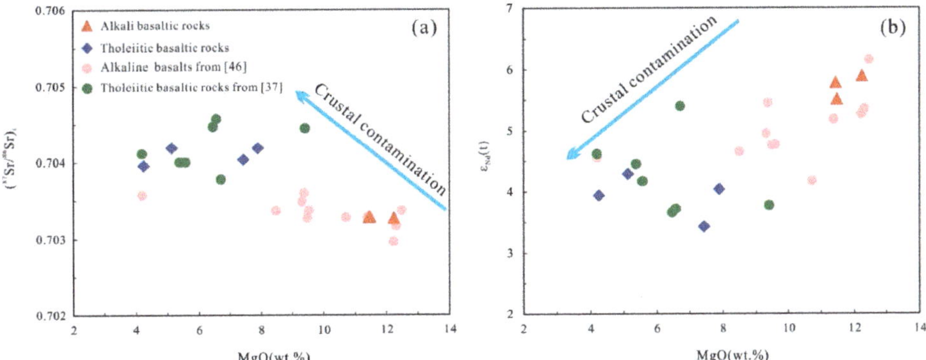

Figure 12. (**a**) Diagram of $(^{87}Sr/^{86}Sr)_i$ versus MgO and (**b**) diagram of $\varepsilon_{Nd}(t)$ versus MgO for the identification of continental contamination.

The low and variable Ni content (63.45–547 ppm) and Mg# values (42.58–65.86) (Table S2), indicating the basaltic samples, have suffered variable degrees of fractional crystallization (Figure 7). All samples show positive correlations of Ni and Co vs. MgO, indicating fractional olivine crystallization (Figure 7j,k). This is consistent with co-variations of NiO with Fo contents in the olivine phenocrysts (Figure 11a). Both AB and TB groups show no obvious Eu anomalies (δEu = 0.81–1.04), indicating that plagioclase fractionation was insignificant. The positive correlation between CaO and MgO content when MgO ≥ 7.0 wt.% and negative linear relation of CaO vs. MgO at MgO < 7.0 wt.% (Figure 7f,i) indicates that olivine was a major crystallizing phase at MgO ≥ 7.0 wt.% but that clinopyroxene started to fractionate at MgO < 7.0 wt.%.

5.2. Mantle Source of the Qianjiadian Mafic Rocks

Continental flood basalts are widely considered as mantle derived. However, it is still unclear whether they are generated from metasomatized lithospheric mantle or partial melting of asthenospheric mantle [68]. In the Sr-Nd diagram (Figure 9a,b), the Qianjiadian mafic rocks have lower $^{87}Sr/^{86}Sr$ and higher $^{143}Nd/^{144}Nd$ ratios than the bulk silicate Earth, indicating a relatively depleted asthenospheric source. Furthermore, the samples from Qianjiadian area are characterized by positive Nb and Ta anomalies and high Nb/La (1.15–1.55), Nb/U (23.0–45.6), and Ce/Pb (3.93–34.9) ratios (those with low Ce/Pb sample may reflect fluid mobile element Pb was affected by post-magmatic processes), with OIB compositional signature (Figure 13a,b), providing another evidence of an asthenospheric origin [69].

However, the enrichment in incompatible elements such as LILEs and LREEs strongly reflecting a fertile composition have been involved in their mantle source. Moreover, Nb/Yb and Ta/Nb ratios overlap with the OIB field and are obviously different from that of MORB (Figure 13c,d). Hence, an additional process must be taken into account for the chemical variation of the Qianjiadian mafic rocks.

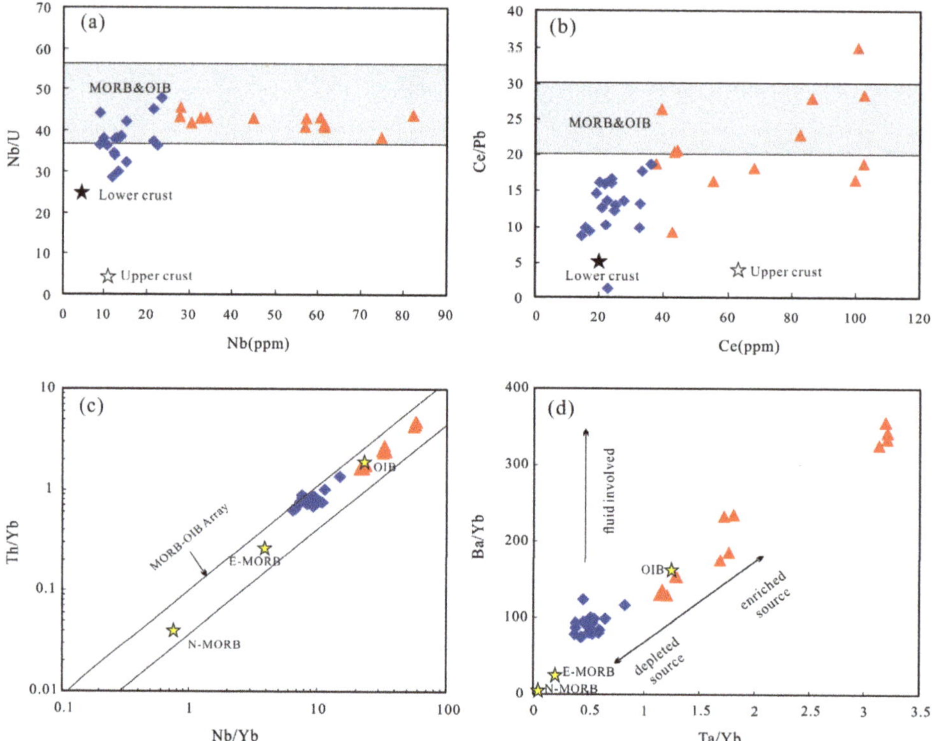

Figure 13. (**a**) Nb/U vs. Nb, (**b**) Ce/Pb vs. Ce, (**c**) Th/Yb vs. Nb/Yb, and (**d**) Ba/Yb vs. Ta/Yb diagrams for the Cenozoic Qianjiadian mafic rocks. Data for MORB and OIB are from [70], data for the upper and lower crusts are calculated after Rudnick and Gao [71], and the field of the MORB-OIB mantle array is from [72].

Ancient subducted continental sediments in the mantle transition zone (MTZ) may comprise K-hollandite with low $^{238}U/^{204}Pb$ value and can eventually evolve to EM 1-type with low $^{206}Pb/^{204}Pb$ after longtime (> 1.5 Ga) separation [73,74], which can be considered as the EM 1 component captured by basaltic magmas in eastern China [75]. However, high $^{206}Pb/^{204}Pb$ for the Qianjiadian mafic rocks cannot be the result of mixing between DM and EM 1 end members (Figure 9c,d). We propose that components of young subducting sediments with high $^{206}Pb/^{204}Pb$ may have been recycled into their mantle source because recent subducting sediments (including those from the Japanese island arc system) have pronounced high radiogenic $^{206}Pb/^{204}Pb$ (18.459–19.452) [76].

The cores of olivine phenocrysts both in the AB and TB group have higher Fe contents but lower Mn/Fe and Ca/Fe ratios than those crystallizing from peridotite melts, suggesting pyroxenite in their mantle source (Figure 11c,d). The low $\delta^{26}Mg$ anomaly and low $\delta^{18}O$ values of the clinopyroxene phenocrysts of Cenozoic continental basalts from the nearby Shuangliao indicate the contribution of water-rich sedimentary carbonate recycled into the upper mantle [77,78]. Recent seismic tomography studies also suggest a stagnant oceanic slab in the MTZ under eastern China [79,80], which can be the result of subduction of the Pacific plate. Some scholars even propose that pyroxenite was formed through reaction of the mantle peridotite with the carbonate-rich melts derived from the subducting oceanic crust [81].

As a whole, recent subducted slab-derived sediments may have an effect on the high radiogenic $^{206}Pb/^{204}Pb$ of the Qianjiadian mafic rocks, which can also be indicated by significant coherence between the mafic rocks and the triangle field determined by Pacific oceanic crust (silica-deficient pyroxenite), depleted mantle (peridotite), and recent subducted sediments (Figure 9c).

5.3. Tectono-Magmatic Significance

The Early Cenozoic magmatic events involved both alkaline and tholeiitic magmatism (AB group and TB group). However, the genetic relationship between this alkaline and tholeiitic magmatism is unclear; for example, it is unclear whether the compositional difference for AB and TB group is the results of crystallization differentiation, variable melting conditions from a similar source, or formed independently.

The AB group displays major and trace element composition different from those of the TB group, especially the lower abundances of most trace elements, such as Rb, Ba, Nb, Ta, K, and Ti of TB group in the primitive mantle-normalized spider diagrams (Figure 8b,d). These features likely indicate these two groups originated from independent sources. However, all the studied samples exhibit a limited range in Sr-Nd isotopic composition (Figure 9) regardless of their petrographic type. Therefore, the analyzed mafic rocks can be regarded as cogenetic, derived from a similar mantle source. The higher total REE contents of AB group than TB group argue against the viewpoint that variable degrees of fractional crystallization caused the compositional differences between the two groups. Therefore, different melting conditions are likely responsible for the compositional variation between AB and TB groups.

It is widely recognized that the melting pressure has an impact on the degree of silica saturation of the magmas [50,82,83]. Particularly, low degrees of partial melting at high pressure condition yield magmas with more normative nepheline (Ne), while partial melting with large degree at lower pressure condition generates magmas with normative hypersthene (Hy) and quartz (Q) [84]. In this study, the AB group has relatively high Ne-normative contents (1.53% to 13.94%), while most samples from the TB group have no Ne content (Table S2).

The plot of La/Yb vs. Sm/Yb has widely been used to distinguish between melting of spinel and garnet peridotite. Considering that Yb is more compatible than La and Sm in garnet, rocks with low degree of partial melting will exhibit high La/Yb and Sm/Yb ratios, whereas rocks with high degrees of batch melting or derived from the spinel stability field will generate magmas with low La/Yb and Sm/Yb ratios [50]. Figure 14a thus reveals that various degrees (1%–20%) of batch melting can produce alkali and tholeiitic mafic rocks in the Qianjiadian U deposit. Specifically, samples of TB group show higher degree of melting (15%–20%) than these AB group (1%–10%). The lower degree of melting in the AB group is consistent with the conclusion investigated using the $(Tb/Yb)_{PM}$–$(Yb/Sm)_{PM}$ diagram (Figure 14b), given that melting degree shows negative correlation with melting depth [85], suggesting that the AB group may have been generated at deeper depth than the TB group.

Zircon U-Pb and previous Ar-Ar dating indicates that the alkali basaltic rocks formed earlier (51~47 Ma) than the tholeiitic basaltic rocks (42–39 Ma). Lithospheric thinning, therefore, can reveal the decreasing alkalinity of the Qianjiadian mafic rocks with time. It is likely that at ~50 Ma the lithosphere was relatively thick so that only the pyroxenite at deeper depths started to melt, forming basanites and alkali basalts (Figure 15a). Following the lithospheric thinning, at ~40 Ma the melting column moved to a relatively shallow level. Consequently, relatively large melting degree of pyroxenite occured to generate the tholeiites (Figure 15b).

Thus, we infer that that the Cenozoic mafic rocks in the Qianjiadian area were generated from a similar mantle source, and variable degrees of partial melting caused by lithospheric thinning may have controlled the genesis of the Qianjiadian mafic rocks.

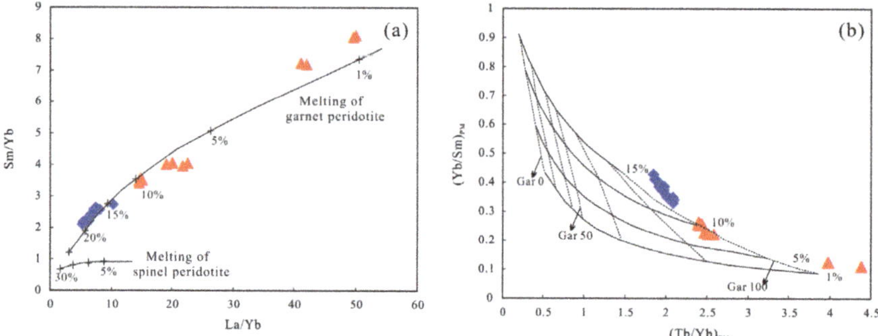

Figure 14. (**a**) La/Yb vs. Sm/Yb for the mafic rocks. Also shown are batch melting curves calculated for garnet peridotite and spinel peridotite. Partition coefficients are taken from [86]. (**b**) $(Tb/Yb)_{PM}$ versus $(Yb/Sm)_{PM}$ for the for the samples from the Qianjiadian region. The black thick lines represent 1%, 5%, 10%, and 15% of aggregated fractional melting of peridotite and the dotted curves suggest the percentage of melt contribution from a garnet-facies mantle (Gar); e.g., Gar 100 corresponds to the melt from garnet peridotite and Gar 0 corresponds to the melt from spinel peridotite. The inverse modeling used here follows [87].

5.4. Constraints on U Mineralization

Uranium is a typical redox sensitive element. It is 'fluid-mobile' when present as oxidized state (U^{6+}) but largely immobile when present as reduced species (U^{4+}) [88–93]. Sandstone-hosted U deposits are epigenetic deposits in which U minerals are present as disseminations and mineral replacements primarily in fluvial, lacustrine, and deltaic sandstones [94]. There are few important prerequisites for the formation of sandstone-hosted U deposits, which include (1) formation of reservoir sand bodies [95], (2) interlayer oxidized zone, (3) a leachable U source, and (4) a reducing environment, where the U is precipitated and concentrated [32,96]. In addition, tectonic inversion also exerts a significant role in the infiltration of meteoric oxidizing fluids and the migration directions of U-bearing groundwaters [1–6]. However, due to the incomplete stratigraphic record during the Cenozoic in the Songliao Basin, the timing of this tectonic inversion remains controversial. Based on the fold and thrust system and the continuous depocenter migration to the northwest Songliao basin, Wang et al. [41] proposed that tectonic inversion began at 79.1 Ma and ended at 64 Ma, whereas Cheng et al. [21] argued that this inversion characterized by the folds and the thrust faults mostly occurred during the Oligo–Miocene (~40–10 Ma), as indicated by apatite and zircon fission-track data. Combined with the research on basalts from the Shuangliao area, the Cenozoic mafic magmatic activity is characterized by alkali basaltic rocks formed at 50~47 Ma, followed by the emplacement of tholeiitic basaltic rocks at 42–40 Ma, indicating lithospheric thinning and crustal extension related to the subduction of the Pacific plate. Therefore, the Eocene (~50–40Ma) mafic rocks in southern Songliao Basin were emplaced prior to the Oligocene inversion (~40 Ma), described in [6,21], which marks the onset of a tectonic uplift period in the Qianjiadian area, concomitant with the main mineralization stage (Figure 15c).

Eocene mafic rocks are widespread in sandstones of the Yaojia Formation hosting the U mineralization in the Qianjiadian deposit (Figure 1). Previous studies proposed that the intrusion of these mafic rocks triggered subsequent circulation of mantle-derived CO_2-rich hydrothermal fluids, which could have been responsible for U leaching from the surrounding sandstones [34]. In addition, it has also been proposed that the genesis of the U mineralization in the Qianjiadian area was controlled by BSR processes within organic matter (OM)-bearing sandstones [28–33]. A recent study on carbonaceous debris from the Qianjiadian U deposit showed the increase of the vitrinite reflectance of carbonaceous debris, hence, revealing the heat effect of diabase dikes and related fluids [20] and suggesting the close spatial relationship between the dikes and the U mineralization. Nevertheless, recent EPMA chemical dating indicated that most of the U mineralization ages are younger than 40 Ma, with a peak in the Miocene or later (<20 Ma) [36], which is in good agreement with the tectonic evolution model proposed in the previous paragraph. Moreover, these ages postdate the emplacement of the mafic rocks, hence, arguing against the model proposed by Nie et al. [34] involving mantle-derived CO_2-rich hydrothermal fluids in the U ore genesis.

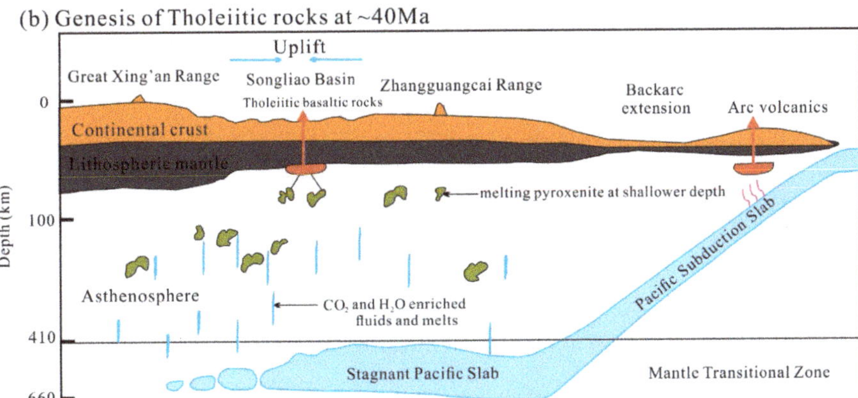

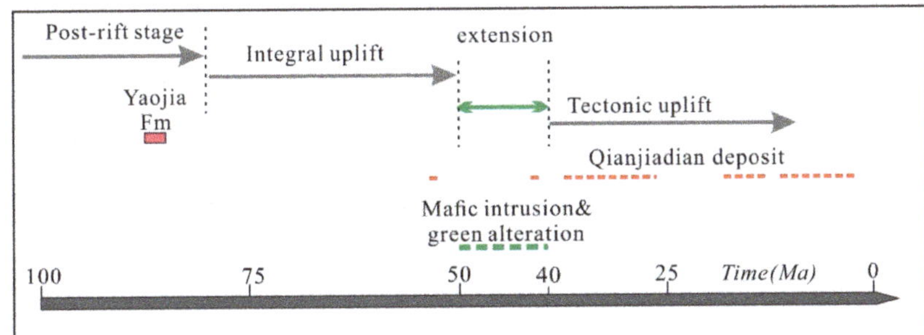

Figure 15. (**a**,**b**) Petrogenetic model of the Early Cenozoic mafic rocks in Qianjiadian area caused by lithospheric thinning attributed to the subduction of the Pacific Plate, modified from [97]. (**c**) Basin evolution associated with sandstone-type uranium mineralization, modified from [27].

Alternatively, in addition to BSR-mediated processes [27,33] occurring in OM-bearing primary reduced sandstone, the mafic rocks may also have provided favorable physicochemical environment (chemical trap), promoting the reduction and deposition of U from oxidizing fluids in the Qianjiadian deposit. The fact that some mafic rocks show evidence for a secondary oxidation demonstrates their interaction with the U-bearing groundwater and also supports their role as reducing barrier for the fluids. Besides, the mafic rocks characterized in the Qianjiadian area have relatively high contents of Fe(II)-bearing minerals such as pyroxene and magnetite (Figure 3d) that could promote U precipitation from oxidizing fluids [98]. For instance, chemical reaction of uranium mineralization could be as follow:

$$UO_2(CO_3)_2{}^{2-} + 2Fe_3O_4 + 2Ca^{2+} + 2H^+ = UO_2 + 3Fe_2O_3 + 2CaCO_3 + H_2O.$$

Moreover, the observations of carbonation in the mafic rocks (Figure 3a) and the common coexistence of uraninite and hematite in ores [16,18,19] would also corroborate this model. Finally, as discussed above, the hydrothermal alteration related to the mafic intrusions locally resulted in the secondary reduction of the host sediments mainly characterized by bleached and green-altered sandstones. As shown in Figure 3a and also described in [27] for the Baxingtu deposit, this alteration characterized by newly precipitated chlorite, epidote, and carbonate as sandstone cement has likely formed a secondary reducing barrier in oxidized sandstones or strengthened the reducing character of the primary reduced sandstones.

Therefore, the mafic rocks and their related hydrothermal alteration in the Qianjiadian area may have played a major role as reducing agent in the genesis of the U mineralization, as suggested by the close spatial relation between ore bodies and the mafic dikes shown in Figure 1c.

5.5. Comparison with the Ordos Basin

Sandstone-hosted uranium deposits in the Ordos Basin are mainly distributed in the northern margin of the basin, including the Dongsheng, Nalinggou, and Daying deposits. The U mineralization of these deposits is located in the contact zone between primary reduced grey sandstones and secondary reduced green sandstones within the Zhiluo Formation [31]. Similar to the Qianjiadian deposit, altered green sandstones are spatially closely related to the U mineralization [99]. Based on the study of clay minerals from ore-bearing sandstone, Ding et al. [100] proposed that these altered green sandstones were the result of the basalt-related alkaline hydrothermal alteration. In the Hangjingqi area basalts were emplaced at 126.2 ± 0.4 Ma [101], prior to the uranium mineralization. The uranium mineralization ages are spatiotemporally related to Miocene tectonics' uplift [102], which is similar to that of the Qianjiadian deposit.

In summary, we propose that the organic compounds (detrital OM or migrated hydrocarbon) are required to allow BSR processes, but the mafic rocks may have also played an important role as reducing barrier promoting the precipitation of U from groundwater oxidizing fluids.

6. Conclusions

Based on the results of this study, we have drawn the following conclusions:

(1) Zircon U-Pb and previous Ar-Ar dating indicate that the alkali mafic rocks formed earlier (51~47 Ma) than the tholeiitic mafic rocks (42–40 Ma).

(2) Cenozoic mafic rocks in the Qianjiadian area were of asthenospheric origin and relatively more radiogenic $^{206}Pb/^{204}Pb$ ratios suggest the metasomatism from subducted Pacific materials prior to melting.

(3) Tholeiites generally equilibrate at shallower depths in the mantle than alkali olivine mafic rocks and variable degrees of partial melting may have an effect on the genesis of the Qianjiadian mafic rocks.

(4) The tectonic inversion from extension to tectonic uplift attributed to the subduction of the Pacific Plate occurred at ~40 Ma. The Eocene (~50–40Ma) mafic rocks from the Qianjiadian U deposit mark the extensional period that preceded the Oligo–Miocene (~40–10 Ma) tectonic uplift and represent reducing environment that may have played a significant role in the U ore genesis during the Oligo–Miocene.

Supplementary Materials: The following are available online at http://www.mdpi.com/2075-163X/10/11/1014/s1, **Table S1**. Zircon U–Pb isotopic data from the mafic rocks in the Qianjiadian U deposit, NE China. **Table S2**. Major element concentrations (in wt.%) and trace element concentrations (in ppm) of the Qianjiadian mafic rocks. **Table S3**. Sr–Nd–Pb isotopic composition. **Table S4**. EPMA data of olivine from Qianjiadian mafic rocks.

Author Contributions: Conceptualization, J.-H.W. and F.-J.N.; funding acquisition, F.-J.N. and F.X.; writing-original draft preparation, D.-G.Y.; writing-review and editing, C.B.; investigation and sampling, F.X., Z.-B.Y., J.-F.C., C.-D.W., H.-T.W. and J.-H.W. All authors have read and agreed to the published version of the manuscript.

Funding: This work was supported by the National Natural Science Foundation of China (grant no. 41372071, 41772068, and 41862010) and National Program on Key Basic Research Project (973 Program) (grant no. 2015CB453002).

Acknowledgments: We thank Geological team No. 243 of the China National Nuclear Corporation in Chifeng, Inner Mongolia, for help with the field expedition. We also thank Yujie Hao of the Key Laboratory of Mineral Resources Evaluation in Northeast Asia, Ministry of Land and Resources of China, Jilin University, China, for help and support during the LA-ICP-MS U–Pb analyses.

Conflicts of Interest: The authors declare no conflict of interest.

References

1. Huston, D.L.; Mernagh, T.P.; Hagemann, S.G.; Doublier, M.P.; Fiorentini, M.; Champion, D.C.; Jaques, A.L.; Czarnota, K.; Cayley, R.; Skirrow, R.; et al. Tectono-metallogenic systems–The place of mineral systems within tectonic evolution, with an emphasis on Australian examples. *Ore Geol. Rev.* **2016**, *76*, 168–210. [CrossRef]
2. Jin, R.S.; Miao, P.S.; SiMa, X.Z.; Li, J.G.; Zhao, H.L.; Zhao, F.Q.; Feng, X.X.; Chen, Y.; Chen, L.L.; Zhao, L.J.; et al. Structure styles of Mesozoic-Cenozoic U-bearing rock series in Northern China. *Acta Geol. Sin.* **2016**, *90*, 2104–2116. [CrossRef]
3. Wang, F.F.; Liu, C.Y.; Qiu, X.W.; Guo, P.; Zhang, S.H.; Cheng, X.H. Characteristics and distribution of world's identified sandstone-type uranium resources. *Acta Geol. Sin.* **2017**, *91*, 2021–2046, (In Chinese with English Abstract).
4. Zhao, Z.H.; Bai, J.P.; Lai, T.G. Reversal structure and its relation to metallization of sandstone-type uranium deposit in Northern Songliao Basin. *Uranium Geol.* **2018**, *34*, 274–279, (In Chinese with English Abstract).
5. Cheng, Y.H.; Wang, S.Y.; Jin, R.S.; Li, J.G.; Ao, C.; Teng, X.M. Global Miocene tectonics and regional sandstone-style uranium mineralization. *Ore Geol. Rev.* **2019**, *106*, 238–250. [CrossRef]
6. Cheng, Y.H.; Wang, S.Y.; Zhang, T.F.; Teng, X.M.; Ao, C.; Jin, R.S.; Li, H.L. Regional sandstone-type uranium mineralization rooted in Oligo–Miocene tectonic inversion in the Songliao Basin, NE China. *Gondwana Res.* **2020**, *88*, 88–105. [CrossRef]
7. Jin, R.S.; Teng, X.M.; Li, X.G.; Si, Q.H.; Wang, W. Genesis of sandstone-type uranium deposits along the northern margin of the Ordos Basin, China. *Geosci. Front.* **2020**, *11*, 215–227. [CrossRef]
8. Leroy, J. The Margnac and Fanay uranium deposits of the La Crouzille district (western Massif Central, France): Geologic and fluid inclusion studies. *Econ. Geol.* **1978**, *73*, 1611–1634. [CrossRef]
9. Marignac, C.; Cuney, M. Ore deposits of the French Massif Central, Insight into the metallogenesis of the Variscan belt. *Miner. Depos.* **1999**, *34*, 472–504. [CrossRef]
10. Ling, H.F.; Shen, W.Z.; Deng, P.; Jiang, S.Y.; Ye, H.M.; Pu, W.; Tan, Z.Z. Geochemical characteristics and genesis of Luxi-Xiazhuang diabase dikes in Xiazhuang uranium orefield, northern Guangdong province. *Acta Geol. Sin.* **2005**, *79*, 497–506.
11. Hu, R.Z.; Bi, X.W.; Zhou, M.F.; Peng, J.T.; Su, W.C.; Liu, S.; Qi, H.W. Uranium metallogenesis in South China and its relationship to crustal extension during the Cretaceous to Tertiary. *Econ. Geol.* **2008**, *103*, 583–598. [CrossRef]
12. Wang, L.X.; Ma, C.Q.; Lai, Z.X.; Marks, M.A.W.; Zhang, C.; Zhong, Y.F. Early Jurassic mafic dykes from the Xiazhuang ore district (South China): Implications for tectonic evolution and uranium metallogenesis. *Lithos* **2015**, *239*, 71–85. [CrossRef]
13. Polito, P.A.; Kyser, T.K.; Rheinberger, G.; Southgate, P.N. A paragenetic and isotopic study of the Proterozoic Westmoreland uranium deposits, Southern McArthur Basin, Northern Territory, Australia. *Econ. Geol.* **2005**, *100*, 1243–1260. [CrossRef]

14. Xia, Y.L.; Lin, J.R.; Li, Z.Y.; Li, S.X.; Liu, H.B.; Wang, Z.M.; Fan, G.; Zheng, J.W.; Li, Z.J.; Zhang, M.Y. Perspective and resource evaluation and metallogenic studies on sandstone-type uranium deposit in Qianjiadian depression of Songliao Basin. *China Nucl. Sci. Technol. Rep.* **2003**, *3*, 105–117, (In Chinese with English Abstract).
15. Xia, Y.L.; Zheng, J.W.; Li, Z.Y.; Li, L.Q.; Tian, S.F. Metallogenic characteristics and model of the Qianjiadian uranium deposit in the Songliao basin. *Miner. Depos.* **2010**, *29*, 154–155, (In Chinese with English Abstract).
16. Jiao, Y.Q.; Wu, L.Q.; Peng, Y.B.; Rong, H.; Ji, D.M.; Miao, A.S.; Li, H.L. Sedimentary-tectonic setting of the deposition-type uranium deposits forming in the Paleo-Asian tectonic domain, North China. *Earth Sci. Front.* **2015**, *22*, 189–205, (In Chinese with English Abstract).
17. Rong, H.; Jiao, Y.; Wu, L.; Jia, J.; Cao, M. Effects of basic intrusions on REE mobility of sandstones and their geological significance: A case study from the Qianjiadian sandstone-hosted uranium deposit in the Songliao Basin. *Appl. Geochem.* **2020**, *120*, 104665. [CrossRef]
18. Luo, Y.; Ma, H.F.; Xia, Y.L.; Zhang, Z.G. Geologic characteristics and metallogenic model of Qianjiadian uranium deposit in Songliao Basin. *Uranium Geol.* **2007**, *23*, 193–201, (In Chinese with English Abstract).
19. Rong, H.; Jiao, Y.; Wu, L.; Ji, D.; Li, H.; Zhu, Q.; Cao, M.; Wang, X.; Li, Q.; Xie, H. Epigenetic alteration and its constraints on uranium mineralization from the Qianjiadian uranium deposit, southern Songliao Basin. *Earth Sci.* **2016**, *41*, 153–166, (In Chinese with English Abstract).
20. Rong, H.; Jiao, Y.Q.; Wu, L.Q.; Wan, D.; Cui, Z.J.; Guo, X.J.; Jia, J.M. Origin of the carbonaceous debris and its implication for mineralization within the Qianjiadian uranium deposit, southern Songliao Basin. *Ore Geol. Rev.* **2019**, *107*, 336–352. [CrossRef]
21. Cheng, Y.H.; Wang, S.Y.; Li, Y.; Ao, C.; Li, Y.F.; Li, J.G.; Li, H.L.; Zhang, T.F. Late Cretaceous–Cenozoic thermochronology in the southern Songliao Basin, NE China: New insights from apatite and zircon fission track analysis. *J. Asian Earth Sci.* **2018**, *160*, 95–106. [CrossRef]
22. Min, M.; Xu, H.F.; Chen, J.; Mostafa, F. Evidence of uranium biomineralization in sandstone-hosted roll-front uranium deposits, northwestern China. *Ore Geol. Rev.* **2005**, *26*, 198–206. [CrossRef]
23. Cai, C.F.; Dong, H.L.; Li, H.T.; Xiao, X.J.; Ou, G.X.; Zhang, C.M. Mineralogical and geochemical evidence for coupled bacterial uranium mineralization and hydrocarbon oxidation in the Shashagetai deposit, NW China. *Chem. Geol.* **2007**, *236*, 167–179. [CrossRef]
24. Grant, B.D.; Charles, R.M.B.; David, J.G. Geology, geochemistry and mineralogy of the lignite-hosted ambassador palaeochannel uranium and multi-element deposit, Gunbarrel Basin, Western Australia. *Miner. Depos.* **2011**, *46*, 761–787.
25. Abzalov, M.Z.; Paulson, O. Sandstone hosted uranium deposits of the great divide basin, Wyoming, USA. *Appl. Earth Sci.* **2012**, *121*, 76–83. [CrossRef]
26. Jiang, L.; Cai, C.F.; Zhang, Y.D.; Mao, S.Y.; Sun, Y.G.; Li, K.K.; Xiang, L.; Zhang, C.M. Lipids of sulfate reducing bacteria and sulfur-oxidizing bacteria found in the Dongsheng uranium deposit. *Chin. Sci. Bull.* **2012**, *57*, 1311–1319. [CrossRef]
27. Bonnetti, C.; Liu, X.; Yan, Z.; Cuney, M.; Michels, R.; Malartre, F.; Mercadier, J.; Cai, J. Coupled uranium mineralisation and bacterial sulphate reduction for the genesis of the Baxingtu sandstone-hosted U deposit, SW Songliao Basin, NE China. *Ore Geol. Rev.* **2017**, *82*, 108–129. [CrossRef]
28. Meunier, J.D.; Trouiller, A.; Brulhet, J.; Pagel, M. Uranium and organic matter in a paleodeltaic environment: The Coutras Deposit (Gironde, France). *Econ. Geol.* **1989**, *84*, 1541–1556. [CrossRef]
29. Cai, C.F.; Zhao, L. Thermochemical sulfate reduction and its effects on petroleum composition and reservoir quality: Advances and problems. *Bull. Mineral. Petrol. Geochem.* **2016**, *35*, 851–859, (In Chinese with English Abstract).
30. Sangély, L.; Chaussidon, M.; Michels, R.; Brouand, M.; Cuney, M.; Huault, V.; Landais, P. Micrometer scale carbon isotopic study of bitumen associated with Athabasca uranium deposits: Constraints on the genetic relationship with petroleum sourcerocks and the abiogenic origin hypothesis. *Earth Planet. Sci. Lett.* **2007**, *258*, 378–396. [CrossRef]
31. Jiao, Y.Q.; Wu, L.Q.; Rong, H.; Peng, Y.B.; Miao, A.S.; Wang, X.M. The relationship between Jurassic coal measures and sandstone-type uranium deposits in the northeastern Ordos Basin, China. *Acta Geol. Sin.* **2016**, *90*, 2117–2132. [CrossRef]

32. Jiao, Y.Q.; Wu, L.Q.; Rong, H. Model of inner and outer reductive media within uranium reservoir sandstone of Sandstone-type Uranium deposits and its ore-controlling mechanism: Case studies in Daying and Qianjiadian uranium deposit. *Earth Sci.* **2018**, *43*, 459–474, (In Chinese with English Abstract).
33. Bonnetti, C.; Zhou, L.; Riegler, T.; Brugger, J.; Fairclough, M. Large S isotope and trace element fractionations in pyrite of uranium roll front systems result from internally-driven biogeochemical cycle. *Geochim. Cosmochim. Acta* **2020**. [CrossRef]
34. Nie, F.J.; Yan, Z.B.; Xia, F.; Li, M.G.; Feng, Z.B.; Lu, Y.Y.; Cai, J.F. Mineralisation from hot fluid flows in the sandstone-type uranium deposit in the Kailu Basin, Northeast China. *Appl. Earth Sci.* **2018**, *127*, 2–14. [CrossRef]
35. Feng, Z.Q.; Jia, C.Z.; Xie, X.N.; Zhang, S.; Feng, Z.H.; Timothy, A.C. Tectonostratigraphic units and stratigraphic sequences of the nonmarine Songliao basin, northeast China. *Basin Res.* **2010**, *22*, 79–95.
36. Zhao, L.; Cai, C.F.; Jin, R.S.; Li, J.G.; Li, H.L.; Wei, J.L.; Guo, F.; Zhang, B. Mineralogical and geochemical evidence for biogenic and petroleum-related uranium mineralization in the Qianjiadian deposit, NE China. *Ore Geol. Rev.* **2018**, *101*, 273–292. [CrossRef]
37. Cheng, Y.H.; Li, Y.; Wang, S.Y.; Li, Y.F.; Ao, C.; Li, J.G.; Sun, L.X.; Li, H.L.; Zhang, T.F. Late Cretaceous tectono-magmatic event in Songliao Basin, NE China: New insights from mafic dyke geochronology and geochemistry analysis. *Geol. J.* **2018**, *53*, 2991–3008. [CrossRef]
38. Wang, P.J.; Xie, X.A.; Mattern, F.; Ren, Y.G.; Zhu, D.F.; Sun, X.M. The Cretaceous Songliao Basin: Volcanogenic succession, sedimentary sequence and tectonic evolution, NE China. *Acta Pet. Sin.* **2007**, *81*, 1002–1011.
39. Wu, F.Y.; Sun, D.Y.; Ge, W.C.; Zhang, Y.B.; Grant, M.L.; Wilde, S.A.; Jahn, B.M. Geochronology of the Phanerozoic granitoids in northeastern China. *J. Asian Earth Sci.* **2011**, *41*, 1–30. [CrossRef]
40. Wu, F.Y.; Sun, D.Y.; Li, H.M.; Wang, X.L. The nature of basement beneath the Songliao Basin in NE China: Geochemical and isotopic constraints. *Phys. Chem. Earth Part A* **2001**, *26*, 793–803. [CrossRef]
41. Wang, P.J.; Mattern, F.; Didenko, A.; Zhu, D.F.; Singer, B.; Sun, X.M. Tectonics and cycle system of the Cretaceous Songliao Basin: An inverted active continental margin basin. *Earth Sci. Rev.* **2016**, *159*, 82–102. [CrossRef]
42. Zhang, W.L.; Su, X.B.; Zhang, B. Discussion on mining utilization coefficient for in situ leaching sandstone type uranium resource: A case study of qianjiadian uranium deposit. *Uranium Minine Metall.* **2017**, *36*, 19–22, (In Chinese with English Abstract).
43. Liu, X.M.; Gao, S.; Diwu, C.R.; Yuan, H.L.; Hu, Z.C. Simultaneous in situ determination of U–Pb age and trace elements in zircon by LA–ICP–MS in 20 μm spot size. *Chin. Sci. Bull.* **2007**, *52*, 1257–1264. [CrossRef]
44. Hoskin, W.O.; Schaltegger, U. The composition of zircon and igneous and metamorphic petrogenesis. *Rev. Mineral. Geochem.* **2003**, *53*, 27–62. [CrossRef]
45. Yang, F.P.; Chen, F.J.; Wang, Y.H.; Li, J.K. Apatite fission track analysis in the central depression, Songliao Basin. *Pet. Explor. Dev.* **1995**, *22*, 20–25, (In Chinese with English Abstract).
46. Xu, Y.G.; Zhang, H.H.; Qiu, H.N.; Ge, W.C.; Wu, F.Y. Oceanic crust components in continental basalts from Shuangliao, Northeast China: Derived from the mantle transition zone. *Chem. Geol.* **2012**, *328*, 168–184. [CrossRef]
47. Irvine, T.H.; Baragar, W.R.A. A guide to the chemical classification of the common volcanic rocks. *Can. J. Earth Sci.* **1971**, *8*, 523–548. [CrossRef]
48. Winchester, J.A.; Floyd, P.A. Geochemical discrimination of different magma series and their differentiation products using immobile elements. *Chem. Geol.* **1977**, *20*, 325–343. [CrossRef]
49. Liu, J.Q.; Chen, L.H.; Wang, X.J.; Zhong, Y.; Yu, X.; Zeng, G.; Erdmann, S. The role of melt-rock interaction in the formation of Quaternary high-MgO potassic basalt from the Greater Khingan Range, Northeast China. *J. Geophys. Res. Solid Earth.* **2017**, *122*, 262–280. [CrossRef]
50. Xu, Y.G.; Ma, J.L.; Frey, F.A.; Feigenson, M.D.; Liu, J.F. Role of lithosphere–asthenosphere interaction in the genesis of Quaternary alkali and tholeiitic basalts from Datong, western North China Craton. *Chem. Geol.* **2005**, *224*, 247–271. [CrossRef]
51. Wang, X.J.; Chen, L.H.; Hofmann, A.W.; Mao, F.G.; Liu, J.Q.; Zhong, Y.; Xie, L.W.; Yang, Y.H. Mantle transition zone-derived EM1 component beneath NE China: Geochemical evidence from Cenozoic potassic basalts. *Earth. Planet. Sci. Lett.* **2017**, *465*, 16–28. [CrossRef]

52. Zou, H.B.; Zindler, A.; Xu, X.S.; Qi, Q. Major, trace element, and Nd, Sr and Pb isotopic studies of Cenozoic basalts in SE China: Mantle sources, regional variations, and tectonic significance. *Chem. Geol.* **2000**, *17*, 33–47. [CrossRef]
53. DePaolo, D.J.; Wasserburg, G.J. The sources of island arcs as indicated by Nd and Sr isotopic studies. *Geophys. Res. Lett.* **1977**, *4*, 465–468. [CrossRef]
54. Boynton, W.V. Geochemistry of the rare earth elements: Meteorite studies. In *Rare Earth Element Geochemistry*; Elsevier: Amsterdam, The Netherlands, 1984; pp. 63–114.
55. Sun, S.S.; McDonough, W.F. Chemical and isotopic systematics of oceanic basalts: Implications for mantle composition and processes. In *Magmatism in the Ocean Basins*; Saunders, A.D., Norry, M.J., Eds.; Geological Society Publications: London, UK, 1989; pp. 313–345.
56. Zindler, A.; Hart, S. Chemical geodynamics. *Annu. Rev. Earth Planet. Sci.* **1986**, *14*, 493–571. [CrossRef]
57. Kuritani, T.; Ohtani, E.; Kimura, J.-I. Intensive hydration of the mantle transition zone beneath China caused by ancient slab stagnation. *Nat. Geosci.* **2011**, *4*, 713–716. [CrossRef]
58. Chung, S.L.; Yang, T.F.; Chen, S.J.; Chen, C.H.; Lee, T.; Chen, C.H. Sr—Nd isotope compositions of high-pressure megacrysts and a lherzite inclusion in alkali basalts from western Taiwan. *J. Geol. Soc. China.* **1995**, *38*, 15–24.
59. Barry, T.L.; Kent, R.W. Cenozoic magmatism in Mongolia and the origin of Central and East Asian basalts. In *Mantle Dynamics and Plate Interactions in East Asia*; Flower, M.F.J., Chung, S.L., Lo, C.H., Lee, T.Y., Eds.; American Geophysical Union: Washington, DC, USA, 1998; pp. 347–364.
60. Hart, S.R. A large isotope anomaly in the southern hemisphere mantle. *Nature* **1984**, *309*, 753–757. [CrossRef]
61. Guo, P.; Ionov, D.A.; Xu, W.L.; Wang, C.G.; Luan, J.P. Mantle and recycled oceanic crustal components inmantle xenoliths from northeastern China and their mantle sources. *J. Geophys. Res. Solid Earth* **2020**, *125*, e2019JB018232. [CrossRef]
62. Danyushevsky, L.V.; Della-Pasqua, F.N.; Sokolov, S. Re-equilibration of melt inclusions trapped by magnesian olivine phenocrysts from subduction-related magmas: Petrological implications. *Contrib. Mineral. Petrol.* **2000**, *138*, 68–83. [CrossRef]
63. Danyushevsky, L.V.; Sokolov, S.; Falloon, T.J. Melt inclusions in olivine phenocrysts: Using diffusive re-equilibration to determine the cooling history of a crystal, with implications for the origin of olivine-phyric volcanic rocks. *J. Petrol.* **2002**, *43*, 1651–1671. [CrossRef]
64. Wang, C.S.; Scott, R.W.; Wan, X.Q.; Graham, S.A.; Huang, Y.J.; Wang, P.J.; Wu, H.C.; Dean, W.E.; Zhang, L.M. Late Cretaceous climate changes recorded in Eastern Asian lacustrine deposits and North American Epieric sea strata. *Earth Sci. Rev.* **2013**, *126*, 275–299. [CrossRef]
65. Sato, H. Nickel content of basaltic magmas: Identification of primary magmas and a measure of the degree of olivine fractionation. *Lithos* **1977**, *10*, 113–120. [CrossRef]
66. Thompson, R.N.; Gibson, S.A. Transient high temperatures in mantle plume heads inferred from magnesian olivines in Phanerozoic picrites. *Nature* **2000**, *407*, 502–506. [CrossRef] [PubMed]
67. Sobolev, A.V.; Hofmann, A.W.; Kuzmin, D.V.; Yaxley, G.M.; Arndt, N.T.; Chung, S.L.; Danyushevsky, L.V.; Elliott, T.; Frey, F.A.; Garcia, M.O. The amount of recycled crust in sources of mantle-derived melts. *Science* **2007**, *316*, 412. [CrossRef]
68. McDonough, W.F. Constraints on the composition of the continental lithospheric mantle. *Earth Planet. Sci. Lett.* **1990**, *101*, 1–18. [CrossRef]
69. Smith, E.I.; Sanchez, A.; Walker, J.D.; Wang, K. Geochemistry of maficmagmas in the Hurricane Volcanic field, Utah: Implications for small-and large-scale chemical variability of the lithospheric mantle. *J. Geol.* **1999**, *107*, 433–448. [CrossRef]
70. Hofmann, A.W.; Jochum, K.P.; Seufert, M.; White, W.M. Nb and Pb in oceanic basalts: New constraints on mantle evolution. *Earth Planet. Sci. Lett.* **1986**, *79*, 33–45. [CrossRef]
71. Rudnick, R.L.; Gao, S. Composition of the continental crust. *Treatise Geochem.* **2003**, *3*, 1–64.
72. Pearce, J.A. Geochemical fingerprinting of oceanic basalts with applications to ophiolite classification and the search for Archean oceanic crust. *Lithos* **2008**, *100*, 14–48. [CrossRef]
73. Murphy, D.T.; Collerson, K.D.; Kamber, B.S. Lamproites from Gaussberg, Antarctica: Possible transition zonemelts of Archaean subducted sediments. *J. Petrol.* **2002**, *43*, 981–1001. [CrossRef]

74. Zhang, M.L.; Guo, Z.F. Origin of Late Cenozoic Abaga–Dalinuoer basalts, eastern China: Implications for a mixed pyroxenite–peridotite source related with deep subduction of the Pacific slab. *Gondwana Res.* **2016**, *37*, 130–151. [CrossRef]
75. Kuritani, T.; Kimura, J.I.; Ohtani, E.; Miyamoto, H.; Furuyama, K. Transition zone origin of potassic basalts from Wudalianchi volcano, northeast China. *Lithos* **2013**, *156–159*, 1–12. [CrossRef]
76. Plank, T. The chemical composition of subducting sediments. In *Treatise on Geochemistry*; Holland, H.D., Turekian, K.K., Eds.; Elsevier: Oxford, UK, 2014; pp. 607–629.
77. Li, S.G.; Yang, W.; Ke, S.; Meng, X.N.; Tian, H.C.; Xu, L.J.; He, Y.; Huang, J.; Wang, X.C.; Xia, Q.; et al. Deep carbon cycles constrained by a large-scale mantle Mg isotope anomaly in eastern China. *Natl. Sci. Rev.* **2017**, *4*, 111–120. [CrossRef]
78. Chen, H.; Xia, Q.K.; Deloule, E.; Ingrin, J. Typical oxygen isotope profile of altered oceanic crust recorded in continental intraplate basalts. *J. Earth Sci.* **2017**, *28*, 578–587. [CrossRef]
79. Huang, J.L.; Zhao, D.P. High-resolution mantle tomography of China and surrounding regions. *J. Geophys. Res.* **2006**, *111*, B09305. [CrossRef]
80. Wei, W.; Zhao, D.; Xu, J.; Wei, F.; Liu, G. P and S wave tomography and anisotropy in Northwest Pacific and East Asia: Constraints on stagnant slab and intraplate volcanism. *J. Geophys. Res.* **2015**, *120*, 1642–1666. [CrossRef]
81. Xu, R.; Liu, Y.S.; Wang, X.H.; Zong, K.Q.; Hu, Z.C.; Chen, H.H.; Zhou, L. Crust recycling induced compositional-temporal-spatial variations of Cenozoic basalts in the Trans-North China Orogen. *Lithos* **2017**, *274*, 383–396. [CrossRef]
82. Takahashi, E.; Kushiro, I. Melting of a dry peridotite at high pressures and basalt magma genesis. *Am. Mineral.* **1983**, *68*, 859–879.
83. Kushiro, I. Partial melting experiments on peridotite and origin of mid-ocean ridge basalt. *Annu. Rev. Earth Planet. Sci.* **2001**, *29*, 71–107. [CrossRef]
84. DePaolo, D.J.; Daley, E.E. Neodymium isotopes in basalts of the southwest basin and range and lithospheric thinning during continental extension. *Chem. Geol.* **2000**, *169*, 157–185. [CrossRef]
85. Langmuir, C.H.; Klein, E.M.; Plank, T. Petrological systematics of mid-ocean ridge basalts: Constraints on melt generation beneath ocean ridges. In *Mantle Flow and Melt Generation at Mid-Ocean Ridges*; Morgan, J.P., Blackman, D.K., Sinton, J.M., Eds.; Geophysical Monograph Series; AGU: Washington, DC, USA, 1992.
86. Johnson, K.T.M.; Dick, H.J.B.; Shimizu, N. Melting in the oceanic upper mantle: An ion microprobe study of diopsides in abyssal peridotites. *J. Geophys. Res.* **1990**, *95*, 2661–2678. [CrossRef]
87. Zhang, Z.C.; Mahoney, J.J.; Mao, J.W.; Wang, F.S. Geochemistry of picritic and associated basalt flows of the western Emeishan flood basalt province, China. *J. Petrol.* **2006**, *47*, 1997–2019. [CrossRef]
88. Langmuir, D. Uranium solution-mineral equilibria at low temperatures with applications to sedimentary ore deposits. *Geochim. Cosmochim. Acta* **1978**, *42*, 547–569. [CrossRef]
89. Wilde, A.; Otto, A.; Jory, J.; MacRae, C.; Pownceby, M.; Wilson, N.; Torpy, A. Geology and mineralogy of uranium deposits from Mount Isa, Australia: Implications for albitite uranium deposit models. *Minerals* **2013**, *3*, 258–283. [CrossRef]
90. Peng, X.; Min, M.; Qiao, H.; Wang, J.; Fayek, M. Uranium-series disequilibria in the groundwater of the Shihongtan sandstone-hosted uranium deposit, NW China. *Minerals* **2016**, *6*, 3. [CrossRef]
91. René, M.; Dolníček, Z.; Sejkora, J.; Škácha, P.; Šrein, V. Uraninite, coffinite and ningyoite from vein-type uranium deposits of the Bohemian Massif (Central European Variscan Belt). *Minerals* **2019**, *9*, 123. [CrossRef]
92. Gigon, J.; Skirrow, R.G.; Harlaux, M.; Richard, A.; Mercadier, J.; Annesley, I.R.; Villeneuve, J. Insights into B-Mg-metasomatism at the Ranger U deposit (NT, Australia) and comparison with Canadian unconformity-related U deposits. *Minerals* **2019**, *9*, 432. [CrossRef]
93. Wu, D.; Pan, J.; Xia, F.; Huang, G.; Lai, J. The mineral chemistry of chlorites and its relationship with uranium mineralization from Huangsha uranium mining area in the middle Nanling range, SE China. *Minerals* **2019**, *9*, 199. [CrossRef]
94. Finch, W.I.; Davis, J.F. Sandstone-type uranium deposits-an introduction. In *Geological Environments of Sandstone-Type Uranium Deposits*; TECDOC-328; IAEA: Vienna, Austria, 1985; pp. 11–20.
95. Jiao, Y.Q.; Wu, L.Q.; Yang, S.K.; Lü, X.B.; Yang, Q.; Wang, Z.H.; Wang, M.F. Sedimentology of uranium reservoir. In *The Exploration and Production Base of Sandstone-Type Uranium Deposits*; Geology Publishing House: Beijing, China, 2006; p. 331. (In Chinese)

96. Rallakis, D.; Michels, R.; Brouand, M.; Parize, O.; Cathelineau, M. The role of organic matter on uranium precipitation in Zoovch Ovoo, Mongolia. *Minerals* **2019**, *9*, 310. [CrossRef]
97. Xu, Y.G.; Li, H.Y.; Hong, L.B.; Ma, L.; Ma, Q.; Sun, M.D. Generation of Cenozoic intraplate basalts in the big mantle wedge under eastern Asia. *Sci. China Earth Sci.* **2018**, *61*, 869–886. [CrossRef]
98. Tappa, M.J.; Ayuso, R.A.; Bodnar, R.J.; Aylor, J.G.; Beard, J.; Henika, W.S.; Vazquez, J.A.; Wooden, J.L. Age of host rocks at the Coles Hill uranium deposit, Pittsylvania county, Virginia, based on zircon U-Pb geochronology. *Econ. Geol.* **2013**, *109*, 513–530. [CrossRef]
99. Li, Z.Y.; Fang, X.Y.; Chen, A.P.; Ou, G.X.; Xiao, X.J.; Sun, Y.; Liu, C.Y.; Wang, Y. Origin of gray-green sandstone in ore bed of sandstone type uranium deposit in north Ordos Basin. *Sci. China Ser. D Earth Sci.* **2007**, *50*, 165–173, (In Chinese with English Abstract). [CrossRef]
100. Ding, B.; Liu, H.X.; Zhang, B.; Yi, C.; Wang, G.; Li, P. Formation mechanism of tabular orebody in Nalinggou uranium deposit, Ordos Basin: Constraints on study of clay minerals from ore-bearing sandstone. *Miner. Depos.* **2020**, *39*, 184–195, (In Chinese with English Abstract).
101. Zhou, H.P.; Zhang, K.; Li, G. Early Cretaceous tectonic-thermal event of the Ordos Basin: Evidence from the Ar-Ar dating of the Hangjinqi basalts. *Geotecton. Metallog.* **2008**, *3*, 360–364, (In Chinese with English Abstract).
102. Zhang, C.; Yi, C.; Dong, Q.; Cai, Y.Q.; Liu, H.X. Geological and geochronological evidence for the effect of Paleogene and Miocene uplift of the Northern Ordos Basin on the formation of the Dongsheng uranium district, China. *J. Geodyn.* **2018**, *14*, 1–18. [CrossRef]

Publisher's Note: MDPI stays neutral with regard to jurisdictional claims in published maps and institutional affiliations.

Article

pXRF Measurements on Soil Samples for the Exploration of an Antimony Deposit: Example from the Vendean Antimony District (France)

Bruno Lemière [1,*], Jeremie Melleton [1], Pascal Auger [1], Virginie Derycke [1], Eric Gloaguen [1], Loïc Bouat [2], Dominika Mikšová [3], Peter Filzmoser [3] and Maarit Middleton [4]

1 BRGM, F-45060 Orléans, France; j.melleton@brgm.fr (J.M.); p.auger@brgm.fr (P.A.); v.derycke@brgm.fr (V.D.); e.gloaguen@brgm.fr (E.G.)

2 Université du Maine, Géosciences, F-72000 Le Mans, France; loic.bouat@etu.univ-orleans.fr

3 CSTAT—Computational Statistics, Vienna University of Technology, 1040 Wien, Austria; miksovadominika1@gmail.com (D.M.); peter.filzmoser@tuwien.ac.at (P.F.)

4 GTK (Geological Survey of Finland), Environmental Solutions, 96101 Rovaniemi, Finland; maarit.middleton@gtk.fi

* Correspondence: b.lemiere@brgm.fr

Received: 24 July 2020; Accepted: 8 August 2020; Published: 18 August 2020

Abstract: Mineral exploration is increasingly challenging in inhabited areas. To evaluate the potential of soil analysis by pXRF (portable X-ray fluorescence) as a low-footprint exploration technique, we revisited a historic Sb district in an agricultural area and performed shallow-soil sampling (Ah and B horizons) along profiles across known veins to capture the endogenic geochemical anomaly signals. Despite an expected bias between pXRF measurements and laboratory analyses, the former effectively located the Sb veins, especially when using their multi-element capabilities. Composition data processing (CoDa) and horizon-selective sampling significantly improved the method's efficiency. On-site measurements allow dynamic sampling and mapping, helping with faster, cost-effective sample selection for further laboratory investigations. Based on this case study, where similar geochemical patterns were obtained for both horizons, application of an on-site approach to a humic horizon can increase survey efficiency and decrease impacts.

Keywords: pXRF; antimony; mineral exploration; Vendean antimony district

---

1. Introduction

For companies, the competitiveness of mineral exploration is based on reducing costs and capital intensity, improving dynamics and shortening delays between target testing and feasibility analysis. The exploration industry faces several constraints depending on the geographical location of projects, often developed far from infrastructure and analytical laboratories. Establishing field laboratories in remote areas and the shipping of samples involve significant logistics challenges and create data-transmission bottlenecks caused by distance between experts. While drill-core geochemistry for resource and reserve estimates for feasibility studies still requires traditional laboratory analysis as an essential step in regional exploration, commodity detection, target investigation and ranking can become significantly cheaper and quicker with the use of field analysers. This is because the goal is not the absolute accuracy of the measurements, but rather the relative ranking of the element concentrations and their anomaly to background contrast. Portable X-ray fluorescence (pXRF) technology allows dynamic decision-making and agile exploration management and facilitates cost-effective exploration [1].

Minimising sample preparation and shipping greatly reduces the environmental footprint of exploration. Case studies of pXRF applications on exploring different metals need to be presented to increase its acceptance.

1.1. Objectives

This study was carried out as part of the European EIT Raw Materials project UpDeep [2]. It is part of a demonstration of surface geochemical methods in southern European conditions, while similar studies are being carried out also in Finland and Greenland to demonstrate the applicability of the methods in northern Europe. Surface geochemistry is tested to provide means for geochemical exploration across Europe besides deep geochemistry [3,4]. The aim of surface geochemistry is to prioritise potential exploration targets by reducing time and cost while improving reliability in target detection. The key benefit is insignificant or non-existing environmental impact in the sampling phase that allows sampling in environmentally sensitive terrains. Low-footprint exploration strategies are necessary in Europe, where social acceptance of surface mining activities is particularly low.

Several methods based on wet chemistry are currently being tested, including partial-extraction soil analysis and plant analysis. Surface geochemical exploration is based on analysing trace amounts of metals or other elements and soil hydrocarbons in plants and soil horizons to discover deep-buried mineralisations. [3,4]. pXRF technology has been tested to greatly reduce impacts even further by diminishing sample shipping and the use of sample digestions. In addition, pXRF would also allow dynamic sampling design and on-site decision making of sample selection for further wet chemistry and survey orientation. To test the applicability of pXRF, two French sites were studied: the Echassières Li-Ta-Sn-W deposit [5] and the Les Brouzils Sb deposit (this study).

Antimony is considered a critical metal and is on the EU's list of critical substances [6] due to its very high dependence (87% in 2017) on its import from China. Antimony is required for a variety of industrial materials such as flame-retardants, plastics, paint pigments, glassware and ceramics, ammunition alloys and batteries [7]. As a result, global Sb consumption has increased to more than 140 kt each year [8,9]. Its most promising use may be in producing rechargeable lithium-ion and sodium-ion batteries, since Sb-based materials are promising ultra-fast high-capacity anodes [10]. On the other hand, monitoring Sb concentrations in soils is also a key aspect of constraining its environmental and possible health [11] impact, in particular on former mining areas [12].

In this context, Sb is a desirable target commodity element for exploration companies. Syngenetic antimony deposits are known in young orogenic belts (Sedex or epithermal, [13]). The most important Sb deposits are orogenic veins of the slate-belt type, like Les Brouzils. In older belts, most Sb occurs as epizonal remobilisation through hydrothermal, fracture-bound circulations [14–16]. In such settings, the surficial footprint of mineralisation is usually elongated and thin (from 1 m to 50 m wide), while the irregular shape of orebodies makes them challenging targets for drilling. Trenching and soil surveys are, therefore, the traditional preferred exploration methods [17]. Reducing the footprint of high-density soil profiles and vegetation surveys is desirable. High-density soil sampling transects are required to detect the veins in addition to laboratory confirmation analyses on a subset of samples, usually selected on site. We try here to demonstrate how pXRF, definitely a low-footprint technique, can be effective for the precise delineation of Sb anomalies if adequate statistical treatment is applied.

1.2. Site and Geology

The study area (Figure 1) belongs to the Vendée antimony district located in western France, southeast of Brittany. The Vendée district has been known for a long time, with the first mention of mining activities on antimony ore deposits during the 18th century at La Ramée [18]. Operations started in the area during this period on several mineralised structures. At the beginning of the 19th century, the discovery of a rich vein at Rochetrejoux led to new activities until 1925. After the discovery of mineralised quartz during the 1950–1970s, the French Geological Survey (BRGM) conducted a stream-sediment geochemical survey [19] followed by soil sampling focusing on sediment anomalies, which led to the discovery of around 20 new prospects distributed on a 50 × 20 km area [18], in particular at Les Brouzils, La Télachère and La Copechagnière (Figure 1). Mining operations started again until the mid-1990s at the Les Brouzils mine.

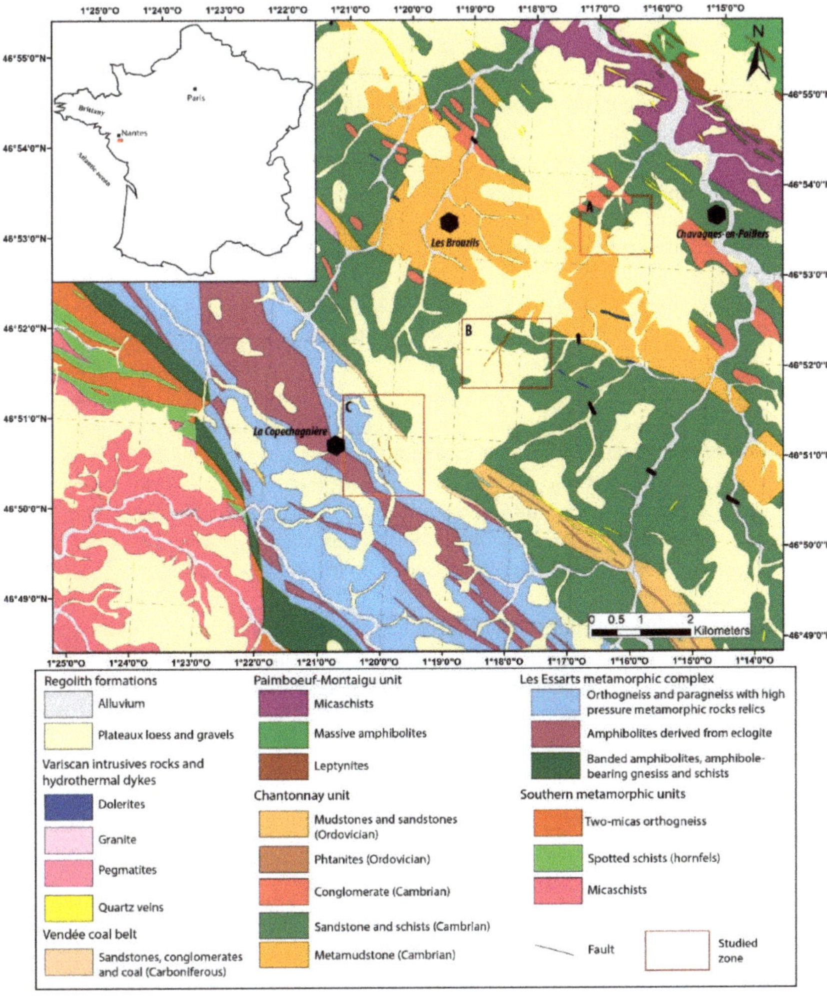

Figure 1. Geological map of the study area (modified from the geological maps at 1/50,000 scale of the Vendée district, [20]). The study areas are: A: La Télachère. B: Les Brouzils, C: La Copechagnière. Size of insert map is 1000 × 1000 km.

The geological framework of the area is Variscan metamorphic rocks (gneiss and amphibolites) and slightly metamorphised sedimentary rocks. The western part of the Vendée Sb district is located on the northern side of the Chantonnay syncline. It is made of slightly metamorphised sedimentary rocks (sandstones, conglomerates and schists), locally crosscut by hectometric dolerite dykes. The edge of the syncline is affected by a regional-scale 120–140° N shear-zone, which is believed to be the structural control of the antimony deposits. Indeed, thrust and shear generated a network of conjugated tension fractures, controlling Sb mineralisations. The Copechagnière mineralised veins occur in the Les Essarts metamorphic complex, which comprises orthogneiss and paragneiss with high-pressure metamorphic relics, amphibolite derived from eclogite and banded amphibolite.

The Les Brouzils ore deposits consists of a principal lode system dipping at 70° towards the southeast.

It extends over at least 800 m horizontally and is recognised up to 100 m vertically [19]. It is a cataclastic zone 6–7 m thick, with a progressive transition towards host rocks. Brecciation intensively developed at the vicinity of the mylonites. Richer veins are localised at the border or the lode systems, but some satellite veins cement breccia elements. Distribution of mineralisation in the lode system is heterogeneous, and five richly mineralised columns were described, with horizontal extension of 30–120 m. The principal characteristic of this ore deposit is the presence of large blades of stibnite. Berthierite, pyrite and arsenopyrite complete the paragenesis. Formation of the deposit is polyphased, with vein infill from several successive hydrothermal episodes [21–23]. The Les Brouzils ore-deposit resources were estimated in the 1990s at 9250 t of metal Sb, with 4800 t of proven reserves with 7.5% ore [19].

In the La Télachère prospects, trenches and drill holes on the two principal anomalies determined the presence of a quartz lode system with associated stibnite. The best results in the area yielded 0.20 m at 9% Sb, and 0.30 m at 12.60% Sb. Paragenesis comprises stibnite, arsenopyrite, galena, sphalerite, berthierite, chalcopyrite, tetrahedrite and pyrrhotite in a quartz gangue. Gold was also observed. In La Télachère, several other Sb anomalies have not yet been investigated further.

In the southern part of the study area, near La Copechanière, two other vein systems were discovered by soil geochemistry and VLF (Very low frequency) investigations during the 1980s. They are subvertical quartz NW-SE veins with stibnite, and with a thickness of 0.2–0.3 m.

The entire area is partially covered by plateaux silts and gravels of mixed allochtonous (eolian) and autochtonous origin. Thickness of the plateau loess can be comprised in the range 0.5–2 m. Its age is assumed to be Würm [24].

Most of the geochemical patterns in soil are endogenic, driven by the underlying bedrock. The aeolian component behaves as a homogeneous dilutant. Soils in the area are well developed, with thickness often exceeding 1 m [25]. Weathering of underlying bedrock can lead to alteration profiles reaching 15 m thick. The typical profile starts, from top to bottom, with a relatively thin humic horizon (Ah, maximum 10 cm), passing to a thick horizon of clay accumulation (B, maximum 1.5 m) and a thick C horizon (up to 10 m).

2. Materials and Methods

2.1. Sampling

At the study area, 73 sampling points were spread over six profiles and three Sb prospects (Figure 2). After a first survey to observe typical soil profiles of the area, samples from Ah horizon were carefully collected after removing surficial vegetation and pebbles. The B horizon was sampled by hand auger at a constant 30 to 60 cm sampling depth (Figure 3).

Raw soil samples collected in the field (200–300 g) in dry conditions were carried in kraft paper bags closed with wire, and were received with some clay agglomerates. They were subsequently disaggregated and further air dried, and then sieved to 2 mm for removal of gravel and vegetation. A riffle splitter was used to prepare subsamples for pXRF and laboratory work. This ensured homogenisation through splitting. Soil material was placed in a Niton sample cup without a film cover, hand pressed and shot from above using a Niton mini-field stand.

Samples collected from Ah and B horizons were processed as independent data sets in order to assess the efficiency of the horizons for Sb exploration by identifying spatial anomaly patterns.

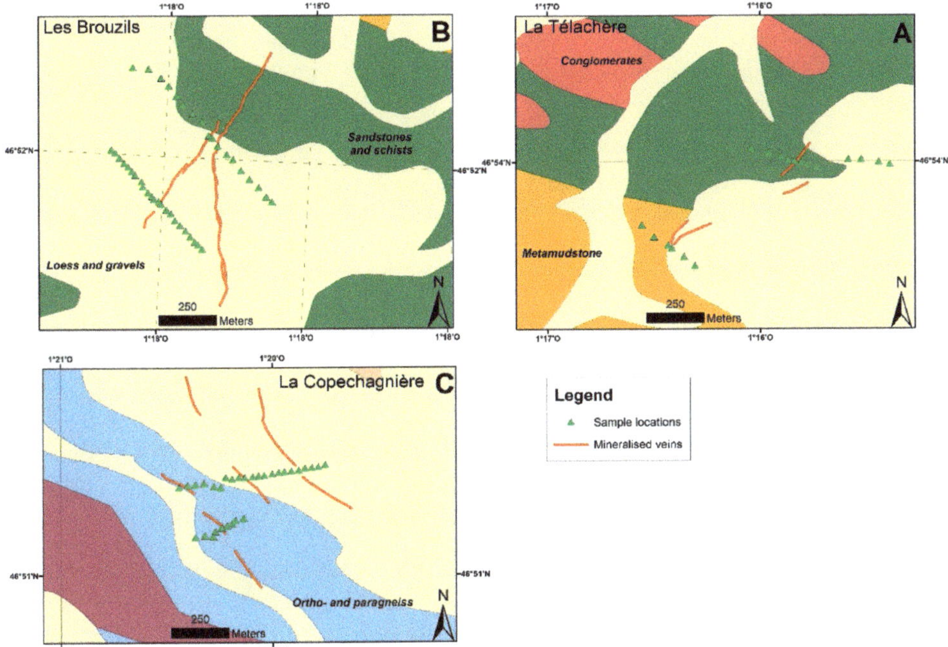

Figure 2. Map of sample locations on geological background map [26]. (**A**): La Télachère; (**B**): Les Brouzils; (**C**): La Copechagnière.

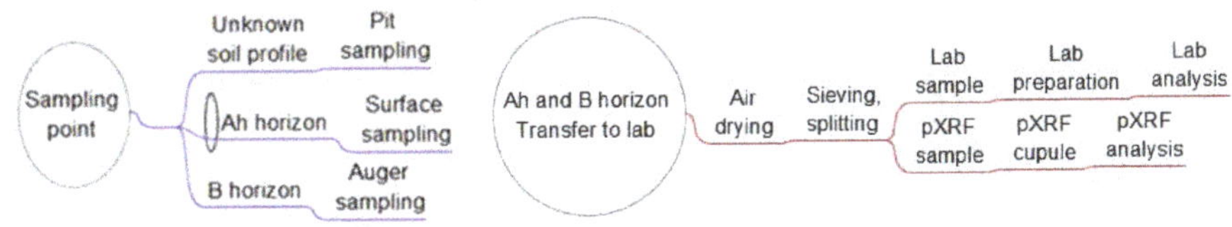

Figure 3. Flow chart showing sample collection and sample preparation.

The sampling transects were designed to be perpendicular to the known mineralised structures (Figure 2). At the time of sampling, the Les Brouzils site was grassland. For La Télachère and La Copechagnière, northern profiles were cultivated (maize), whereas southern profiles are located in wild forest environments. To minimise possible anthropic contaminations, samples were collected at hedges between agricultural fields. The southern profiles were located in the forested woodland.

2.2. pXRF Analyses

To fit best the objectives of the project, the pXRF survey should have been conducted on-site during sampling. However, this was not possible for logistical reasons, so samples were analysed by pXRF in the laboratory one week after the campaign, using almost the same method as usually performed on-site (on-site sieving being performed without drying). This allowed simulation of the on-site selection of samples for laboratory analyses [1]. It is well-known [27–30] that moisture and sample preparation significantly affect raw pXRF measurements, and that a laboratory-type sample preparation offers results closer to laboratory analyses, but at the expense of a slower and more difficult workflow on the field. For the needs of our demonstration, it was chosen to perform measurements the same way as they would have been on site, i.e., on roughly homogenised samples without drying, and with moderate porosity reduction. No matrix-specific calibration was attempted. For the same reason, only the usual certified reference materials (CRMs) were used for quality monitoring, without the acquisition of matrix-matching CRMs, as would have been the case on-site.

Measurements were performed in soil mode with a Niton XL3t-980 pXRF spectrometer (Thermo Fischer, Billerica, MA, USA) and a lab stand. The spectrometer uses a 50-kV tube with an Ag anode, a large silicon drift detector (SDD) and a set of three filters. Soil mode (Compton) was used and counting time was 90 s (30 s per filter). Sb measurements are made using the Kα and Lα lines according to the instrument's software. The LOD was defined as the 3σ standard deviation of the blank, and was reported for each sample according to its matrix composition.

Samples were analysed as air dried but not fully dry. The impact of residual moisture is limited under 20%, especially as linearity is more important than absolute accuracy.

The spectrometer's built-in analysis program reports results for Ag, As, Au, Ba, Ca, Cd, Co, Cr, Cs, Cu, Fe, Hg, K, Mn, Mo, Ni, Pb, Pd, Rb, S, Sb, Sc, Se, Sn, Sr, Te, Th, Ti, U, V, W, Zn and Zr. Among these, Au, Hg, Pd and Te were not kept, as they are often affected by interferences, and because their distribution is too heterogeneous and prone to nugget effects in most samples. Some elements were found to be below the analytical limit of the spectrometer for most samples and do not appear in the Results Section 3.

2.3. *QA/QC*

Quality assurance (QA) and quality control (QC) were performed similar to any type of laboratory analysis, using blanks at regular intervals, certified reference materials (CRMs), internal standards and duplicates. QA/QC methods are fully applicable to pXRF [27].

2.4. *Laboratory Analyses*

Laboratory analyses were performed by ACME Bureau Veritas (Vancouver, BC, Canada) using the AQ250 method, which includes aqua regia digestion of 0.5 g of sample at 95 °C and ultra-trace analysis by ICP/AES (Inductively coupled plasma atomic emission spectroscopy) and ICP/MS (Inductively Coupled Plasma Mass Spectrometry) for Ag, Al, As, Au, B, Ba, Be, Bi, Ca, Cd, Ce, Co, Cr, Cs, Cu, Dy, Er, Eu, Fe, Ga, Gd, Ge, Hf, Hg, Ho, In, K, La, Li, Lu, Mg, Mn, Mo, Na, Nb, Nd, Ni, P, Pb, Pd, Pr, Pt, Rb, Re, S, Sb, Sc, Se, Sm, Sn, Sr, Ta, Tb, Te, Th, Ti, Tl, Tm, U, V, W, Y, Yb, Zn and Zr.

The aqua regia digestion method is not total, while pXRF is a total analysis technique. This may introduce biases for refractory elements [1,31]. The elements discussed in Section 3 are not refractory enough to lead to significant biases.

2.5. *Data Processing*

Classical statistical analysis requires that the data be normal or log-normal and represent one population. With typical geochemical data, the samples are compositions of chemical element contributions, and this interdependence already reflects that the relevant information is not contained in the absolute concentration values but rather in their (log-)ratios. Compositional data analysis (CoDa) [32,33] allows analysis of these log-ratios, and thus, the relative information rather than the absolute information.

Therefore, concentrations do not vary independently and should not be plotted in the Euclidean geometry. This is particularly true for pXRF analyses, which are multielement data, and include both major and trace elements. Element concentrations should, thus, be treated as relative information. Multivariate techniques such as PCA (Principal Component factor Analysis) and correlation analysis [32] are particularly affected by compositional data structure. Hence, absolute concentrations must be first opened by pairwise log ratios of transform techniques, such as the centred log ratio (clr) [33–35].

To avoid the biases mentioned by [33,34], the most significant variables and their correlations were not identified from correlation analysis on raw element concentrations, but were deduced from PCA (Principal Component factor Analysis) and CA (correspondence analysis). PCA was used to understand the relationships and possible dependencies between variables, while CA allowed us to understand the relationships between observations and variables. The Cochran C test was then used to identify the variables with an estimate of variance significantly larger than others. Statistics were provided by the XLSTAT package version 2019.3.2 (Addinsoft, Bordeaux, France), which offers most of the classical functions through an Excel add-on, and a window on R for more elaborate features. Further processing is currently being performed using R, with a dedicated interface currently being developed by the UpDeep project, which should allow processing in the field.

3. Results

3.1. Exploratory Data Analysis

Data analysis is conducted here with the pXRF results alone, in order to simulate what exploration geologists would do without laboratory results, the latter being used only for later quality assessment. Our objective is to make full use of the multi-element capabilities of the pXRF, which need no extra time or cost, rather than using it as a single-element analyser. Beyond this, we also wish to demonstrate the benefit of CoDa analysis of multielement data over raw data. Both raw and clr data sets were processed. It might seem that such data processing would affect the real time benefits of pXRF, but it can be achieved by mobile software in field conditions [2].

According to [33,34], the most significant variables and their correlations were identified by PCA (Principal Component factor Analysis) and CA (correspondence analysis) [35] on raw values first (Figures S1–S11 in Supplementary File), and then on clr-transformed values (Figures 4–9 below). Among the 29 elements measured by pXRF (Ag, As, Ba, Ca, Cd, Co, Cr, Cs, Cu, Fe, K, Mn, Mo, Ni, Pb, Rb, S, Sb, Sc, Se, Sn, Sr, Th, Ti, U, V, W, Zn and Zr), nine elements (Ag, Cd, Co, Cs, Sc, Se, Sn, U and W) had no or little values above the lower analytical limit (LOD) reported by the instrument. Descriptive statistics for the other 20 elements are provided in Tables 1 and 3 for elements with over 20% measurements above LOD.

	Rb	K	Ti	Sr	Th	Zr	S	Pb	Ba	V	Ca	Cu	Zn	Cr	Fe	Mn	Sb	As
Rb	1.000	0.813	0.777	0.525	0.774	0.800	0.377	0.684	0.251	0.204	-0.046	0.134	-0.158	-0.094	-0.531	-0.742	-0.428	-0.705
K	0.813	1.000	0.690	0.506	0.484	0.577	0.416	0.358	0.419	0.170	0.298	0.131	-0.109	-0.155	-0.413	-0.661	-0.501	-0.560
Ti	0.777	0.690	1.000	0.727	0.602	0.883	0.441	0.587	0.163	0.224	0.064	0.152	-0.235	-0.288	-0.602	-0.649	-0.433	-0.650
Sr	0.525	0.506	0.727	1.000	0.476	0.626	0.477	0.277	0.247	0.209	0.217	0.175	-0.123	-0.267	-0.290	-0.452	-0.478	-0.673
Th	0.774	0.484	0.602	0.476	1.000	0.735	0.264	0.674	0.143	0.132	-0.257	0.091	-0.162	-0.023	-0.426	-0.529	-0.354	-0.711
Zr	0.800	0.577	0.883	0.626	0.735	1.000	0.325	0.681	0.067	0.060	-0.178	0.144	-0.293	-0.275	-0.712	-0.655	-0.227	-0.650
S	0.377	0.416	0.441	0.477	0.264	0.325	1	0.226	-0.015	0.105	0.432	0.003	0.096	-0.240	-0.088	-0.338	-0.398	-0.470
Pb	0.684	0.358	0.587	0.277	0.674	0.681	0.226	1.000	-0.078	0.073	-0.200	0.090	-0.095	-0.117	-0.442	-0.517	-0.197	-0.593
Ba	0.251	0.419	0.163	0.247	0.143	0.067	-0.015	-0.078	1.000	0.268	-0.030	-0.025	-0.145	0.236	0.204	-0.126	-0.380	-0.327
V	0.204	0.170	0.224	0.209	0.132	0.060	0.105	0.073	0.268	1.000	0.039	-0.153	-0.004	0.072	0.144	-0.125	-0.395	-0.236
Ca	-0.046	0.298	0.064	0.217	-0.257	-0.178	0.432	-0.200	-0.030	0.039	1.000	0.102	0.178	-0.251	-0.008	-0.169	-0.313	0.006
Cu	0.134	0.131	0.152	0.175	0.091	0.144	0.003	0.090	-0.025	-0.153	0.102	1.000	-0.236	-0.154	-0.166	-0.131	-0.207	-0.204
Zn	-0.158	-0.109	-0.235	-0.123	-0.162	-0.293	0.096	-0.095	-0.145	-0.004	0.178	-0.236	1.000	0.119	0.266	0.068	-0.208	0.059
Cr	-0.094	-0.155	-0.288	-0.267	-0.023	-0.275	-0.240	-0.117	0.236	0.072	-0.251	-0.154	0.119	1.000	0.309	0.184	-0.218	-0.029
Fe	-0.531	-0.413	-0.602	-0.290	-0.426	-0.712	-0.088	-0.442	0.204	0.144	-0.008	-0.166	0.266	0.309	1.000	0.523	-0.119	0.292
Mn	-0.742	-0.661	-0.649	-0.452	-0.529	-0.655	-0.338	-0.517	-0.126	-0.125	-0.169	-0.131	0.068	0.184	0.523	1.000	0.122	0.395
Sb	-0.428	-0.501	-0.433	-0.478	-0.354	-0.227	-0.398	-0.197	-0.380	-0.395	-0.313	-0.207	-0.208	-0.218	-0.119	0.122	1.000	0.502
As	-0.705	-0.560	-0.650	-0.673	-0.711	-0.650	-0.470	-0.593	-0.327	-0.236	0.006	-0.204	0.059	-0.029	0.292	0.395	0.502	1

Figure 4. Pearson correlations on clr data for B horizon Group B observations, pXRF measurements. The use of colour shading and its interpretation is described in [35].

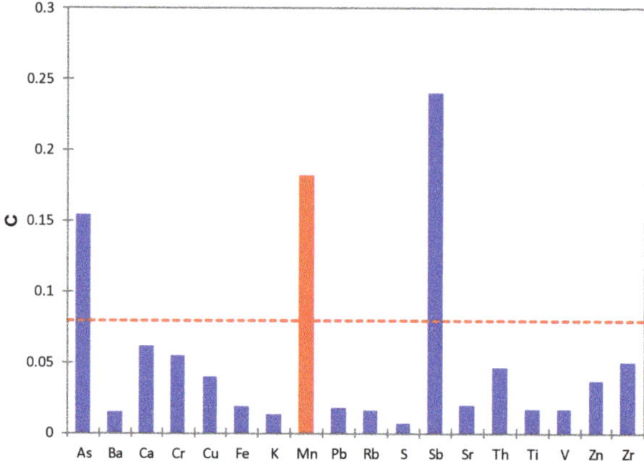

Figure 5. Cochran's C test. Samples from B horizon, pXRF data. The C test detects one exceptionally large variance value at a time.

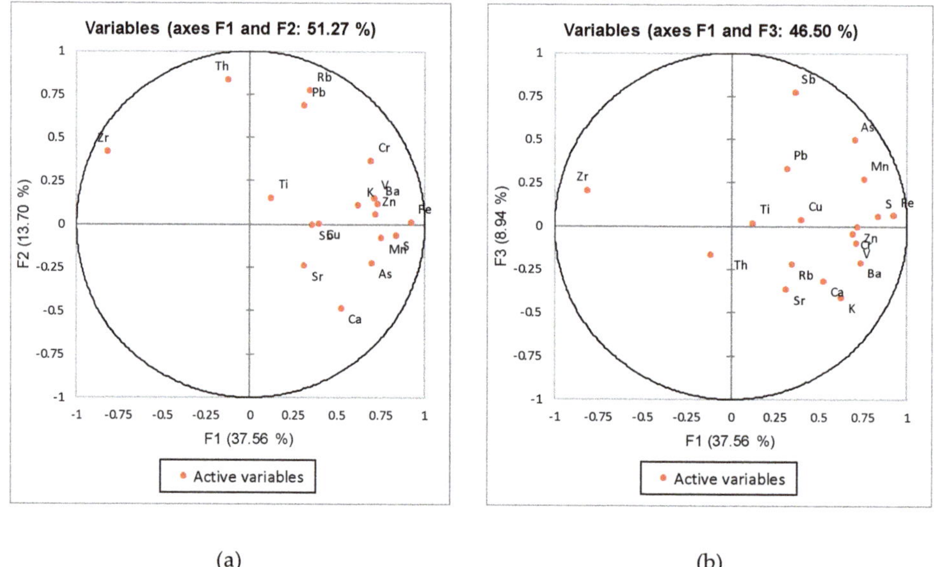

(a) (b)

Figure 6. PCA factor diagrams for F1-F2 (**a**) and F1-F3 (**b**). Samples from B horizon, pXRF clr data.

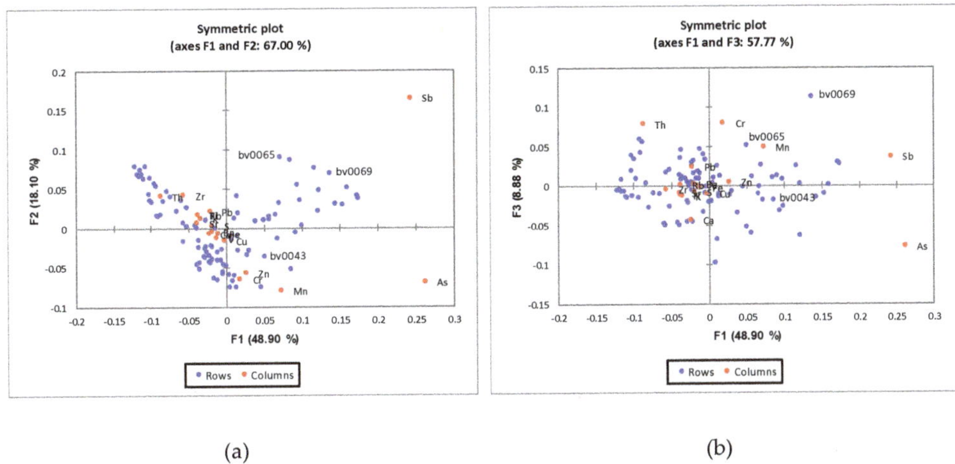

(a) (b)

Figure 7. CA symmetric plots for F1-F2 (**a**) and F1-F3 (**b**). Samples from B horizon, pXRF clr data.

	Rb	Ti	K	Zr	Pb	Sr	V	Th	Fe	Cr	Ba	S	Cu	Mn	Zn	Ca	As	Sb
Rb	1.000	0.801	0.788	0.761	0.609	0.438	0.484	0.670	0.045	0.040	0.049	0.020	-0.242	-0.555	-0.534	-0.636	-0.585	-0.418
Ti	0.801	1.000	0.699	0.866	0.682	0.632	0.538	0.587	0.005	-0.044	0.016	-0.095	-0.321	-0.447	-0.716	-0.686	-0.451	-0.378
K	0.788	0.699	1.000	0.597	0.394	0.420	0.448	0.422	0.129	0.075	0.164	-0.032	-0.325	-0.481	-0.522	-0.364	-0.404	-0.513
Zr	0.761	0.866	0.597	1.000	0.575	0.608	0.232	0.659	-0.242	-0.219	-0.055	0.019	-0.071	-0.585	-0.601	-0.635	-0.485	-0.19
Pb	0.609	0.682	0.394	0.575	1.000	0.385	0.453	0.405	0.079	-0.115	-0.212	0.097	-0.127	-0.367	-0.366	-0.583	-0.381	-0.340
Sr	0.438	0.632	0.420	0.608	0.385	1.000	0.126	0.267	0.014	-0.124	-0.107	0.115	0.081	-0.353	-0.288	-0.229	-0.500	-0.401
V	0.484	0.538	0.448	0.232	0.453	0.126	1.000	0.293	0.490	0.289	0.135	-0.227	-0.556	-0.092	-0.468	-0.516	-0.175	-0.458
Th	0.670	0.587	0.422	0.659	0.405	0.267	0.293	1.000	-0.058	-0.014	0.101	-0.122	-0.225	-0.370	-0.470	-0.599	-0.467	-0.326
Fe	0.045	0.005	0.129	-0.242	0.079	0.014	0.490	-0.058	1.000	0.399	0.307	-0.305	-0.388	0.207	-0.077	-0.204	0.066	-0.478
Cr	0.040	-0.044	0.075	-0.219	-0.115	-0.124	0.289	-0.014	0.399	1.000	0.340	-0.217	-0.363	0.170	-0.106	-0.016	-0.152	-0.376
Ba	0.049	0.016	0.164	-0.055	-0.212	-0.107	0.135	0.101	0.307	0.340	1.000	-0.709	-0.510	0.161	-0.282	-0.147	0.005	-0.174
S	0.020	-0.095	-0.032	0.019	0.097	0.115	-0.227	-0.122	-0.305	-0.217	-0.709	1.000	0.477	-0.335	0.340	0.283	-0.235	-0.017
Cu	-0.242	-0.321	-0.325	-0.071	-0.127	0.081	-0.556	-0.225	-0.388	-0.363	-0.510	0.477	1.000	-0.135	0.453	0.340	-0.141	0.280
Mn	-0.555	-0.447	-0.481	-0.585	-0.367	-0.353	-0.092	-0.370	0.207	0.170	0.161	-0.335	-0.135	1.000	0.312	0.270	0.174	-0.042
Zn	-0.534	-0.716	-0.522	-0.601	-0.366	-0.288	-0.468	-0.470	-0.077	-0.106	-0.282	0.340	0.453	0.312	1.000	0.618	0.012	0.092
Ca	-0.636	-0.686	-0.364	-0.635	-0.583	-0.229	-0.516	-0.599	-0.204	-0.016	-0.147	0.283	0.340	0.270	0.618	1.000	0.168	0.088
As	-0.585	-0.451	-0.404	-0.485	-0.381	-0.500	-0.175	-0.467	0.066	-0.152	0.005	-0.235	-0.141	0.174	0.012	0.168	1.000	0.527
Sb	-0.418	-0.378	-0.513	-0.187	-0.340	-0.401	-0.458	-0.326	-0.478	-0.376	-0.174	-0.017	0.280	-0.042	0.092	0.088	0.527	1.000

Figure 8. Pearson correlations on clr data for Group Ah, pXRF measurements (as heat map). The use of colour shading and its interpretation is described in [35].

The accuracy of pXRF analyses is often debated because pXRF measurements on raw or roughly prepared loose samples often differ from laboratory analyses of the same samples. There are many possible reasons for discrepancies, apart from instrumental ones: a more thorough sample preparation in the laboratory, the use of a wet chemical analysis method instead of XRF and the type of digestion used. Comparisons should be based on pressed pellet XRF in the laboratory, but this is seldom used, and most laboratory work nowadays is based on ICP/AES and ICP/MS.

However, measurement data sets are consistent and usually exhibit quasi-linear relationships with laboratory analyses, provided that a total digestion method is used [1]. Censored data (under the LOD) were replaced by the maximum 3σ value reported by the instrument. This value is greater than the most probable value under the pXRF calibration, but is not necessarily an accurate value. The only important feature is that such values are consistent with the regular (>LOD) data. Element associations were investigated through the PCA and CA analyses. They were used to identify the elements carrying most of the variance in the data set. This was performed separately for the Ah and B data sets.

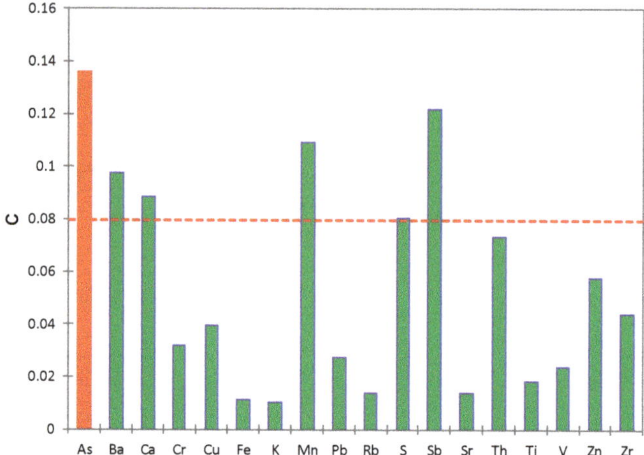

Figure 9. Cochran's C test. Samples from Ah horizon, pXRF raw data.

3.1.1. B Horizon

Samples from the B horizon (mineral subsoil) were usually collected at a depth of 30 to 60 cm. Descriptive statistics are provided in Table 1.

Table 1. pXRF measurements with over 20% observations above LOD, B horizon soils (pXRF, in mg/kg).

Measurements	As	Ba	Ca	Cr	Cu	Fe	K	Mn	Mo	Ni
number	79	96	96	60	27	96	96	87	17	5
min	8	135	958	28	21	5188	9266	68	7	46
max	117	497	9471	98	66	49,077	22,665	4177	9	71
avg	28	306	2696	46	28	20,367	15,440	474	7	57
med	19	307	2180	43	27	19,517	15,048	386	7	49
Measurements	**Pb**	**Rb**	**S**	**Sb**	**Sr**	**Th**	**Ti**	**V**	**Zn**	**Zr**
number	96	96	1	33	96	94	96	94	92	96
min	9	42		19	49	5	4852	53	11	181
max	32	93	879	515	134	15	9184	170	67	433
avg	20	67		107	81	9	5637	93	27	310
med	20	67		52	79	9	5613	89	24	302

Pearson correlations on raw data (Figures S2 and S9 in Supplementary File) show no significant meaningful trend in connection with Sb and chalcophile elements, apart from Mn. Most trends observed are related with parent lithology. The data set was then converted to clr—centred log ratios [32]. The Pearson correlation matrix clearly separates one lithology-related group of elements (Rb, K, Ti, Sr, Th, Zr) and another mineralisation-related group of elements (Sb, As), which includes Mn and Fe in heat map format (Figure 4). A second group of elements (S, Pb, Ba) seems to be unrelated with (Sb, As, Mn) and closer to the lithology group.

The Cochran C test was then used to identify the variables with an estimate of variance significantly larger than others. Among these elements, only Sb, As and Mn show significant anomalies (Figure 5), and Fe is also controlled by lithology.

To reduce compositional data bias, we examined element relationships through PCA. The correlation matrix was used, 18 eigenvalues were calculated and graphs were selected from six eigenvectors. Rotation was not applied. An Sb-As-Mn association appears more clearly for Factor F3 (Figure 6 and Table 2).

Table 2. Contribution of each element to the six main factors for B horizon observations, pXRF clr data.

Contribution	F1	F2	F3	F4	F5	F6
As	0.273	−0.147	**0.389** [1]	−0.141	−0.116	0.035
Ba	0.286	0.070	−0.168	−0.019	−0.177	−0.402
Ca	0.206	−0.312	−0.254	−0.128	0.152	0.446
Cr	0.268	0.229	−0.039	0.054	−0.159	−0.217
Cu	0.157	−0.002	0.025	0.048	0.637	0.150
Fe	0.359	0.005	0.044	0.086	0.024	−0.166
K	0.243	0.066	−0.329	−0.451	−0.112	0.064
Mn	0.293	−0.054	**0.210** [1]	0.208	0.041	−0.209
Pb	0.125	0.433	0.257	0.047	0.166	0.311
Rb	0.136	0.488	−0.178	−0.269	0.006	0.105
S	0.325	−0.043	0.038	0.026	0.171	0.121
Sb	0.143	−0.005	**0.604** [1]	−0.139	0.061	−0.088
Sr	0.125	−0.157	−0.294	0.502	0.328	−0.234
Th	−0.043	0.527	−0.132	0.106	0.239	−0.105
Ti	0.051	0.094	0.011	0.533	−0.404	0.507
V	0.277	0.092	−0.080	0.215	−0.309	0.022
Zn	0.280	0.031	−0.008	0.007	0.018	0.196
Zr	−0.309	0.262	0.158	0.125	0.051	−0.036

[1] Bold figures indicate meaningful positive contributions.

A Pb-Ba association is noted for F5, but with less than 6.5% of total variance. Correspondence analysis (CA) was used to relate these associations with specific samples and possibly with mineralisation signatures or alteration phenomena.

The association between Sb and As is a major driver for F1 (Figure 7), while Mn is less clearly supported by F1 and appears clearly on F2. When reporting samples along these factors, samples BV0065 and BV0069 are highlighted, and sample BV0043 is opposed on F2. Samples BV0065 and BV0069 are high-Mn, with only BV0069 with high Sb, while sample BV0043 has low Mn and high Sb. The positive driver for F1 is a Zr-Th-K-Rb association, most likely related with lithology (heavy minerals, micas). BV0043 differs also by its contents in Ca, while all other samples are extremely depleted.

3.1.2. Ah Horizon

Samples from the Ah horizon (humic topsoil) were collected usually between 2–7 cm deep. Elementary statistics are provided in Table 3.

Table 3. Elementary statistics for Ah horizon observations, pXRF measurements (in mg/kg).

Measurement	As	Ba	Ca	Cr	Cu	Fe	K	Mn	Mo	Ni
number	81	70	96	38	19	96	96	91	13	0
min	7	62	1297	27	19	6837	8811	78	7	
max	486	409	69,043	86	183	32,732	21,678	1992	8	
avg	29	169	5511	42	41	17,982	14,865	409	7	
med	16	163	3329	40	24	16,921	14,442	334	7	
Measurement	**Pb**	**Rb**	**S**	**Sb**	**Sr**	**Th**	**Ti**	**V**	**Zn**	**Zr**
number	96	96	27	27	96	84	96	93	94	96
min	12	43	523	20	52	5	2672	49	13	170
max	60	113	3252	436	137	12	6667	156	330	390
avg	23	64	1204	80	79	8	5326	87	38	284
med	21	64	979	44	76	8	5282	83	26	280

Pearson correlations on raw data are given in Figure 8. We observed Sb-As and Cu-Zn correlations separately.

We then examined element relationships through PCA. An Sb-As association appears more clearly (Table 4, Figures 8 and 9). The Sb-Pb association is carried by factor F4 (Table 4). S is in relation with both groups.

Table 4. Contribution of each element to the six main factors for Ah horizon, clr pXRF measurements.

Contribution	F1	F2	F3	F4	F5	F6
As	−0.627	−0.035	−0.401	**0.516** [1]	−0.140	−0.084
Ba	−0.425	0.573	−0.449	−0.232	−0.145	0.133
Ca	−0.807	−0.371	−0.017	−0.198	−0.040	−0.028
Cr	−0.682	0.418	0.045	−0.170	0.015	−0.004
Cu	−0.520	−0.645	0.179	0.025	−0.179	0.291
Fe	−0.824	0.367	0.016	0.045	0.137	0.003
K	−0.582	0.461	−0.006	−0.285	−0.300	−0.402
Mn	−0.763	0.182	−0.096	0.075	0.185	0.283
Pb	−0.017	0.292	0.574	**0.536** [1]	0.286	0.085
Rb	−0.251	0.635	0.413	−0.101	−0.454	0.029
S	−0.311	−0.581	0.586	−0.008	−0.092	−0.131
Sb	−0.419	−0.323	−0.312	**0.533** [1]	−0.430	0.231
Sr	−0.431	−0.158	−0.079	−0.368	0.516	0.460
Th	0.277	0.512	0.280	−0.132	−0.405	0.482
Ti	0.201	0.788	−0.039	0.223	0.230	0.172
V	−0.433	0.687	0.173	0.233	0.179	−0.157
Zn	−0.811	−0.315	0.250	−0.045	−0.071	0.177
Zr	0.834	0.125	−0.068	−0.008	−0.235	0.341

[1] Bold figures indicate meaningful positive contributions.

The Pearson correlation matrix clearly separates one lithology-related group of elements (Rb, K, Ti, Sr, Th, Zr) and one mineralisation-related group of elements (Sb, As, Mn, Figure 10), which includes Fe in heat map format (Figure 8). The lithology-related group of elements (Rb, K, Ti, Sr, Th, Zr, Pb) also includes Pb in the heat-map format, and the mineralisation-related group of elements includes Ca and Fe in this format (Sb, As, Mn, Fe, Ca), suggesting that Ca and Fe are controlled at least partly by mineralisation processes. Among the elements, As, Sb and Mn show significant anomalies (Figure 9), as do Ba and Ca.

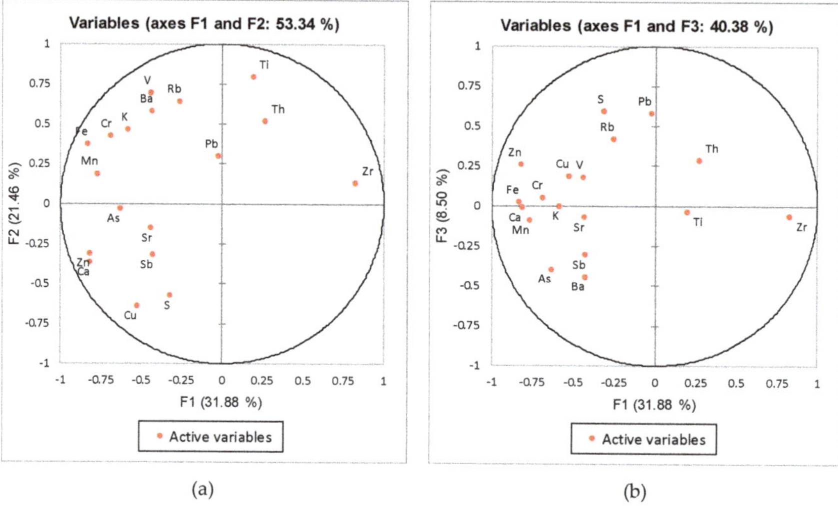

(a) (b)

Figure 10. PCA factor diagrams for F1-F2 (**a**) and F1-F3 (**b**). Samples from group Ah, pXRF clr data.

3.1.3. Variations Between Ah and B Horizons

The concentration ranges do not differ significantly between B and Ah horizons (Tables 2 and 3). Differences in geochemical behaviour are observed in the geochemical signatures (Sections 3.1.1 and 3.1.2) and spatial distribution (Section 3.2). Variations were, therefore, analysed as ratios and enrichment factors (Figure 11). Average and median values were similar between B and Ah for As, Cr, Cu, K, Mn, Rb and Ti. Higher values were observed in the B horizon for Ba, Fe, Sb, V and Zr compared to the Ah horizon. Lower values were observed for Ca, Pb, S and Zn. Sb and As seem to display a moderate depletion in the Ah horizon (Figure 11).

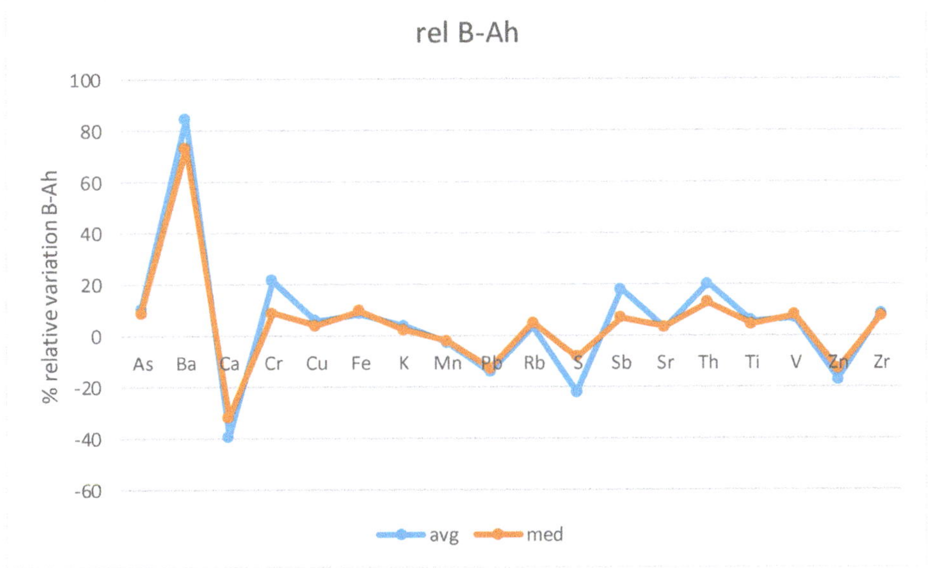

Figure 11. B vs. Ah enrichment factors by element (B concentration/Ah concentration, averaged pXRF raw data for all points).

3.2. Spatial Anomaly Mapping

To better understand the possible benefits of pXRF soil surveys in exploration, samples from the Ah and B horizons were processed and mapped as independent data sets. Maps were also drawn using PCA and CA factor scores for samples. They do not show much more information than single-element maps, but might be useful for other data sets.

On the element maps (Figures 12–17), the classes boundaries for each element were selected according to quantiles. Unlike distribution breaks, this allows the lower values of each element's concentration to be put forward in order to show low-level anomalies best.

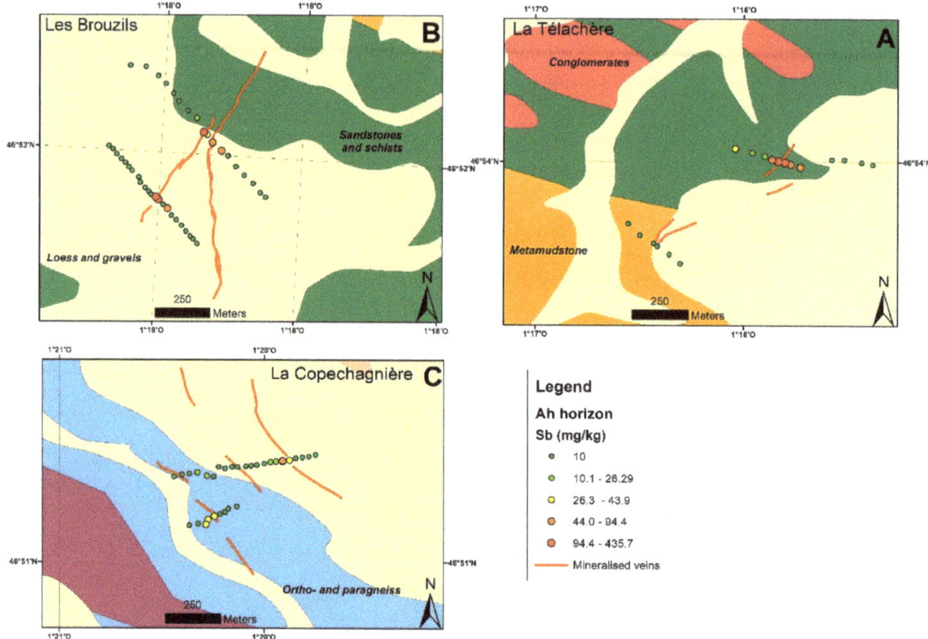

Figure 12. Sb map of Ah horizon samples, (pXRF measurements in mg/kg). (**A**): La Télachère; (**B**): Les Brouzils; (**C**): La Copechagnière.

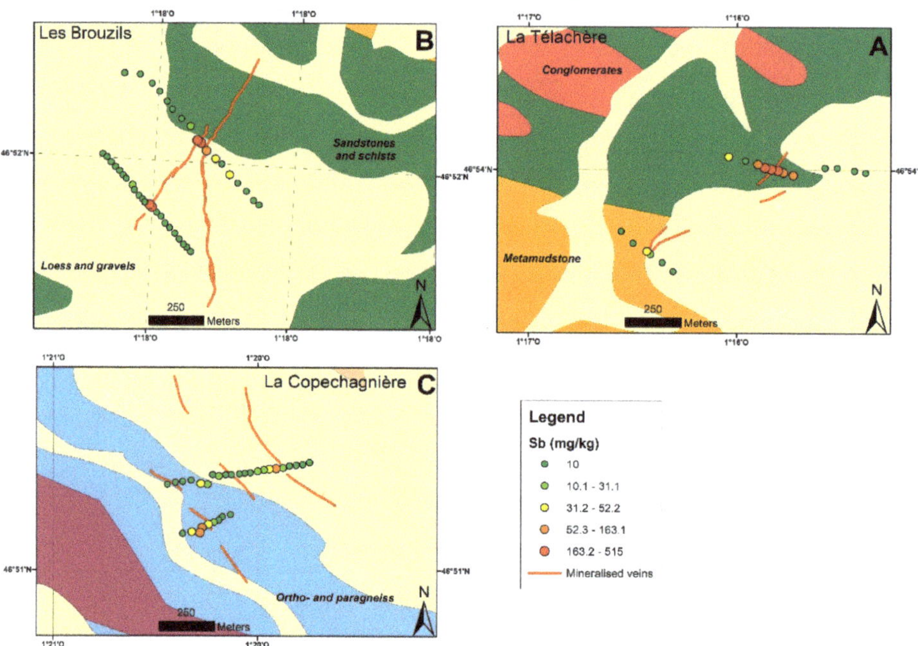

Figure 13. Sb map of B horizon samples (pXRF measurements in mg/kg). (**A**): La Télachère; (**B**): Les Brouzils; (**C**): La Copechagnière.

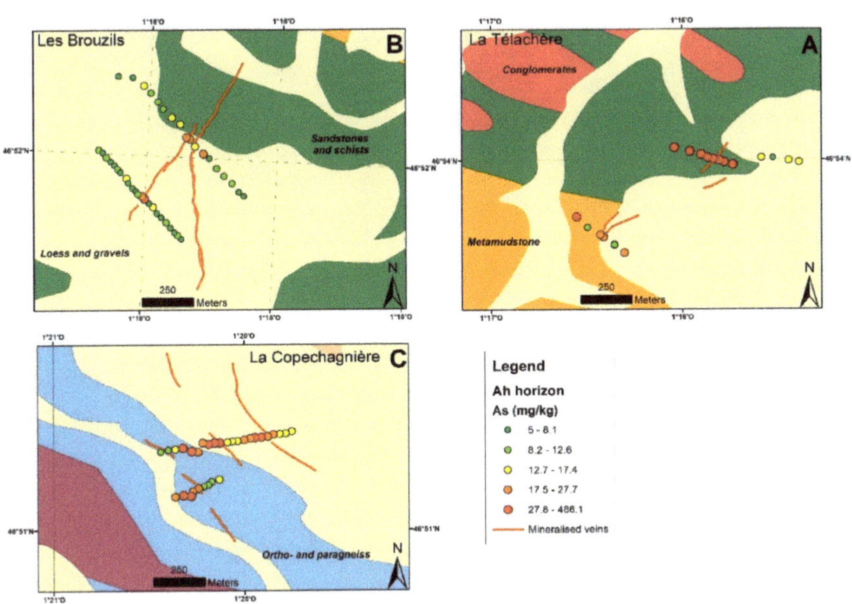

Figure 14. As map of Ah horizon samples (pXRF measurements in mg/kg). (**A**): La Télachère; (**B**): Les Brouzils; (**C**): La Copechagnière.

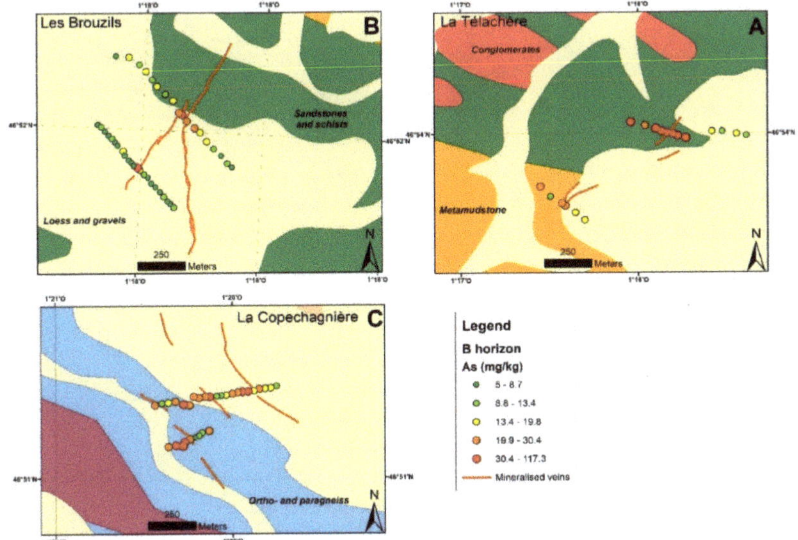

Figure 15. As map of B horizon samples (pXRF measurements in mg/kg). (**A**): La Télachère; (**B**): Les Brouzils; (**C**): La Copechagnière.

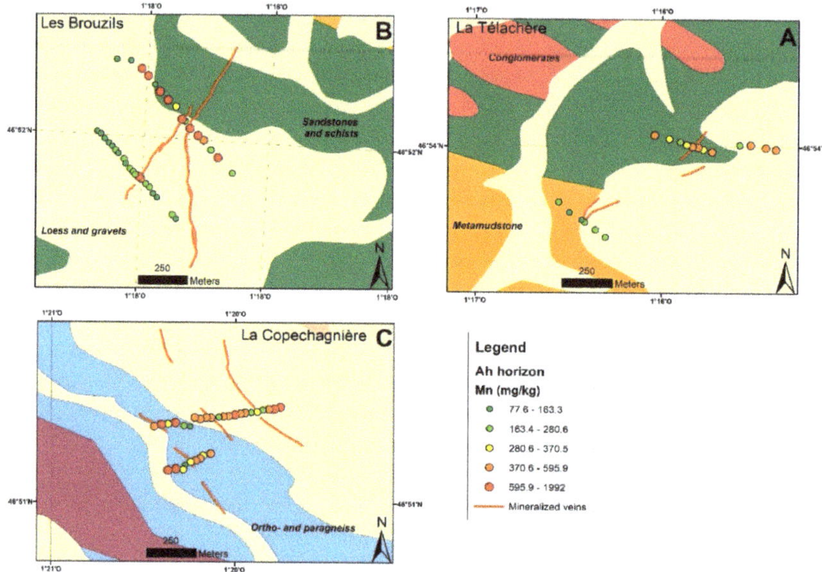

Figure 16. Mn map of Ah horizon samples (pXRF measurements in mg/kg). (**A**): La Télachère; (**B**): Les Brouzils; (**C**): La Copechagnière.

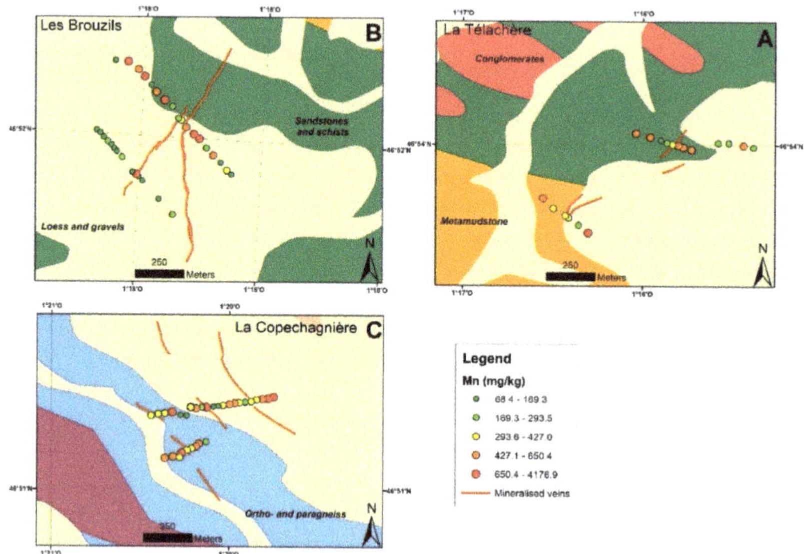

Figure 17. Mn map of B horizon samples (pXRF measurements in mg/kg). (**A**): La Télachère; (**B**): Les Brouzils; (**C**): La Copechagnière.

3.2.1. Sb Spatial Anomaly Patterns

The profiles sampled in the three areas illustrated in Figure 2 are shown in Figures 10–16, separately for the horizons Ah and B.

In the La Télachère area, a moderate but conspicuous Sb anomaly was observed both on Ah (Figure 12) and B (Figure 13) data, with higher Sb concentrations in the latter. A similar pattern was observed for As (Figures 14 and 15), while the Mn anomaly was weaker, which was mainly observable in the topsoil (Figure 16) and slightly shifted to the east.

In La Copéchagnière, the Sb anomaly was weak, close to the LOD of the pXRF instrument. It would require a careful comparison with laboratory analyses to determine whether Sb variations by pXRF are meaningful or not, and properly located. We observed arsenic anomalies in the B horizon, but they are less obvious in the Ah horizon. A weak Mn signal was observed in the Ah horizon data.

The most prominent anomalies were observed in the Les Brouzils area near the abandoned mine works. Sb anomalies were located near the expected position of the quartz-Sb vein on both profiles, in both horizons (Figures 12 and 13). The distance to reported mineralisation was no more than 50 m. Additional single sample anomalies were observed further away, but their position near roads or tracks might be related to the past mining activity.

3.2.2. As Spatial Anomaly Patterns

Sharp anomalies were observed in the Les Brouzils area, south profile; La Télachère, north profile; and la Copechagnière, north profile. Weaker As anomalies are observable on the other profiles, especially on the B horizon (Figures 14 and 15). Similar to Sb, the distance to reported mineralisation was no more than 50 m.

3.2.3. Mn Spatial Anomaly Patterns

The anomalies for Mn are observed on a slightly wider area (Figures 16 and 17), though still centred on known mineralisation, with a slight drift towards the southeast. This may reflect the larger contrast between Mn's lower analytical limit and observed Mn concentrations. Sb has more intense gradients than Mn along the profiles.

Maps drawn with PCA factor scores (Figure 18) did not bring a significant advantage. They are similar to Mn maps, but represent the La Télachère orebody better.

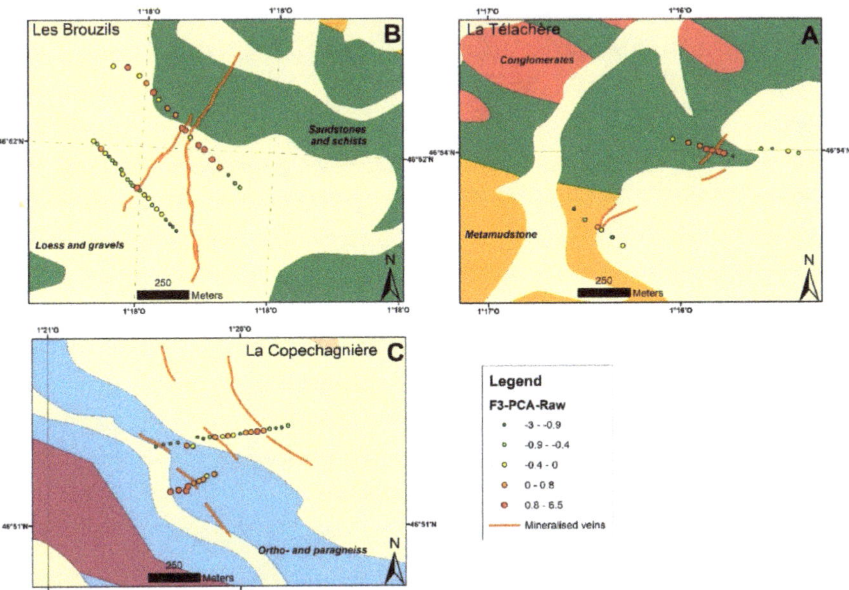

Figure 18. PCA F3 map of B horizon samples (pXRF measurements in mg/kg). (**A**): La Télachère; (**B**): Les Brouzils; (**C**): La Copechagnière.

3.3. *Quality Control*

3.3.1. pXRF QA/QC

On the basis of regular pXRF measurements of CRMs, all reported elements showed satisfactory agreement for at least one CRM (Table 5). Discrepancies may reflect matrix incompatibility. Blank sample measurements are satisfactory.

To assess reproducibility, triplicate measurements were made on a subset, shooting on slightly different locations (Table 6). Thus, the reproducibility also includes the sample heterogeneity. This explains higher (10 to 20%) variation rates for As, Sb, S, Cr, Mo, Ni and V, which are controlled by mineralised grains. Variability is less than 10% for the other elements.

3.3.2. pXRF Quality Control by Laboratory Analyses

Usually, external control of pXRF measurements is performed on a subset of the analysed samples by submitting this subset to an external laboratory analysis, operating under its own QA/QC scheme [27]. It would be ideal to perform this by comparing pXRF results with laboratory XRF analyses to avoid possible biases related to digestion. This was not possible within the budget and time frame, but the comparison of pXRF measurements with aqua regia ICP/AES and ICP/MS analyses provided confirmation of consistency and LODs. This was possible because the main elements under investigation (Sb, As, Mn, Pb and Zn, see the Exploratory data analysis section) are readily soluble in aqua regia for the mineral paragenesis observed. Another possible source for discrepancies lies in sample preparation, as pXRF measurements were carried on dried and sieved but not milled soil samples, instead of lab-ready pulps. This was chosen to better simulate on-site measurements.

Although laboratory analyses were performed on lab-ready pulps after aqua regia digestion, and our measurements were on dried raw samples, an acceptable linear correlation was observed (Figure 19, and similar patterns for Mn and Zn). Both Sb and As showed a positive bias for pXRF (27 and 14%), which did not affect sample ranking.

Overestimation for Sb and As cannot be caused by porosity. Underestimation due to partial dissolution by aqua regia is possible, but unlikely, because most Sb and As minerals are not refractory. In other projects, negative bias for these elements may be observed. We believe that standard calibration in soil mode is not optimal for Sb below 200 mg/kg. One possible explanation is that the factory calibration and standards did not closely match the samples' ranges. Having a custom calibration with matching standards, or better, with spiked SRMs from the same matrix, might improve this bias, but would need repeated measurements. We did not investigate this in detail, as the ranking of the samples was not affected. The bias for As, combined with good linearity ($R^2 = 0.955$), is acceptable for exploration. Good correlations were also observed for Mn and Zn, with minor bias (−5% and +3%).

Table 5. Blank and CRM (soil and waste) pXRF measurements (mg/kg). na: not available, nc: not calculated

Reference	As	Ba	Ca	Cr	Cu	Fe	K	Mn	Mo	Ni	Pb	Rb	S	Sb	Sr	Th	Ti	V	Zn	Zr
Blank																				
average	<LD	<LD	668	<LD	<LD	281	446	<LD	<LD	<LD	<LD	<LD	<LD	<LD	102	<LD	99	<LD	<LD	20
std deviation			15			26									5		14			3
NIST 2709																				
average	16	875	18,892	113	37	33,878	18,818	492	5	74	17	90	<LOD	16	221	11	3474	117	86	136
std deviation	4	40	883	25	9	507	387	77	2	8	3	2			5	1	110	19	8	6
recommended	18	968	18,900	130	35	35,000	20,300	538	2	88	19	96	890	8	na	11	3420	112	106	160
+/−	1	40	500	4	1	1100	600	17	nc	5	1	nc	20	1	na	nc	240	5	3	nc
NIST 2710																				
average	17	90	<LOD	16	221	11	3474	117	86	136	5548	126	2525	41	316	33	2659	75	6894	115
std deviation	3	2			5	1	110	19	8	6	193	4	1472	13	7	7	257	16	275	5
recommended	19	96	890	8	na	11	3420	112	106	160	5532	120	2400	38	330	13	2830	77	6952	na
+/−	1	nc	20	1	na	nc	240	5	3	nc	80	nc	60	3	nc	nc	100	2	91	na
NIST 2710a																				
average	1689	877	8221	60	3395	47,587	22,104	2273	9	50	5572	112	11,926	53	250	49	3052	90	4363	207
std deviation	99	83	385		49	8440	900	408	3		253	6		18	6		120	40	196	1
recommended	1540	792	9640	23	3420	43,200	21,700	2140	na	8	5520	117	na	53	255	18	3110	82	4180	na
+/−	100	36	450	6	50	800	1300	60	na	1	30	3	na	2	7	0	70	9	150	na
NIST 2780																				
average	<LOD	1106	2394	36	184	28,619	33,977	499	12	43	5152	183	12,262	175	233	34	6859	241	2167	175
std deviation		29	461	15	18	1157	2236	53	2	3	186	3	710	3	4	8	355	17	181	1
recommended	49	993	1950	44	216	27,840	33,800	462	11	12	5770	175	12,630	160	217	12	6990	268	2570	176
+/−	3	71	200	nc	8	800	2600	21	nc	nc	410	nc	420	nc	18	nc	190	13	160	nc

Table 6. Replicate pXRF measurement statistics (in triplicates, in mg/kg).

Measurements	As	Ba	Ca	Cr	Cu	Fe	K	Mn	Mo	Ni	Pb	Rb	S	Sb	Sr	Th	Ti	V	Zn	Zr
average	43	1298	10,070	77	447	65,349	30,653	1769	37	106	588	227	7958	60	78	39	3961	113	209	240
median of averages	38	541	2543	70	74	26,952	30,803	640	15	81	98	233	7868	42	69	27	3921	100	102	242
average of standard deviations	6	42	151	11	20	698	437	83	5	17	10	6	1646	19	3	5	123	16	13	5
median of standard deviations	5	29	129	10	12	310	391	55	5	13	7	5	1986	13	3	4	109	12	10	5
sdev/average	13%	3%	1%	15%	4%	1%	1%	5%	12%	16%	2%	2%	21%	32%	3%	11%	3%	15%	6%	2%

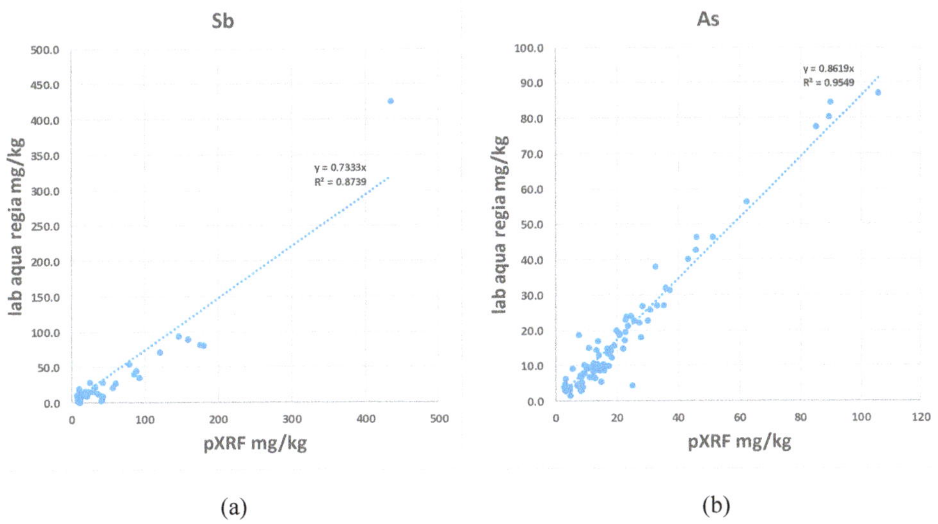

Figure 19. pXRF measurements vs. ICP/AES analyses, Ah horizon soils, in mg/kg: (**a**) Sb; (**b**) As.

4. Discussion

4.1. CoDa Processing

Raw data processing results are less efficient than clr data processing results, demonstrating that CoDa interpretation is more efficient. Raw data are what the geochemist will see while sampling and analysing. CoDa processing may seem complex but can be performed on the same day or the next, using an ordinary laptop. On-site selection and dynamic sampling are still possible as the geochemist is still on site.

4.2. Soil Horizons

No systematic variations were observed between horizons Ah and B, apart from Ca and Ba. Variations of Sb and pathfinder elements (As, Mn) are of the same order of magnitude. Arsenic anomalies observed in the B horizon are less obvious than in the Ah horizon. The weak Mn signal in the Ah horizon is shifted, and thus, we suspect that it reflects surficial enrichment.

The strong Ca enrichment in the Ah horizon (Figure 11) may reflect pedogenetic processes or lime addition for agricultural practices in acid soils. Ba enrichment could not be explained.

4.3. Spatial Anomaly Mapping

The richer profiles at area B (Les Brouzils) showed small but conspicuous Sb anomalies (Figures 12 and 13), while area A (La Télachère) and area C (La Copéchagnière) profiles showed weaker anomalies. Using a lower concentration profile for Sb at area A (La Télachère) improved the resolution, despite the poor absolute accuracy, as the numbers were very close to the LOD.

Using quantiles, Sb classes boundaries were selected preferentially in the lower range of the element concentration, in order to show better low-level anomalies (Figures 12 and 13).

It seems that Mn anomalies are less sharp, from a spatial point of view, or more diffuse than Sb or As anomalies. Mn is known for scavenging Sb during Sb mobility under reducing conditions [36]. However, this does not explain the spatial distribution of Mn. Other possibilities would be an earlier Mn deposition during mineralisation events, or a wider diffusion of Mn in the vein selvages due to higher mobility.

4.4. Application to Exploration

We were able, using pXRF measurements, to detect shallow Sb mineralisation based on single-element Sb patterns, but this did not give positive results for all occurrences. The mineralisation was missed at the La Télachère south profile and the La Copechagnière north profile. Detection was improved using Sb and pathfinder elements (As, Mn) patterns where single-element Sb patterns failed. At least one pathfinder could be detected at profiles where Sb was not detected. Threshold values can be estimated for these elements (As: 30 mg/kg, Mn: 400 mg/kg). The larger contrast between the lower Mn analytical limit and observed Mn concentrations makes Mn a more sensitive pathfinder than As. We cannot provide mineralogical evidence for ore-forming or hydrothermal processes explaining this, as we did not have access to any vein outcrop or to historic drill cores, but the association is confirmed statistically and spatially. It is not often cited [36,37] and it might be related to the aptitude of Mn hydroxides to scavenge Sb by changing its redox state [12,38]. However, this is not applicable to geochemical dispersion under surface conditions, as the cartographic distribution of Mn is linked with vein position. It is more likely that Mn was part of the mineralising event, either by deposition or by selvage alteration.

We did not attempt deeper ore detection from soil profiles, as previous prospection did not target deeper ore. The profiles we tested were set across known anomalies and were too short to have a chance to cross other anomalies. Longer profiles or grid patterns would be needed.

On the Les Brouzils mineral occurrences, soil anomalies appeared no further than 50 m from the known position of antimony veins. Single-sample anomalies were observed further away, but their position near roads or tracks might be caused by the past mining activity.

The representativeness of pXRF data alone is satisfactory according to two criteria:

- Ranking samples according to Sb and pathfinder concentrations, as a linear relationship is observed between pXRF measurements and laboratory analyses, even with a bias affecting absolute accuracy.
- Delineating precise anomalies, as the spatial consistency of anomalies with known mineralisation location is good.

Beyond the absolute accuracy issues, exploration decisions based on the pXRF geochemical data set are reliable, in terms of the dynamic sampling plan. Sampling uncertainties and matrix heterogeneity issues in laboratory analyses can be reduced by using multiple measurements by pXRF on site, prior to shipping, allowing optimised sample selection.

This selection can be based on pXRF results due to linearity [1,39]. Decisions with financial consequences, such as drilling programs or detailed investigations, will require this laboratory confirmation. However, decisions on further reconnaissance investigations, such as higher density sampling or area extensions, can be made on the basis of pXRF measurements and geology alone.

The lower accuracy of field analyses is compensated for by the much larger number of analyses made possible by on-site methods. An added benefit of pXRF analyses is the quantified uncertainty for each sample, allowing direct integration into geostatistics.

More detailed multielement processing based on the CoDa toolbox has yet to be completed in the UpDeep project. It may be tested later on promising structures without known mineralisation.

The B horizon is generally the preferred medium target for geochemical exploration, as it concentrates most indicator elements. Similar geochemical contrasts between the B and humic horizon (Ah) have already been demonstrated [40]. Earlier, [41] demonstrated that it could also be representative of the C horizon. For the Vendée case study, our results agree well with these statements. Ah can be sampled quickly and easily.

5. Conclusions: Prospecting for Sb with a pXRF

This work contributes to evaluating whether pXRF is an effective tool for focusing investigations at the early stage, defining geochemical patterns and allowing fast vectorisation based on quasi-real time measurements. It also allows for the more effective selection of samples for laboratory analyses, focusing on critical samples and reducing the number of routine ones. Further benefits would be more efficient field campaigns and reducing sampling and drilling needs in Sb exploration. To achieve that, data quality are ensured by denser measurements and by QA/QC monitoring.

Samples were lain densely (10 to 30 m) on predefined profiles, so narrow and shallow veins were located effectively. It would have been beneficial to further reduce the spacing between sampling points. From multielement pXRF measurements, we conclude that:

- Based on single-element Sb patterns, mineralisation can often be detected, but not for all intercepts.
- Based on Sb, As and Mn patterns, Sb mineralisation can be detected using pathfinders. A composite signature search (Sb, As, Mn) turned out to be more effective for mineralisation detection than single Sb maps. The pathfinder signature needs to be determined prior to the survey. Maps drawn with PCA factor scores did not bring a significant advantage, but this may be due to the rather simple signature and to the lack of lithogeochemical influence. Factor-score maps might be useful at other sites.
- Sb, As and Mn contrast is good, but the background values are not much above the lower analytical limit of the instrument on raw samples.
- Using the same signature (Sb, As, Mn) for deeper ore detection is theoretically possible but more difficult. Based on the thickness of the surficial cover and on the structural control, a weaker signal could be expected.
- Detection of weak anomalies may be hampered by background noise and scatter. No demonstration was made on the site, as previous prospection did not target the deeper ore.

Portable XRF measurements on soil samples from carefully selected horizons Ah and B provide relevant information while exploring for vein-type or structure-hosted antimony mineralisations. The low cost and fast execution of pXRF measurements allow high spatial density on profiles or grids, therefore reducing the potential uncertainty of the significance of measurements close to the lower analytical limit.

The execution of pXRF profiles by auger does not require heavy field work and may be conducted at high spatial density. In terms of footprint, analysing soil for selected horizons is similar to the collection of surficial pedological samples. It is easily accepted by landowners and farmers, as it causes very limited damage to the vegetation cover. This damage can be further minimised by careful sampling procedures with cover restoration, and by focusing on the Ah horizon. Analyses with portable XRF of the Ah horizon are a cost-effective and efficient method to target such Sb ore deposits.

It might look difficult to conduct multivariate analysis or result mapping using GIS (Geographic information systems) while on site, but the computing capabilities of laptops, tablets and even smartphones are now sufficient to allow proper data interpretation on site, within minutes. The objective of the UpDeep project [2] is to provide a high level of data integration and processing, in a smooth and robust way, in order to overcome limitations on interpretation.

Elemental signatures and spatial patterns can be determined while sampling and can be used for planning further sampling points. The immediate availability of results allows sampling plans, profiles and grids to be refined according to observed anomalies. This approach, based on dynamic workplans [42] and adaptive sampling (ASAP) strategies [43], is both time- and cost-effective at the early stages. The main limitation is determining absolute concentrations, which may need time-consuming sample preparation or matrix-specific calibration to reduce bias, but bias does not affect spatial patterns or anomaly detection, which are the primary objectives of a survey.

Due to the relatively low cost of pXRF measurements, much higher density sampling is possible than for laboratory analyses, resulting in higher quality data sets [44]. On-site soil analyses by pXRF are effective for outlining anomalies, creating maps and locating mineralisation. However, the analyses need careful confirmation in the laboratory, such as anomaly ranking and deposit pre-evaluation before exploration moves to the next stage.

Field methods provide invaluable help for laboratory sample selection and screening. They improve the cost-effectiveness of analytical programs. Field methods help to control the representativeness of laboratory samples, for instance by assessing heterogeneity by multiple shots while sampling. The availability of a large field data set reduces the risk of overlooking sampling and preparation uncertainties with laboratory results.

The Vendée case study highlights that the humic horizon can increase efficiency of the survey and decrease its impacts, with faster access to the sample.

Supplementary Materials: The following are available online at http://www.mdpi.com/2075-163X/10/8/724/s1, Figure S1: pXRF measurements with over 20% observations above LOD, B horizon soils (pXRF, in mg/kg), Figure S2: Pearson correlations on pXRF raw data for B horizon soils (as heat map). The use of colour shading and its interpretation is described in [1], Figure S3: Contribution of each element to the 6 main factors for B horizon observations, pXRF raw data. Coloured cells indicate meaningful positive contributions, Figure S4: PCA factor diagrams for F1, F2 (a) and F3 (b). Samples from B horizon, pXRF raw data, Figure S5: CA symmetric plots for F1, F2 (a) and F3 (b). Samples from B horizon, pXRF raw data, Figure S6: Pearson correlations on clr data for B horizon Group B observations, pXRF measurements. The use of colour shading and its interpretation is described in [1], Figure S7: Cochran's C test. Samples from B horizon, pXRF data. The C test detects one exceptionally large variance value at a time, Figure S8: Elementary statistics for Ah horizon observations, pXRF measurements (in mg/kg), Figure S9: Pearson correlations on raw data for Ah horizon, pXRF raw measurements (as heat map). The use of colour shading and its interpretation is described in [1], Figure S10: Contribution of each element to the 6 main factors for Ah horizon, pXRF measurements. Coloured cells indicate meaningful positive contributions, Figure S11: PCA factor diagrams for F1, F2 (a) and F3 (b). Samples from group Ah, pXRF raw data.

Author Contributions: Conceptualization methodology and validation, B.L. and J.M.; sampling, V.D., E.G., L.B.; investigation, P.A., V.D.; data curation, B.L., D.M., P.F.; writing—original draft preparation, review and editing, B.L., J.M., P.F., M.M.; visualization, J.M.; supervision and project administration, J.M., M.M. All authors have read and agreed to the published version of the manuscript.

Funding: This research is part of an upscaling project UpDeep, Upscaling deep buried geochemical exploration techniques into European business, funded by the European EIT Raw Materials. This activity received funding from the European Institute of Innovation and Technology (EIT), a body of the European Union, under the Horizon 2020, the EU Framework Programme for Research and Innovation.

Acknowledgments: Karen M. Tkaczyk, professional editor, proofread the manuscript and checked the language.

Conflicts of Interest: The authors declare no conflict of interest.

References

1. Lemiere, B. A review of pXRF (field portable X-ray fluorescence) applications for applied geochemistry. *J. Geochem. Explor.* **2018**, *188*, 350–363. [CrossRef]
2. Middleton, M.S.; Nykänen, V.; Melleton, J.; Lemiere, B.; Sarala, P.; Filzmoser, P.; Järvinen, P.; Rinkkala, M.; Rönnqvist, J.; Thaarup, S. Upscaling deep buried geochemical exploration techniques into European business—UpDeep. In Proceedings of the Resources for Future Generations—RFG2018, Vancouver, BC, Canada, 16–21 June 2018; p. 1227.
3. Cameron, E.M.; Hamilton, S.M.; Leybourne, M.I.; Hall, G.E.M.; McClenaghan, M.B. Finding deeply buried deposits using geochemistry. *Geochem. Explor. Environ. Anal.* **2004**, *4*, 7–32. [CrossRef]
4. Heberlein, D.; Dunn, C. Sweat, Sap And Emanations—What Trees and Snow Can Reveal About Hidden Mineralization. In Proceedings of the Resources for Future Generations—RFG2018, Vancouver, BC, Canada, 16–21 June 2018; p. 1941.
5. Melleton, J.; Lemière, B.; Derycke, V.; Serrand, A.S.; Fournier, E.; Gloaguen, E.; Lacquement, F.; Auger, P.; Middleton, M.; Nykänen, V. Exploration geochemistry: Comparison between classic trace elements geochemistry, soil partial leaches, portable XRF, on soils and biogeochemistry in Western Europe Environment. Example from Li-Ta-Sn and W deposits. In Proceedings of the Resources for Future Generations—RFG2018, Vancouver, BC, Canada, 16–21 June 2018; p. 1395.
6. European Commission. Communication from the Commission to the European Parliament, the Council, the European Economic and Social Committee and the Committee of the Regions Tackling the Challenges in Commodity Markets and on Raw Materials (COM/2011/0025 Final). 2011. Available online: https://eur-lex.europa.eu/legal-content/EN/TXT/?uri=CELEX:52011DC0025 (accessed on 20 December 2018).

7. USGS 2015 Minerals Yearbook—Antimony. Available online: https://minerals.usgs.gov/minerals/pubs/commodity/antimony/myb1-2015-antim.pdf (accessed on 20 December 2018).
8. Guo, X.; Wu, Z.; He, M.; Meng, X.; Jin, X.; Qiu, N.; Zhang, J. Adsorption of antimony onto iron oxyhydroxides: Adsportion behavior and surface structure. *J. Hazard. Mater.* **2014**, *276*, 339–345. [CrossRef]
9. Henckens, M.L.C.M.; Driessen, P.P.J.; Worrell, E. How can we adapt to geological scarcity of antimony? Investigation of antimony's substitutability and of other measures to achieve a sustainable use. *Resour. Conserv. Recycl.* **2016**, *108*, 54–62. [CrossRef]
10. He, J.; Wei, Y.; Zhai, T.; Li, H. Antimony-based materials as promising anodes for rechargeable lithium-ion and sodium-ion batteries. *Mater. Chem. Front.* **2018**, *3*, 437–455. [CrossRef]
11. ATSDR, Public Health Statement for Antimony, in Toxic Substances Portal. Available online: https://www.atsdr.cdc.gov/phs/phs.asp?id=330&tid=58 (accessed on 31 January 2020).
12. Herath, I.; Vithanage, M.; Bundschuh, J. Antimony as a global dilemma: Geochemistry, mobility, fate and transport. *Environ. Pollut.* **2017**, 545–559. [CrossRef] [PubMed]
13. Pohl, W. *Economic Geology: Principles and Practice: Metals, Minerals, Coal and Hydrocarbons—Introduction to Formation and Sustainable Exploitation of Mineral Deposits*; Wiley-Blackwell: Chichester, UK, 2011.
14. Munoz, M.; Courjault-Rade, P.; Tollon, F. The massive stibnite veins of the French Palaeozoic basement: A metallogenic marker of Late Variscan brittle extension. *Terra Nova* **2007**. [CrossRef]
15. Pochon, A.; Gloaguen, E.; Branquet, Y.; Poujol, M.; Ruffet, G.; Boiron, M.C.; Boulvais, P.; Gumiaux, C.; Cagnard, F.; Gouazou, F.; et al. Variscan Sb-Au mineralization in Central Brittany (France): A new metallogenic model derived from the Le Semnon district. *Ore Geol. Rev.* **2018**, *97*, 109–142. [CrossRef]
16. Scratch, R.B.; Watson, G.P.; Kerrich, R.; Hutchinson, R.W. Fracture-controlled antimony-quartz mineralization, Lake George Deposit, New Brunswick; mineralogy, geochemistry, alteration, and hydrothermal regimes. *Econ. Geol.* **1984**, *79*, 1159–1186. [CrossRef]
17. Sainsbury, C.L. Geochemical exploration for antimony in southeastern Alaska. *USGS Open-File Rep.* **1955**, 55–158. [CrossRef]
18. Marcoux, E.; Serment, R.; Allon, A. Les gites d'antimoine de Vendée (Massif armoricain, France); historique des recherches et synthèse métallogénique. *Chron. Rech. Min.* **1984**, *476*, 3–30.
19. Le Fur, Y.; Allon, A.; Biron, R.; Lequertier, M.; Roussel, M. La découverte du gisement d'antimoine des Brouzils en Vendée (Massif Armoricain, France). Historique des travaux, description du gisement et projet d'exploitation. *Chron. Rech. Min.* **1988**, *492*, 5–18.
20. Godard, G.; Bouton, P.; Poncet, D.; Carlier, G.; Chevallier, M. *Geological Map 1:50,000, Montaigu*; BRGM Editions: Orleans, France, 2007; p. 536.
21. Bailly, L.; Bouchot, V.; Beny, C.; Milesi, J.-P. Fluid inclusion study of stibnite using infrared microscopy: An example from the Brouzils antimony deposit (Vendee, Armorican massif, France). *Econ. Geol.* **2000**, *95*, 221–226. [CrossRef]
22. Pochon, A.; Gapais, D.; Gloaguen, E.; Gumiaux, C.; Branquet, Y.; Cagnard, F.; Martelet, G. Antimony deposits in the Variscan Armorican belt, a link with mafic intrusives? *Terra Nova* **2016**, *28*, 138–145. [CrossRef]
23. Pochon, A.; Branquet, Y.; Gloaguen, E.; Ruffet, G.; Poujol, M.; Boulvais, P.; Gumiaux, C.; Cagnard, F.; Baele, J.-M.; Kéré, I.; et al. A Sb ± Au mineralizing peak at 360 Ma in the Variscan belt, BSGF. *Earth Sci. Bull.* **2019**, *190*, 4. [CrossRef]
24. Ters, M. Action morphologique des phénomènes périglaciaires dans la region littorale vendéenne. *Bull. Assoc. Géogr. Fr.* **1953**, *232–233*, 78–87. [CrossRef]
25. INRA. Base de Données Géographique des Sols de France à 1/1,000,000. 1998. Available online: www.gissol.fr (accessed on 24 July 2020).
26. Béchennec, F. Carte Géologique Harmonisée du Département de Loire-Atlantique. BRGM Report RP-55703-FR. 2007. 369p. Available online: http://infoterre.brgm.fr/rapports/RP-55703-FR.pdf (accessed on 18 August 2020).
27. Hall, G.; Buchar, A.; Bonham-Carter, G. Quality Control Assessment of Portable XRF Analysers: Development of Standard Operating Procedures, Performance on Variable Media and Recommended Uses. Canadian Mining Industry Research Organization (Camiro) Exploration Division, Project 10E01 Phase I Report. 2012. Available online: https://www.appliedgeochemists.org/index.php/publications/other-publications/2-uncategorised/106-portable-xrf-for-the-exploration-and-mining-industry (accessed on 24 July 2020).

28. Gray, A. Form, Distribution, and Genesis of Precious Metal Mineralization within the Bald Hill Antimony Deposit, South-Central New Brunswick, Canada. Master's Thesis, University of New Brunswick, Fredericton and Saint John, NB, Canada, 2019.
29. Bastos, R.O.; Melquiades, F.L.; Biasi, G.E.V. Correction for the effect of soil moisture on in situ XRF analysis using low-energy background. *X-Ray Spectrom.* **2012**, *41*, 304–307. [CrossRef]
30. Ge, L.; Lai, W.; Lin, Y. Influence of and correction for moisture in rocks, soils and sediments on in situ XRF analysis. *X-Ray Spectrom* **2005**, *34*, 28–34. [CrossRef]
31. Caporale, A.G.; Adamo, P.; Capozzi, F.; Langella, G.; Terribile, F.; Vingiani, S. Monitoring metal pollution in soils using portable-XRF and conventional laboratory-based techniques: Evaluation of the performance and limitations according to metal properties and sources. *Sci. Total Environ.* **2018**, *643*, 516–526. [CrossRef]
32. Aitchison, J. *The Statistical Analysis of Compositional Data. Monographs on Statistics and Applied Probability*; Chapman & Hall Ltd.: London, UK, 1986; p. 416.
33. Filzmoser, P.; Hron, K.; Reimann, C. The bivariate statistical analysis of environmental (compositional) data. *Sci. Total Environ.* **2010**, *408*, 4230–4238. [CrossRef]
34. Filzmoser, P.; Hron, K.; Reimann, C.; Garrett, R. Robust factor analysis for compositional data. *Comput. Geosci.* **2009**, *35*, 1854–1861. [CrossRef]
35. Reimann, C.; Filzmoser, P.; Hronc, K.; Kynčlová, P.; Garrett, R.G. A new method for correlation analysis of compositional (environmental) data—A worked example. *Sci. Total Environ.* **2017**, *607–608*, 965–971. [CrossRef]
36. Nicholson, K. Contrasting mineralogical-geochemical signatures of manganese oxides; guides to metallogenesis. *Econ. Geol.* **1992**, *87*, 1253–1264. [CrossRef]
37. Ashley, P.M.; Craw, D.; Graham, B.P.; Chappell, D.A. Environmental mobility of antimony around mesothermal stibnite deposits, New South Wales, Australia and southern New Zealand. *J. Geochem. Explor.* **2003**, *7*, 1–14. [CrossRef]
38. Belzile, N.; Chen, Y.; Wang, Z. Oxidation of antimony(III) by amorphous iron and manganese oxyhydroxides. *Chem. Geol.* **2001**, *174*, 379–387. [CrossRef]
39. Young, K.E.; Evans, C.A.; Hodges, K.V.; Bleacher, J.E.; Graff, T.G. A review of the handheld X-ray fluorescence spectrometer as a tool for field geologic investigations on Earth and in planetary surface exploration. *Appl. Geochem.* **2016**, *72*, 77–87. [CrossRef]
40. Cook, S.J.; Dunn, C.E. Final report on Results of the Cordilleran Geochemistry Project: A comparative assessment of soil geochemical methods for detecting buried mineral deposits. *Geosci. BC Pap.* **2007**, *7*, 225.
41. Baker, B.E. An application of soil humix substances to geochemical exploration. *Appl. Geochem.* **1986**, *2*, 307–310. [CrossRef]
42. Robbat, J.R.A. *Dynamic Workplans and Field Analytics: The Keys to Cost-Effective Site Investigations*; Case Study; Tufts University: Medford, MA, USA, 1997.
43. US Department of Energy. *Adaptive Sampling and Analysis Programs (ASAPs)*; Report DOE/EM-0592; US Department of Energy: Washingon, DC, USA, 2001.
44. Ramsey, M.H.; Boon, K.A. Can in situ geochemical measurements be more fit-for-purpose than those made ex situ? *Appl. Geochem.* **2012**, *27*, 969–976. [CrossRef]

MDPI
St. Alban-Anlage 66
4052 Basel
Switzerland
Tel. +41 61 683 77 34
Fax +41 61 302 89 18
www.mdpi.com

Minerals Editorial Office
E-mail: minerals@mdpi.com
www.mdpi.com/journal/minerals

www.ingramcontent.com/pod-product-compliance
Lightning Source LLC
LaVergne TN
LVHW070056170726
843515LV00007B/1546

* 9 7 8 3 0 3 6 5 1 9 2 0 3 *